Current Topics in Behavioral Neurosciences

Volume 48

Current Topics in Behavioral Neurosciences provides critical and comprehensive discussions of the most significant areas of behavioral neuroscience research, written by leading international authorities. Each volume in the series represents the most informative and contemporary account of its subject available, making it an unrivalled reference source. Each volume will be made available in both print and electronic form.

With the development of new methodologies for brain imaging, genetic and genomic analyses, molecular engineering of mutant animals, novel routes for drug delivery, and sophisticated cross-species behavioral assessments, it is now possible to study behavior relevant to psychiatric and neurological diseases and disorders on the physiological level. The Behavioral Neurosciences series focuses on translational medicine and cutting-edge technologies. Preclinical and clinical trials for the development of new diagnostics and therapeutics as well as prevention efforts are covered whenever possible. Special attention is also drawn on epigenetical aspects, especially in psychiatric disorders.

CTBN series is indexed in PubMed and Scopus.

More information about this series at http://www.springer.com/series/7854

Allan H. Young • Mario F. Juruena
Editors

Bipolar Disorder: From Neuroscience to Treatment

Springer

Editors
Allan H. Young
PO72 De Crespigny Park
King's College London
London, UK

Mario F. Juruena
PO72 De Crespigny Park
King's College London
London, UK

ISSN 1866-3370 ISSN 1866-3389 (electronic)
Current Topics in Behavioral Neurosciences
ISBN 978-3-030-72142-8 ISBN 978-3-030-72143-5 (eBook)
https://doi.org/10.1007/978-3-030-72143-5

This Springer imprint is published by the registered company Springer Nature Switzerland AG.
The registered company address is: Gewerbestrasse 11, 6330 Cham, Switzerland

Preface

Bipolar disorder is currently a clinical syndrome, as the present state of evidence is insufficient to allow its characterization based on etiology or pathophysiology. Despite this illness's frequent severity and high prevalence, the underlying neurobiological mechanisms are far from being elucidated, and this limits the development of new treatments substantially. A fundamental limitation is that current diagnostic systems are constructed on symptoms alone with no objective markers or verifying tests. Indeed, within different bipolar clinical subtypes, no biomarker differentiates one from another. Additionally, it is known that a diversity of disorders can display similar clinical manifestations and that the disorder can manifest with different features in different people.

Validated biomarkers, based on knowledge of neurobiology and genetics, would allow the integration of neuroscience into psychiatric diagnostic practice. A great advance would be to integrate reliable neurobiological findings to our diagnoses currently based on the clinical syndrome. A future diagnostic criteria system in which etiology and pathophysiology are thus integrated would refine categorical classification significantly.

Bipolar disorder destabilizes more than just a person's mood, thinking, and physiology. It can weaken one's identity, life, and dreams. It is a disorder that can intensely and skillfully destroy lives and a sense of meaning, not just for the person who has the illness but also their family.

The relationship between stress and bipolar disorders is an excellent example of a field of study that might be better understood from a comprehensive perspective. Emotional stress may modify the internal homeostatic state in a subject. During acute stress, adaptive physiological responses occur, if a critical interruption of this balance occurs an illness may result from the interactions on the nervous, endocrine, and immune systems.

Thus, the causes, development, and outcome for bipolar patients may be the product of the interaction of psychological, social, and cultural factors with neurochemistry and neurophysiology. Neurochemistry and neurophysiology are not separate and apart from our life experiences and/or recent stressful situations.

Neuroscience is replete with studies that report that the brain and its cognitive processes involve these fascinating interactions.

Both psychosocial and physical settings may have a huge impact on our neurobiology and behavior, and they also stimulate the development of adaptation or "allostasis." As our experiences modify our brain and cognition, changing our thoughts, we are changing our neurobiology. The balance of the internal homeostatic state of an individual may determine the vulnerability to diseases related to psychological stress in genetically predisposed individuals.

Currently, available treatment options for bipolar disorder are often insufficient to manage the acute episodes, relapses, cyclicity, suicide attempts, and recurrences that are key features of this disorder or for restoring premorbid functioning. In the last few years, we have witnessed a more wide-ranging understanding of the neural circuits and the various mechanisms of synaptic and neural plasticity, the molecular mechanisms of receptors, and the process by which genes code for specific functional proteins.

Translational research is one of the approaches that is being used with increasing frequency in neuroscience mainly because it gives us the tools to integrate different findings to enable new treatment developments. Translational research is developed to create a bridge between basic science and clinical developments and between clinical development and practice. Translational research broadly aims at the discovery of novel therapeutic agents.

To this end, the chapters included in this volume show how much has already been added to our knowledge of bipolar disorders. This volume presents essential contributions by leading international experts in key areas of research who have advanced our understanding of the causes and treatment of bipolar disorder. Moreover, it is an invaluable compendium of current knowledge and a concise guide to the integrative management of bipolar disorder. The potential therapeutic implications of new research are emphasized throughout the book. We hope you benefit from reading it as much as we have from preparing it.

London, UK
2021

Allan H. Young
Mario F. Juruena

Contents

The Neurobiology of Bipolar Disorder

Allan H. Young and Mario F. Juruena

Contents

Abstract Bipolar disorders are severe and have a high prevalence; despite this, the neurobiological mechanisms are far from being elucidated, and this limits the development of new treatments. Although the aetiology of bipolar disorders is not yet fully understood, it is accepted that the disorder(s) may result from the interaction between genetic factors that cause susceptibility and predisposing, precipitating and perpetuating environmental factors, such as stress and traumatic events. A pathophysiological formulation of the disease suggests that dysfunctions in intracellular biochemical cascades, oxidative stress and mitochondrial dysfunction impair the processes linked to neuronal plasticity, leading to cell damage and the consequent loss of brain tissue that has been identified in post-mortem and neuroimaging studies. The data we have reviewed suggests that peripheral biomarkers related to hormones, inflammation, oxidative stress and neurotrophins are altered in bipolar disorders, especially during acute mood episodes. Together, these changes have

A. H. Young and M. F. Juruena (✉)
Centre for Affective Disorders, Department of Psychological Medicine, Institute of Psychiatry, Psychology and Neuroscience, King's College London, London, UK
e-mail: allan.young@kcl.ac.uk; mario.juruena@kcl.ac.uk

Curr Topics Behav Neurosci (2021) 48: 1–20
https://doi.org/10.1007/7854_2020_179
Published Online: 14 January 2021

been associated with a systemic toxicity of the disease and the damage resulting from multiple episodes. Systemic toxicity related to recurrent episodes in bipolar disorder may influence brain anatomical changes associated with the progression of stress and neuroplasticity in bipolar disorder and the response to treatment.

Keywords Bipolar disorder · Hormones · Inflammation · Neurobiology · Neuroendocrine · Neurotrophins · Oxidadtive stress

1 Introduction

Bipolar affective disorder is a severe and chronic mental disorder, which often puts the lives of those affected at risk. Despite this severity and high prevalence, the pathophysiological mechanisms are far from being elucidated, which limits the development of new treatments.

Bipolar disorder (BD) is considered a clinical syndrome, as the current state of the evidence is insufficient to allow its characterization based on aetiology or pathophysiology. Also, BD has a complex clinical course, involving manic, depressive and/or mixed episodes, which makes study a challenge for researchers in the field.

Conceptually, it is clear that BD is complex and multifactorial. Although the aetiology of BD is not yet defined (Bobo 2017), it is accepted that the disorder may result from the interaction between genetic factors that cause susceptibility and predisposing, precipitating and perpetuating environmental factors, such as stress and traumatic events (Caspi and Moffitt 2006; Barnett and Smoller 2009).

A research approach that describes reliable neurobiological findings based on psychopathological syndrome will be more solid contrasted to a non-aetiologic system of classification. Integrative approaches to understanding complex health issues can transcend disciplinary and knowledge boundaries and provide opportunities to view phenomena from diverse perspectives. A future diagnostic criteria system in which aetiology and pathophysiology are essential in diagnostic decision-making would bring psychiatry closer to other specialties of medicine. The precision medicine might deconstruct traditional symptom-based categories. Patients with a range of mood disorders may be studied across several platforms to parse current heterogenous syndromes into homogenous clusters (see Fig. 1).

Although bipolar disorder (BD) is not a typical neurodegenerative disease, several studies have associated BD with stress and death of glial cells and neurons (Machado-Vieira et al. 2009). In an attempt to understand the changes present in BD, initially, the focus of the research was the dysregulation of neurotransmitters, and then it became intracellular signalling (Hahn and Friedman 1999; Manji et al. 2011); more recently, mitochondrial energy metabolism, oxidative stress and neurogenesis have been the focus of studies in BD (Berk et al. 2011). The hypothesis was raised that the loss of neurotrophic support, oxidative stress and inflammation are associated with more advanced stages of BD, occurring progressively (Berk et al. 2014). In

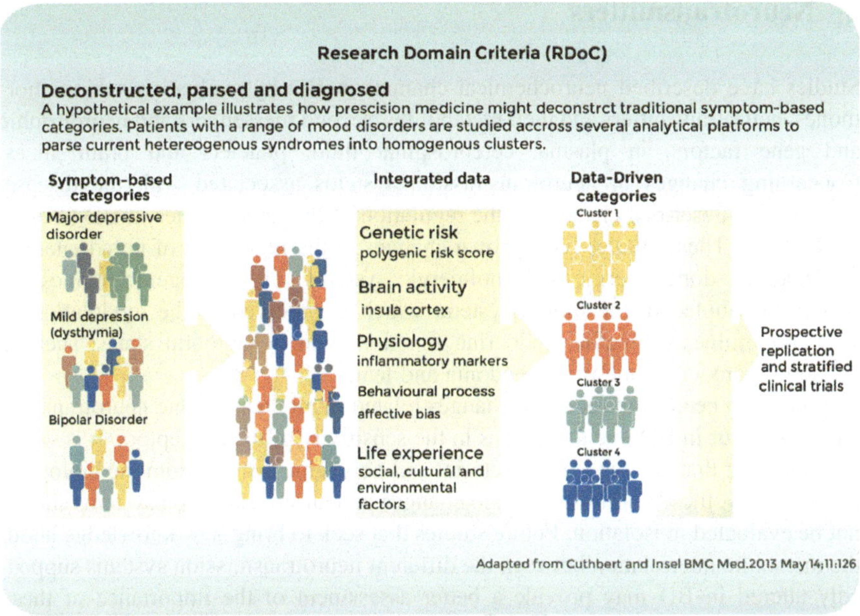

Fig. 1 Precision medicines in mood disorders integrate platforms to parse current heterogenous syndromes into homogenous clusters. Adapted from Cuthbert and Insel (2013). Illustrations courtesy of Romayne Gadelrab

this view, mitochondrial function and neuroprotection may be particularly relevant in BD.

One of the strategies most used by researchers is to study the biochemical targets of therapeutically relevant drugs, verifying changes in these targets in both manic and depressive episodes. Biochemical and molecular targets believed to be involved in the neurobiology/pathophysiology of this disorder are neurotransmitter systems (serotonergic, dopaminergic and glutamatergic), neurotrophic factors, PKC, GSK-3, mitochondria and oxidative stress (Zarate et al. 2006).

To explain the changes in brain cells in BD, different hypotheses have been considered, including the presence of greater oxidative stress in the CNS of patients with BD (Machado-Vieira et al. 2007; Wang et al. 2009). Reactive oxygen species (ROS) are toxic substances produced in all cells of the human body that have the potential to cause molecular damage and impair cell function. It is necessary that the cells have detoxification systems, which, under normal physiology, is accomplished by the action of antioxidant substances. However, there may be situations in which the production of ROS exceeds the competence of the antioxidant system and oxidative degradation of molecules may occur, characterized as a state of oxidative stress. It is possible to investigate the presence of oxidative stress in cells by detecting the final products of oxidative reactions and studying the activity of antioxidant enzymes (Andreazza et al. 2008).

2 Neurotransmitters

Studies have described neurochemical changes in BD by evaluating various hormones, neurotransmitters and their metabolites, second messengers and neurotrophic and gene factors, in plasma, cerebrospinal fluid, platelets and brain slices. Concerning changes in neurotransmission systems associated with the disease, studies have described changes in the regulation of biogenic amines in BD (Young et al. 1994). These studies have shown changes in the regulation of noradrenergic, serotonergic, dopaminergic and cholinergic systems. These biogenic amines are widely distributed in the limbic system, which is involved in the modulation of sleep-wakefulness, appetite, endocrine functions and behavioural states, such as irritability, anxiety, euphoria, anhedonia and fear.

It has also been suggested that changes related in monoaminergic neurotransmitters may occur in BD due to changes in the sensitivity of their receptors. It is worth remembering that, as it is a multifactorial and polygenic disease from the biological point of view, these findings related to changes in neurotransmission systems should not be evaluated in isolation. Future studies that seek to bring new knowledge about the metabolic interrelation between the different neurotransmission systems supposedly altered in BD may provide a better assessment of the importance of these findings. The study of the possible relationship between the abnormalities observed in these neurotransmission systems and the clinical and cognitive manifestation of BD will also assist in the evaluation of the influence of specific biochemical changes in the modulation of both mood and cognitive and neurovegetative functions.

2.1 Serotoninergic System

Serotonin (5-HT) modulates different neuronal activities and several physiological and behavioural functions, such as impulse control, aggressiveness and suicidal tendencies (Shiah and Yatham 2000). Decreased 5-HT release and activity may be associated with some abnormalities such as suicidal ideation, suicide attempts, aggression and sleep disorders, which are frequent findings in bipolar disorders (Ackenheil 2001).

Since the 1970s, many authors (e.g. Prange et al. 1974) have suggested the involvement of 5-HT in the pathophysiology of BD, formulating the permissive hypothesis, in which a deficit in central serotonergic neurotransmission would allow expression of both the manic phase and the depressive; however, such phases would differ to central catecholamine (norepinephrine and dopamine) levels, which would be elevated in mania and decreased in depression. Also, decreased levels of 5-hydroxy indole acetic acid (5-HIAA), the major metabolite of serotonin, have been demonstrated in the CSF of manic and depressed patients compared to healthy controls, suggesting that both mania and depression are associated with a reduction in function central serotoninergic. A post-mortem study of the brains of patients with

BD also found significantly lower levels of 5-HIAA in the frontal and parietal cortex, compared with controls, providing further evidence for the hypothesis of decreased central serotonergic activity in bipolar disorders (Baumann et al. 1999). Neuroendocrine challenge studies (Juruena 2014), when analysed together, suggest that the presynaptic activity of serotonin in the CNS is decreased, whereas the sensitivity of receptors postsynaptic is increased in mania (Shiah and Yatham 2000).

2.2 *Dopaminergic System*

Dopamine is a catecholaminergic neurotransmitter present in the central and peripheral nervous systems. Dopamine is synthesized in the brain from L-dihydroxyphenylalanine (L-Dopa) and has five distinct receptors, which have different affinities for dopamine. Type 1 dopamine receptors include D1 and D5 receptors, while type 2 receptors include D2, D3 and D4 receptors (Beaulieu and Gainetdinov 2011). Type D1 receptors are coupled to the Gαs protein, which activates adenylate cyclase, leading to increase in cyclic adenosine monophosphate (cyclic AMP), while D2 receptors are coupled to the Gαi protein, which inhibits adenylate cyclase, reducing the cytoplasmic concentration of cyclic AMP. Dopamine is involved in physiological functions such as blood pressure, kidney function, glucose homeostasis, voluntary movements, cognition, reward system, sleep and memory. Therefore, some researchers suggest that the dopaminergic system is involved in the pathophysiology of BD and that excessive dopaminergic stimulation is related to and perhaps triggers manic symptoms (Beaulieu and Gainetdinov 2011).

One of the most consistent findings about the role of dopamine in the neurobiology of BD is the fact that dopaminergic agonists, both direct and indirect, simulate episodes of mania or hypomania in patients with bipolar disorder with an underlying or predisposition (Brunello and Tascedda 2003). Ackenheil (2001) suggested that although the results were not consistent, a higher dopaminergic activity induced by increased release, decreased buffering capacity by vesicles synaptic disorders or the higher sensitivity of the dopaminergic may be associated with the development of manic symptoms, while decreased activity dopaminergic is associated with depression.

Pioneering studies on dopaminergic changes in BD had demonstrated high levels of dopamine in the urine of BD patients during manic episodes (Joyce et al. 1995). Also, a clinical study has shown that the administration of amphetamine (AMPH), an agonist substance in the dopaminergic system, triggers manic symptoms in both BD patients and healthy volunteers, with behavioural changes being more intense in those with BD (Anand et al. 2011). Studies suggest that the pathophysiology of BD may be linked to changes in the dopaminergic system and that the therapeutic effects of mood stabilizers may be linked to the modulation of this system (Bunney and Garland 1982). Therefore, the decrease in the expression or activity of the dopamine transporter (DAT) can lead to an increase in extracellular dopamine, suggesting that changes in this protein may contribute to the clinical manifestations of BD (Berk et al. 2007).

Therefore, it is believed that defects in the homeostatic mechanisms that respond to a hyperdopaminergic state, in the manic phase, may result in an excessive reduction in dopaminergic function, leading to a hypodopaminergic state and depression. In contrast, a weak regulatory response to a hypodopaminergic state may lead to a manic state. Dopamine is associated with induction of oxidative stress state in the brain and is metabolized via (a) enzymatic processes by the monoamine oxidase (MAO), which generates H2O2, a ROS, and dihydroxyphenylacetic acid, or via (b) autocatalysis by reaction with Fe2+ and H2O2 forming 6-hydroxydopamine (Obata 2002). Further studies are indeed necessary to understand how MAO contributes to oxidative damage in BD. Such studies would contribute a crucial piece of information – how oxidative stress modulates neurotransmitter levels.

2.3 *Norepinephrinergic System*

Studies have described the function of this system in depressive states. In these states, lower norepinephrine deficits and lower a2 receptor sensitivity have been reported, in contrast to a tendency toward higher norepinephrine activity in manic states (Ackenheil 2001). Following this logic, Baumann et al. (1999) observed that individuals with BD present higher numbers of pigmented cells in the locus coeruleus than do unipolar patients. Also, Shiah and Yatham (2000) suggested that diminished central 5-HT function concomitant with increased noradrenergic function may be involved in the genesis of mania.

Studies – evaluating norepinephrinergic metabolism in manic patients compared to healthy controls – describe that the results for plasma levels of MHPG may reflect the pathophysiology of BD from the manic state to the depressive state more than the plasma levels of HVA or BDNF. These data suggest that peripheral MHPG, which is associated with norepinephrine levels in the brain, could be used as a biomarker for whole mood states in BD. The MHPG levels, reflecting the norepinephrine levels in the brain, are likely to reflect the clinical characteristics of the switching process in BD and to have prognostic significance for the treatment of manic and depressive states (Kurita et al. 2014). When lithium was used, a significant decrease in these markers was observed in manic patients. However, on the other hand, the level of MHPG can be very different from individual to individual. Therefore, a patient's MHPG level must be known because it can change over time (Kurita et al. 2014).

2.4 *GABAergic System*

The monoaminergic hypothesis is not able to explain the pathophysiology of mood disorders fully. Moreover, the balance between excitatory and inhibitory impulses may be essential for processing information and preserving cognitive functions. Furthermore, new therapies have been developed to modulate the neurotransmission of gamma-aminobutyric acid (GABA) and glutamate, the main inhibitory and excitatory neurotransmitters, respectively (Wilkinson and Sanacora 2019). The dysregulation of the levels of these leads to an alteration of activity and of the neural resting state, which, chronically, might result in a maladaptation of these neuronal systems contributing to the appearance of mood symptoms (Fee et al. 2017).

Clinical data indicate that decreased GABAergic activity is associated with manic and depressive states and that GABAergic agonists have mood-stabilizing properties. Low levels of GABAergic activity have been found in the plasma of bipolar patients during both depressive and manic episodes (Petty et al. 1993).

The ratio between GABA and glutamate appears to be altered in patients with MDD (Sanacora et al. 2004). There is still no consensus about either peripheral or central levels of these neurotransmitters in patients with bipolar disorders due to the difficulty of studies with regard to controlling the effects of medication and post-mortem changes (Brady et al. 2013; Sanacora et al. 2008). Even so, patients who present a reduction in cortical GABA appear to have more significant cognitive impairments, especially inhibitory control (Bhagwagar et al. 2008; Luscher et al. 2011).

2.5 *Glutamatergic System*

Glutamate is the most important excitatory neurotransmitter in the brain. The role of this amino acid is to generate excitatory postsynaptic potential. However, its performance in brain functioning is much more diverse and complex (Rahn et al. 2012). In brain development, glutamate acts on migration, differentiation and the neuron's survivability (Luján et al. 2005). In mature brains, it is also inherently involved with neuroplasticity and neurotoxicity (Dong et al. 2009). The involvement of glutamate in excitatory synaptic transmission, neurotoxicity and neuroplasticity processes has led to growing interest in its role in the pathophysiology of psychiatric disorders, particularly in BD.

Investigation of the glutamatergic system in BD has been carried out using different approaches. Studies that measured glutamate in the plasma of patients with BD have found conflicting results and have been done on small samples of both patients and controls. Altamura et al. (1993) and Hoekstra et al. (2006) found increased levels of glutamate in patients in depression and mania, respectively. Although this results from the first line of evidence that the glutamatergic system is altered in BD, plasma glutamate levels may not adequately represent brain levels, given that the glutamate present in the brain is produced locally since that from outside the CNS does not cross the blood-brain barrier.

A greater number of studies have investigated glutamate in BD in post-mortem brain tissue (Law and Deakin 2001; McCullumsmith et al. 2007). Hashimoto et al. (2007) measured glutamate levels in the brain and found increased levels in the frontal cortex of patients with BD. Other studies investigated the protein or RNA expression of receptor subunits or other molecules belonging to the glutamatergic system and found abnormalities in areas of the brain associated with BD, including the dorsolateral prefrontal cortex (DLPFC) (Beneyto and Meador-Woodruff 2006), the hippocampus (Law and Deakin 2001; McCullumsmith et al. 2007), the striatum (Kristiansen and Meador-Woodruff 2005) and the thalamus (Clinton and Meador-Woodruff 2004). The results are consistent, showing a decrease in the expression of several molecules linked to glutamatergic transmission in the brains of patients with BD. These findings have supported the hypothesis of hyperglutamatergic state in the brain of patients with BD, leading to a subsensitization of receptors, as a compensation mechanism to modulate excessive glutamatergic activity (Rao et al. 2007). In fact, chronic administration of N-methyl-D-aspartate (NMDA), a glutamate analogue from which the NMDA receptor is named, causes a decrease in the expression of NR-1 and NR-3A subunits of this receptor in cell cultures neurons (Rao et al. 2007).

Further evidence that the glutamatergic system is altered in BD comes from studies demonstrating the action of mood stabilizers in this system. Chronic treatment with lithium or valproic acid decreases the synaptic levels of glutamate through several mechanisms (Sanacora et al. 2008). Both drugs have the effect of decreasing the expression of the GluR1 subunit of alpha-amino-3-hydroxy-methyl-5-4-isoxazolpropionic (AMPA) receptors in the hippocampus with prolonged treatment at clinically relevant doses (Du et al. 2003).

More support comes from recent studies indicating that AMPA receptor antagonists attenuate several manifest behaviours produced by the administration of amphetamine to rats (Bäckström and Hyytiä 2003). Lithium appears to increase neuronal excitability in the hippocampal CA1 synapses (Colino et al. 1998), increasing the effectiveness of these pathways. This appears to be due to the capacity of lithium to potentiate currents through AMPA receivers, through the selective increase in the opening of these channels (Gebhardt and Cull-Candy 2010). Lamotrigine, a drug used to prevent and treat depressive episodes in BD, inhibits the release of glutamate in the hippocampus of rats (Leach et al. 1986). Riluzole, another substance that inhibits the release of glutamate at the synapse, has shown antidepressant effect in preliminary studies (Zarate et al. 2005; Sanacora et al. 2007).

The above findings contribute to the hypothesis that glutamate levels are increased in BD and some of the effectiveness of mood stabilizers may come from an effect on glutamatergic neurotransmission.

In summary, the most current studies on the pathophysiology of BD have identified several macroscopic changes in brain areas and circuits, in addition to histopathological and chemical changes, at the tissue, intracellular and synaptic levels associated with the disease. A pathophysiological hypothesis of the disease suggests that dysfunctions in intracellular biochemical cascades, oxidative stress and mitochondrial dysfunction impair the processes linked to neuronal plasticity, leading to cell damage and the consequent loss of brain tissue that has been identified in post-mortem and neuroimaging. The investigation of the glutamatergic system in BD has been carried out using different methodological approaches. Studies that measured glutamate in the plasma of patients with BD have found conflicting results and have been done on small samples from patients and controls. Altamura et al. (1993) and Hoekstra et al. (2006) found increased levels of glutamate in patients in depression and mania, respectively. Although this results from the first line of evidence that the glutamatergic system is altered in BD, plasma glutamate levels may not adequately represent brain levels, given that the glutamate present in the brain is produced locally, since that from outside the CNS does not cross the blood-brain barrier.

A greater number of studies have investigated glutamate in BD in post-mortem brain tissue (Law and Deakin 2001; McCullumsmith et al. 2007). Hashimoto et al. (2007) measured glutamate levels in the brain and found increased levels in the frontal cortex of patients with BD. Other studies investigated the protein or RNA expression of receptor subunits or other molecules belonging to the glutamatergic system and found abnormalities in areas of the brain associated with BD, including the dorsolateral prefrontal cortex (DLPFC) (Beneyto and Meador-Woodruff 2006), the hippocampus (Law and Deakin 2001; McCullumsmith et al. 2007), the striatum (Kristiansen and Meador-Woodruff 2005) and the thalamus (Clinton and Meador-Woodruff 2004). The results are consistent, showing a decrease in the expression of several molecules linked to glutamatergic transmission in the brains of patients with BD. These findings have supported the hypothesis of a hyperglutamatergic state in the brain of patients with BD, leading to a subsensitization of receptors, as a compensatory mechanism to modulate excessive glutamatergic activity (Rao et al. 2007). In fact, chronic administration of N-methyl-D-aspartate (NMDA), a glutamate analogue from which the NMDA receptor is named, causes a decrease in the expression of NR-1 and NR-3A subunits of this receptor in cell cultures neurons (Rao et al. 2007).

Other evidence that the glutamatergic system is altered in BD comes from studies demonstrating the action of mood stabilizers in this system. Chronic treatment with lithium or valproic acid decreases the synaptic levels of glutamate through several mechanisms (Sanacora et al. 2008). Both drugs have the effect of decreasing the expression of the GluR1 subunit of alpha-amino-3-hydroxy-methyl-5-4-isoxazolpropionic (AMPA) receptors in the hippocampus with prolonged treatment and at clinically relevant doses (Du et al. 2003).

Additional support for this evidence comes from recent studies indicating that AMPA receptor antagonists attenuate several manifest behaviours produced by the administration of amphetamine to rats (Bäckström and Hyytiä 2003). Lithium appears to increase neuronal excitability in the hippocampal CA1 synapses (Colino et al. 1998), increasing the effectiveness of these pathways. This appears to be due to the capacity of lithium to potentiate currents through AMPA receivers, through the selective increase in the opening of these channels (Gebhardt and Cull-Candy 2010). Lamotrigine, a drug used to prevent and treat depressive episodes in BD, inhibits the release of glutamate in the hippocampus of rats (Leach et al. 1986). Riluzole, another substance that inhibits the release of glutamate at the synapse, has shown antidepressant effect in preliminary studies (Zarate et al. 2005; Sanacora et al. 2007).

In summary, the above findings contribute to the hypothesis that glutamate levels are increased in BD and some of the effectiveness of mood stabilizers comes from an effect on glutamatergic neurotransmission.

Most current studies on the pathophysiology of BD have identified several macroscopic changes in brain areas and circuits, in addition to histopathological and chemical changes, at the tissue, intracellular and synaptic levels associated with the disease. A pathophysiological hypothesis of the disease suggests that dysfunctions in intracellular biochemical cascades, oxidative stress and mitochondrial dysfunction impair the processes linked to neuronal plasticity, leading to cell damage and the consequent loss of brain tissue that has been identified in post-mortem and neuroimaging. The metabolites studied provide important clues about the processes linked to cellular energy metabolism, the intracellular pathway of phosphatidylinositol, neurotransmission and glutamatergic metabolism, in addition to providing measures of cellular vulnerability.

3 Intracellular Signalling

Most current studies on the pathophysiology of BD have identified several macroscopic changes in brain areas and circuits, in addition to histopathological and chemical changes, at the tissue, intracellular and synaptic levels associated with the disease. A pathophysiological hypothesis of the disease suggests that dysfunctions in intracellular biochemical cascades, oxidative stress and mitochondrial dysfunction impair the processes linked to neuronal plasticity, leading to cell damage and the consequent loss of brain tissue that has been identified in post-mortem and neuroimaging. Therefore, the metabolites studied provide essential clues about the processes linked to cellular energy metabolism, the intracellular pathway of phosphatidylinositol, neurotransmission and glutamatergic metabolism, in addition to providing measures of cellular vulnerability (Zarate et al. 2006).

In the case of oxidative stress, changes in the main antioxidant enzymes have already been described, such as the increased activity of superoxide dismutase (SOD, EC 1.15.1.1) in patients experiencing an episode of mania or depression and decreased catalase (CAT, EC 1.11.1.6) in medicated euthymic patients and an increase in non-medicated manic patients (Steckert et al. 2010). Among several markers of oxidative stress already evaluated in bipolar disorder, a meta-analysis showed that levels of lipid peroxidation, nitric oxide and DNA/RNA damage are significantly elevated in patients when compared to controls (Brown et al. 2014).

Bioenergetic changes in BD have been described in the last decades. Several studies have presented robust evidence to support the involvement of mitochondrial dysfunction in the pathophysiology of BD, through changes involving oxidative phosphorylation, the glycolytic pathway, phospholipid metabolism, decreased total energy production and/or availability of substrates, as well as abnormalities in the morphology and intracellular distribution of mitochondria. In fact, a recent integrative literature review, involving neuroimaging studies, has demonstrated increased levels of lactate in various brain regions and cerebrospinal fluid in patients with the disorder, which indicates an increase in anaerobic and extramitochondrial glucose metabolism, consistent with a weakened mitochondrial metabolism in BD. This fact is supported by data demonstrating that BD patients also have significantly lower brain levels of phosphocreatine (PCr, a high-energy compound), as well as a reduction in N-acetyl-aspartate (NAA) levels, along with correlation negative between NAA/creatine + PCr levels or NAA levels and disease duration. These indicate neurodevelopmental changes and provide indirect evidence that mitochondrial dysfunction may play a role in disease progression (Scaini et al. 2016). Also, it has been shown that patients with this disorder have reduced levels of inorganic phosphate, a regulator of oxidative phosphorylation, and Na+/K+-ATPase activity, as well as a significant decrease in the levels of adenosine diphosphate (ADP), but not in ATP concentrations. These suggest that the reduction may be driven by increased adenylate kinase activity, providing ATP at the cost of ADP. Added to this hypothesis is the fact that, during the process of visual stimulation, healthy individuals show significant reductions in PCr levels, but not in ATP levels, as expected since PCr is being used for the synthesis of ATP from the creatine kinase reaction. However, patients with BD have a different profile, since significant reductions in ATP levels were observed, but not in PCr levels. Also, a recent study demonstrated a significant reduction in the rate of direct reaction effect of creatine kinase in the absence of changes in concentration in ATP and PCr in patients with BD during the first episode. These data corroborating the previously presented data, according to which patients with the disorder have a baseline and normal concentrations of ATP and PCr, but disability to replenish ATP concentrations in the brain during periods of high energy demand (Du et al. 2018).

Corroborating the involvement of mitochondrial dysfunction in the pathophysiology of BD, a recent study demonstrated that patients with the disorder have an imbalance between the processes of fusion and mitochondrial fission, observed by increased levels in fission protein (Fis-1) and decreased levels of protein of fusion proteins (Mfn-2 and Opa-1), suggesting that the process of mitochondrial dynamics in patients with BD is dysfunctional, which may increase mitochondrial fragmentation (Scaini et al. 2017). In fact, morphological abnormalities have been described (more mitochondria in a smaller size) and an abnormal pattern of agglomeration and marginalization in the intracellular distribution of mitochondria in neurons in the prefrontal cortex of the post-mortem brain and peripheral cells of patients with BD (Cataldo et al. 2010). Furthermore, it is known that morphological and process changes in mitochondrial dynamics can directly activate the apoptotic pathway. This fact was evidenced by an increase in the protein levels of the active form of caspase-3, which was shown to be negatively correlated with the protein levels of Mfn-2 and Opa-1, as well as by a decrease in anti-apoptotic factors. In addition, the same study indicated that the changes observed peripherally were directly correlated with functional decline in patients with BD (Scaini et al. 2017).

As previously described, mitochondrial dynamics involves not only the processes of fission and fusion but also the movement of mitochondria through neurons, a process influenced by the concentration of ATP and Ca + 2. In addition to changes in

ATP levels, as already mentioned, studies have pointed out that BD patients have changes in intracellular Ca + 2 signalling, causing an increase in cytosolic Ca + 2 concentrations, related to excitotoxicity, decreased mitochondrial viability and, for the end, cell death (Duchen 2000).

In addition to providing most of the cellular ATP, mitochondria also play a central role in a wide variety of metabolic pathways and cellular functions. Since the brain uses about 20% of the body's total ATP, mitochondrial dysfunction significantly impacts brain functions. Although many questions remain open, many studies have shown that mitochondrial dysfunction, at both peripheral and CNS levels, is related to the pathophysiology of bipolar disorders.

4 Adenylate Cyclase Signalling Pathway

Intracellular ATP plays a central role in energy homeostasis, being the main product of reactions such as photophosphorylation, respiration and fermentation, acting as an energetic donor in most endergonic biosynthetic processes that support cell survival, proliferation and motility. The cytoplasm of most mammalian cells contains an ATP concentration in the range of 5–10 mM, and even higher concentrations are stored in the form of vesicles at synaptic terminals of neurons (Adinolfi et al. 2010).

Once in the extracellular environment, ATP has a signalling role, as in proliferation, mitogenesis and cell differentiation (Burnstock and Verkhratsky 2010), and in high concentrations, ATP extracellular cell causes cytotoxic effects (Lemmens et al. 1996). Specifically, in the central nervous system (CNS), extracellular ATP can act both as a fast-exciting neurotransmitter, by supporting Ca+2 waves between astrocytes, and as a neuromodulator (Fam et al. 2000). Since the extracellular concentration of ATP is in the nanomolar range in physiological situations, even a small release of ATP can generate a very significant signal, due to this low background (Adinolfi et al. 2010). In this context, studies demonstrate that ATP can activate microglial cells to release cytokines (Sanz and Di Virgilio 2000) and, therefore, seems to act in a pro-inflammatory way, while adenosine, a product of the hydrolysis of this nucleotide, may have an anti-inflammatory effect (Bours et al. 2006, Di Virgilio et al. 2009). Adenosine is also a neuromodulator capable of mediating neuroprotection by decreasing membrane excitability, limiting the influx of calcium and exerting modulating effects on glial cells (Ribeiro et al. 2003). Also, it is well established as a presynaptic inhibitor, reducing the release of neurotransmitters such as glutamate, dopamine, serotonin and acetylcholine (Brundege and Dunwiddie 1997).

There is essential adaptive plasticity of the ectonucleotidase pathway, with the main function in the CNS to protect the cells of this noble organ, avoiding accumulations, absence and exacerbated stimuli of these signalling molecules, when necessary. In general, in the sense of a decrease in ATP levels, and an increase in adenosine levels, since this is a neuroprotective molecule (Bonan 2012).

Associated with changes in neurotrophins, other factors linked to the homeostasis of the central nervous system may change during episodes, such as the increase in free radicals (oxidative stress). The imbalance in the production and control of reactive oxygen species, demonstrated, among other results.

5 Neurotrophins and Neurogenesis

Biochemical studies from peripheral samples from patients have been important in clarifying the pathophysiology of bipolar disorder. In this context, a growing body of evidence suggests changes in molecules associated with neuroplasticity (Scola and Andreazza 2015). Among these molecules, the role of neurotrophins stands out, which comprise a class of highly abundant proteins in the nervous system with fundamental functions in neuronal survival, growth and plasticity (Huang and Reichardt 2001).

In this perspective, the neurotrophins are proteins responsible for mediating neuronal survival and differentiation, modulation and transmission of synaptic plasticity. BDNF (brain-derived neurotrophic factor) is the most abundant neurotrophic factor in the central nervous system of adult humans and is highly expressed in brain areas known to regulate complex cognitive functions (Post 2007). It is known that there is evidence of changes in BDNF in neuropsychiatric diseases such as multiple sclerosis, Alzheimer's disease and Parkinson's disease (Duman and Monteggia 2006).

The evidence suggests a role for BDNF in the pathogenesis of BD. Blood levels of BDNF have been described as decreased in BD patients during manic, depressed and even euthymic episodes (Palomino et al. 2006; Machado-Vieira et al. 2007; Monteleone et al. 2008; de Oliveira et al. 2009). But these results have not been replicated in other studies (Yoshimura et al. 2006; Mackin et al. 2007). In one study, it was described that BD patients had lower BDNF levels than healthy controls (Lin 2009). Taking into account the different affective poles, the results showed statistically significant differences in BDNF only among manic patients or depressed state and controls and not between patients in euthymic state and controls. Also, BDNF levels increased significantly after pharmacological treatment of the manic state. These findings indicate that BDNF levels are abnormally reduced in the manic and depressed states of BD and that the reduced level in the manic state increases after pharmacological treatment. This suggests a potential role in the level of BDNF in the blood as a state-dependent BD biomarker (Hashimoto 2010).

Also, studies also indicate changes in other neurotrophins and trophic factors in bipolar patients, including neurotrophin-3, neurotrophin-4/5, glia-derived neurotrophic factor (GDNF) and neural (NGF) and endothelial growth factors vascular (VEGF), reinforcing the hypothesis that impairments in neuroplasticity are involved in the pathophysiology of bipolar disorder (Scola and Andreazza 2015).

6 Neuroendocrine

There is evidence that in bipolar disorder, dysfunction of the hypothalamic-pituitary-adrenal (HPA) axis is associated with neuroprogression, which can induce persistent changes in the HPA axis that lead to depression and more refractory bipolar episodes in adults. It is believed that disturbance of the hormonal stress system can cause problems of thought and memory and worsen depressive symptoms in bipolar disorder (Watson et al. 2004). Hypoactivation of the HPA axis has been observed in depression with atypical characteristics, but also in other depression subtypes, in manic/hypomanic episodes. These abnormalities are related to the feedback of glucocorticoids, on the HPA axis, and their binding to glucocorticoid receptors (GR and MR). These abnormalities seem to be related to changes in the ability to circulate glucocorticoids to exert their negative feedback on the secretion of hormones from the HPA axis by binding to mineralocorticoid (MR) and glucocorticoid (GR) receptors in HPA tissues (Juruena et al. 2015). Bipolar disorder is multifactorial concerning the possible factors that contribute to their aetiology – biological and environmental factors (Juruena 2014, see Fig. 2).

Regarding neurobiological markers, cortisol is considered one of the main mediators of allostatic load (McEwen 2006). Several studies in bipolar disorder – and major depressive disorder – demonstrate dysregulation of the HPA axis, with the maintenance of high levels of cortisol even in stages of remission of the disease, due to possible negative impaired feedback in the HPA axis (Watson et al. 2004; Ellenbogen et al. 2010; Juruena et al. 2010). Also, abnormalities in axis regulation may predict depressive (Juruena et al. 2009) and manic relapses (Maripuu et al. 2016). It can be considered that the reduction in resilience with the repetition of episodes and the increase in reactivity to stressors in the progression of the disease may be related to the deregulation of the HPA axis (Juruena 2014).

Bipolar disorder has also been associated with impaired cell resilience. Changes in cell signalling pathways, that is, in the cell's ability to deal with a given stimulus and the response to it, may be associated with greater vulnerability to stressful events, and this includes changes in neurotransmitters, trophic cascade, calcium, anti-apoptotic factors and pathway survival (Belvederi Murri et al. 2016).

HPA axis and renin-angiotensin-aldosterone system (RAAS) impairment may be some of the intrinsic aetiological factors (Juruena et al. 2015; Heuser et al. 2000; Keller et al. 2017; Murck et al. 2019). MR and GR receptors in the brain are involved in the regulation of stress hormone secretion and complex behaviour, such as emotion, memory and sleep. In the pathophysiology of stress-related psychiatric disorders, these receptors have not been sufficiently characterized, but HPA axis and RAAS interact with each other, and both are mediated by the limbic system (de Kloet and Joëls 2017; Harris et al. 2013; Murck et al. 2012).

Aldosterone and cortisol bind to mineralocorticoid receptors (MR) and glucocorticoid receptors (GR), respectively, to exert their actions on RAAS and HPA axis. MR/GR imbalance and inadequate levels of aldosterone and cortisol may impair resilience capacity of subjects, raising their vulnerable phenotype to disorders related

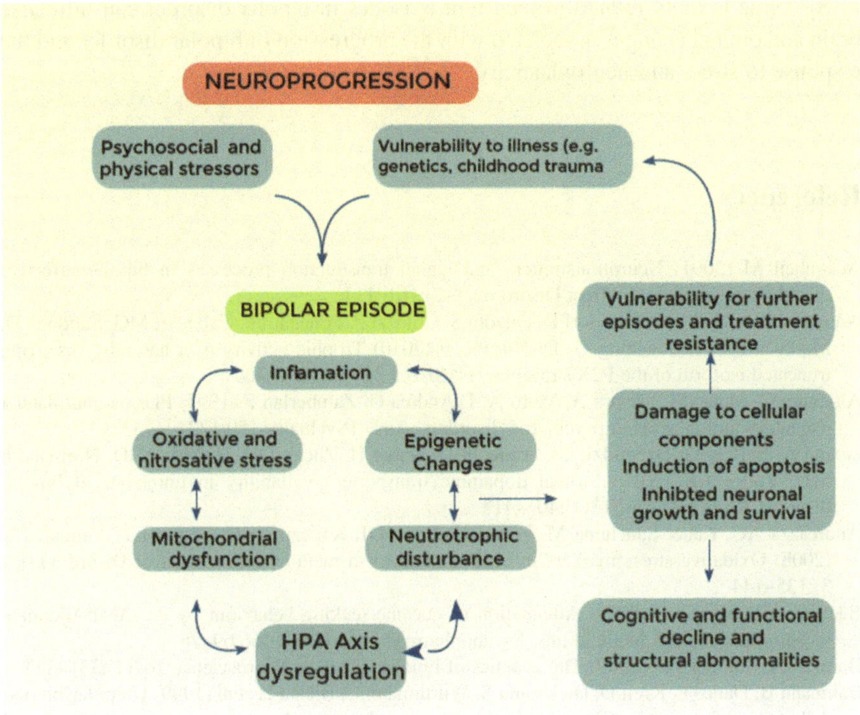

Fig. 2 The impact on neuroprogression, adapted from Moylan et al. (2013). Illustrations courtesy of Romayne Gadelrab

to stress exposition (de Kloet and Joëls 2017; Gomez-Sanchez and Gomez-Sanchez 2014). As HPA axis and RAAS adequate functioning also depend on their receptors' well-functioning, their gene receptors' polymorphisms are important biological factors to be considered in bipolar patients. Some of MR and GR gene polymorphisms have been shown in the literature associated with the imbalanced functioning HPA axis and RAAS (DeRijk et al. 2008; van Leeuwen et al. 2011).

7 Conclusion

The data we have reviewed suggests that peripheral biomarkers related to hormones, inflammation, oxidative stress and neurotrophins are altered in the bipolar disorder, especially during acute mood episodes. Together, these changes have been associated with a systemic toxicity of the disease and the damage resulting from multiple episodes.

Systemic toxicity related to recurrent episodes in bipolar disorder can influence brain anatomical changes associated with the progression of bipolar disorder and the response to stress and neuroplasticity.

References

Ackenheil M (2001) Neurotransmitters and signal transduction processes in bipolar affective disorders: a synopsis. J Affect Disord 62(1–2):101–111

Adinolfi E, Cirillo M, Woltersdorf R, Falzoni S, Chiozzi P, Pellegatti P, Callegari MG, Sandonà D, Markwardt F, Schmalzing G, Di Virgilio F (2010) Trophic activity of a naturally occurring truncated isoform of the P2X7 receptor. FASEB J 24(9):3393–3404

Altamura C, Mauri M, Ferrara A, Moro A, D'Andrea G, Zamberlan F (1993) Plasma and platelet excitatory amino acids in psychiatric disorders. Am J Psychiatry 150:1731–1733

Anand A, Barkay G, Dzemidzic M, Albrecht D, Karne H, Zheng QH, Hutchins GD, Normandin MD, Yoder KK (2011) Striatal dopamine transporter availability in unmedicated bipolar disorder. Bipolar Disord 13(4):406–413

Andreazza AC, Kauer-Sant'anna M, Frey BN, Bond DJ, Kapczinski F, Young LT, Yatham LN (2008) Oxidative stress markers in bipolar disorder: a meta-analysis. J Affect Disord 111(2-3):135–144

Bäckström P, Hyytiä P (2003) Attenuation of cocaine-seeking behaviour by the AMPA/kainate receptor antagonist CNQX in rats. Psychopharmacology (Berl) 166:69–76

Barnett JH, Smoller JW (2009) The genetics of bipolar disorder. Neuroscience 164(1):331–343

Baumann B, Danos P, Krell D, Diekmann S, Wurthmann C, Bielau H et al (1999) Unipolar-bipolar dichotomy of mood disorders is supported by noradrenergic brainstem system morphology. J Affect Disord 54(1–2):217–224

Beaulieu JM, Gainetdinov RR (2011) The physiology, signalling, and pharmacology of dopamine receptors. Pharmacol Rev 63(1):182–217

Belvederi Murri M, Prestia D, Mondelli V et al (2016) The HPA axis in bipolar disorder: systematic review and metaanalysis. Psychoneuroendocrinology 63:327–342

Beneyto M, Meador-Woodruff J (2006) Lamina-specific abnormalities of AMPA receptor trafficking and signaling molecule transcripts in the prefrontal cortex in schizophrenia. Synapse 60:585–598

Berk M, Dodd S, Kauer-Sant'anna M, Malhi GS, Bourin M, Kapczinski F, Norman T (2007) Dopamine dysregulation syndrome: implications for a dopamine hypothesis of bipolar disorder. Acta Psychiatr Scand Suppl 434:41–49

Berk M, Kapczinski F, Andreazza AC, Dean OM, Giorlando F, Maes M, Yücel M, Gama CS, Dodd S, Dean B, Magalhães PV, Amminger P, McGorry P, Malhi GS (2011) Pathways underlying neuroprogression in bipolar disorder: focus on inflammation, oxidative stress and neurotrophic factors. Neurosci Biobehav Rev 35(3):804–817

Berk M, Berk L, Dodd S, Cotton S, Macneil C, Daglas R, Conus P, Bechdolf A, Moylan S, Malhi GS (2014) Stage managing bipolar disorder. Bipolar Disord 16(5):471–477

Bhagwagar Z et al (2008) Low GABA concentrations in occipital cortex and anterior cingulate cortex in medication-free, recovered depressed patients. Int J Neuropsychopharmacol 11 (2):255–260

Bobo WV (2017) The diagnosis and management of bipolar I and II disorders: clinical practice update. Mayo Clin Proc 92(10):1532–1551

Bonan CD (2012) Ectonucleotidases and nucleotide/nucleoside transporters as pharmacological targets for neurological disorders. CNS Neurol Disord Drug Targets 11(6):739–750

Bosaipo NB, Borges VF, Juruena MF (2017) Bipolar disorder: a review of conceptual and clinical aspects. Medicina 50(Suppl 1):72–84

Bours MJ, Swennen EL, Di Virgilio F, Cronstein BN, Dagnelie PC (2006) Adenosine 5-′-triphosphate and adenosine as endogenous signaling molecules in immunity and inflammation. Pharmacol Ther 112(2):358–404
Brady RO et al (2013) Brain gamma-aminobutyric acid (GABA) abnormalities in bipolar disorder. Bipolar Disord 15(4):434–439
Brown NC, Andreazza AC, Young LT (2014) An updated meta-analysis of oxidative stress markers in bipolar disorder. Psychiatry Res 218(1–2):61–68
Brundege JM, Dunwiddie TV (1997) Role of adenosine as a modulator of synaptic activity in the central nervous system. Adv Pharmacol 39:353–391
Brunello N, Tascedda F (2003) Cellular mechanisms and second messengers: relevance to the psychopharmacology of bipolar disorders. Int J Neuropsychopharmacol 6(2):181–189
Bunney WE Jr, Garland BL (1982) A second generation catecholamine hypothesis. Pharmacopsychiatria 15(4):111–115
Burnstock G, Verkhratsky A (2010) Long-term (trophic) purinergic signalling: Purinoceptors control cell proliferation, differentiation and death. Cell Death Dis 1:e9
Caspi A, Moffitt TE (2006) Gene-environment interactions in psychiatry: joining forces with neuroscience. Nat Rev Neurosci 7(7):583–590
Cataldo AM, McPhie DL, Lange NT, Punzell S, Elmiligy S, Ye NZ et al (2010) Abnormalities in mitochondrial structure in cells from patients with bipolar disorder. Am J Pathol 177 (2):575–585
Clinton SM, Meador-Woodruff JH (2004) Abnormalities of the NMDA receptor and associated intracellular molecules in the thalamus in schizophrenia and bipolar disorder. Neuropsychopharmacology 29:1353–1362
Colino A, García-Seoane JJ, Valentín A (1998) Action potential broadening induced by lithium may cause a presynaptic enhancement of excitatory synaptic transmission in neonatal rat hippocampus. Eur J Neurosci 10:2433–2443
Cuthbert BN, Insel TR (2013) Toward the future of psychiatric diagnosis: the seven pillars of RDoC. BMC Med 11:126
de Kloet ER, Joëls M (2017) Brain mineralocorticoid receptor function in control of salt balance and stress-adaptation. Physiol Behav 178:13–20
de Oliveira GS, Ceresér KM, Fernandes BS et al (2009) Decreased brain-derived neurotrophic factor in medicated and drug-free bipolar patients. J Psychiatr Res 43:1171–1174
Derijk RH, van Leeuwen N, Klok MD, Zitman FG (2008) Corticosteroid receptor-gene variants: modulators of the stress-response and implications for mental health. Eur J Pharmacol 585 (2–3):492–501
Di Virgilio F, Ceruti S, Bramanti P, Abbracchio MP (2009) Purinergic signalling in inflammation of the central nervous system. Trends Neurosci 32(2):79–87
Dong XX, Wang Y, Qin ZH (2009) Molecular mechanisms of excitotoxicity and their relevance to pathogenesis of neurodegenerative diseases. Acta Pharmacol Sin 30:379–387
Du J, Gray NA, Falke C, Yuan P, Szabo S, Manji HK (2003) Structurally dissimilar antimanic agents modulate synaptic plasticity by regulating AMPA glutamate receptor subunit GluR1 synaptic expression. Ann N Y Acad Sci 1003:378–380
Du F, Yuksel C, Chouinard VA, Huynh P, Ryan K, Cohen BM, Öngür D (2018) Abnormalities in high-energy phosphate metabolism in first-episode bipolar disorder measured using 31P-magnetic resonance spectroscopy. Biol Psychiatry 84(11):797–802
Duchen MR (2000) Mitochondria and Ca(2+)in cell physiology and pathophysiology. Cell Calcium 28(5–6):339–348
Duman RS, Monteggia LM (2006) A neurotrophic model for stress-related mood disorders. Biol Psychiatry 59(12):1116–1127
Ellenbogen MA, Santo JB, Linnen AM, Walker CD, Hodgins S (2010) High cortisol levels in the offspring of parents with bipolar disorder during two weeks of daily sampling. Bipolar Disord 12:77–86

Fam SR, Gallagher CJ, Salter MW (2000) P2Y(1) purinoceptor-mediated Ca(2+) signaling and Ca (2+) wave propagation in dorsal spinal cord astrocytes. J Neurosci 20(8):2800–2808
Fee C, Banasr M, Sibille E (2017) Somatostatin-positive GABA Interneuron deficits in depression: cortical microcircuit and therapeutic perspectives. Biol Psychiatry 82(8):549–559
Gebhardt C, Cull-Candy SG (2010) Lithium acts as a potentiator of AMPAR currents in hippocampal CA1 cells by selectively increasing channel open probability. J Physiol 588:3933–3941
Gomez-Sanchez E, Gomez-Sanchez CE (2014) The multifaceted mineralocorticoid receptor. Compr Physiol 4(3):965–994
Hahn CG, Friedman E (1999) Abnormalities in protein kinase C signaling and the pathophysiology of bipolar disorder. Bipolar Disord 1(2):81–86
Harris AP, Holmes MC, de Kloet ER, Chapman KE, Seckl JR (2013) Mineralocorticoid and glucocorticoid receptor balance in control of HPA axis and behaviour. Psychoneuroendocrinology 38(5):648–658
Hashimoto K (2010) Brain-derived neurotrophic factor as a biomarker for mood disorders: an historical overview and future directions. Psychiatry Clin Neurosci 64(4):341–357
Hashimoto K, Sawa A, Iyo M (2007) Increased levels of glutamate in brains from patients with mood disorders. Biol Psychiatry 62:1310–1316
Heuser I, Deuschle M, Weber B, Stalla GK, Holsboer F (2000) Increased activity of the hypothalamus-pituitary-adrenal system after treatment with the mineralocorticoid receptor antagonist spironolactone. Psychoneuroendocrinology 25(5):513–518
Hoekstra R, Fekkes D, Loonen A, Pepplinkhuizen L, Tuinier S, Verhoeven W (2006) Bipolar mania and plasma amino acids: increased levels of glycine. Eur Neuropsychopharmacol 16:71–77
Huang EJ, Reichardt LF (2001) Neurotrophins: roles in neuronal development and function. Annu Rev Neurosci 24:677–736. Review
Joyce PR, Fegursson DM, Woollard G, Abbott RM, Horwood LJ, Upton J (1995) Urinary catecholamines and plasma hormones predict mood state in rapid cycling bipolar affective disorder. J Affect Disord 33(4):233–243
Juruena MF, Cleare AJ, Papadopoulos AS, Poon L, Lightman S, Pariante CM (2010) The prednisolone suppression test in depression: dose-response and changes with antidepressant treatment. Psychoneuroendocrinology 35(10):1486–1491
Juruena MF (2014) Early-life stress and HPA axis trigger recurrent adulthood depression. Epilepsy Behav 38:148–159
Juruena MF, Pariante CM, Papadopoulos AS, Poon L, Lightman S, Cleare AJ (2009) Prednisolone suppression test in depression: prospective study of the role of HPA axis dysfunction in treatment resistance. Br J Psychiatry 194:342–349
Juruena MF, Baes CW, Menezes IC, Graeff FG (2015) Early life stress in depressive patients: role of glucocorticoid and mineralocorticoid receptors and of hypothalamic-pituitary-adrenal axis activity. Curr Pharm Des 21(11):1369–1378
Keller J, Gomez R, Williams G, Lembke A, Lazzeroni L, Murphy GM et al (2017) HPA Axis in major depression: cortisol, clinical symptomatology, and genetic variation predict cognition. Mol Psychiatry 22(4):527–536
Kristiansen L, Meador-Woodruff J (2005) Abnormal striatal expression of transcripts encoding NMDA interacting PSD proteins in schizophrenia, bipolar disorder and major depression. Schizophr Res 78:87–93
Kurita M, Nishino S, Numata Y, Okubo Y, Sato T (2014) The noradrenaline metabolite MHPG is a candidate biomarker from the manic to the remission state in bipolar disorder I: a clinical naturalistic study. PLoS One 9(6):e100634
Law AJ, Deakin JF (2001) Asymmetrical reductions of hippocampal NMDAR1 glutamate receptor mRNA in the psychoses. Neuroreport 12:2971–2974
Leach MJ, Marden CM, Miller AA (1986) Pharmacological studies on lamotrigine, a novel potential antiepileptic drug: II. Neurochemical studies on the mechanism of action. Epilepsia 27:490–497

Lemmens R, Vanduffel L, Teuchy H, Culic O (1996) Regulation of proliferation of LLC-MK2 cells by nucleosides and nucleotides: the role of ectoenzymes. Biochem J 316(Pt 2):551–557
Lin PY (2009) State-dependent decrease in levels of brain-derived neurotrophic factor in bipolar disorder: a meta-analytic study. Neurosci Lett 466:139–143
Luján R, Shigemoto R, López-Bendito G (2005) Glutamate and GABA receptor signalling in the developing brain. Neuroscience 130:567–580
Luscher B, Shen Q, Sahir N (2011) The GABAergic deficit hypothesis of major depressive disorder. Mol Psychiatry 16(4):383–406
Machado-Vieira R, Dietrich MO, Leke R et al (2007) Decreased plasma brain derived neurotrophic factor levels in unmedicated bipolar patients during manic episode. Biol Psychiatry 61:142–144
Machado-Vieira R, Manji HK, Zarate CA Jr (2009) The role of lithium in the treatment of bipolar disorder: convergent evidence for neurotrophic effects as a unifying hypothesis. Bipolar Disord 11(Suppl 2):92–109
Mackin P, Gallagher P, Watson S, Young AH, Ferrier IN (2007) Changes in brain-derived neurotrophic factor following treatment with mifepristone in bipolar disorder and schizophrenia. Aust N Z J Psychiatry 41:321–326
Manji HK, Henter ID, Zarate CA Jr (2011) Bipolar disorder: a neurobiological synthesis. Curr Top Behav Neurosci 5:331–340
Maripuu M, Wikgren M, Karling P, Adolfsson R, Norrback KF (2016) Relative hypocortisolism is associated with obesity and the metabolic syndrome in recurrent affective disorders. J Affect Disord 204:187–196
Mazer AK, Cleare AJ, Young AH, Juruena MF (2019) Bipolar affective disorder and borderline personality disorder: differentiation based on the history of early life stress and psychoneuroendocrine measures. Behav Brain Res 357–358:48–56
McCullumsmith R, Kristiansen L, Beneyto M, Scarr E, Dean B, Meador-Woodruff J (2007) Decreased NR1, NR2A, and SAP102 transcript expression in the hippocampus in bipolar disorder. Brain Res 1127:108–118
McEwen B (2006) Dialogues Clin Neurosci 8(4)
Monteleone P, Serritella C, Martiadis V, Maj M (2008) Decreased levels of serum brain-derived neurotrophic factor in both depressed and euthymic patients with unipolar depression and in euthymic patients with bipolar I and II disorders. Bipolar Disord 10:95–100
Moylan S, Maes M, Wray N, Berk M (2013) The neuroprogressive nature of major depressive disorder: pathways to disease evolution and resistance, and therapeutic implications. Mol Psychiatry 18:595–606
Murck H, Schüssler P, Steiger A (2012) Renin-angiotensin-aldosterone system: the forgotten stress hormone system: relationship to depression and sleep. Pharmacopsychiatry 45(3):83–95
Murck H, Braunisch MC, Konrad C, Jezova D, Kircher T (2019) Markers of mineralocorticoid receptor function: changes over time and relationship to response in patients with major depression. Int Clin Psychopharmacol 34(1):18–26
Obata T (2002) Dopamine efflux by MPTP and hydroxyl radical generation. J Neural Transm (Vienna) 109(9):1159–1180
Palomino A, Vallejo-Illarramendi A, González-Pinto A et al (2006) Decreased levels of plasma BDNF in first-episode schizophrenia and bipolar disorder patients. Schizophr Res 86:321–322
Petty F, Kramer GL, Fulton M, Moeller FG, Rush AJ (1993) Low plasma GABA is a marker of a characteristic similar to bipolar disorder. Neuropsychopharmacology 9(2):125–132
Post RM (2007) Role of BDNF in bipolar and unipolar disorder: clinical and theoretical implications. J Psychiatr Res 41(12):979–990
Prange AJ Jr, Wilson IC, Lynn CW, Alltop LB, Stikeleather RA (1974) L-tryptophan in mania. Contribution to a permissive hypothesis of affective disorders. Arch Gen Psychiatry 30 (1):56–62
Rahn KA, Slusher BS, Kaplin AI (2012) Glutamate in CNS neurodegeneration and cognition and its regulation by GCPII inhibition. Curr Med Chem 19:1335–1345

Rao J, Ertley R, Rapoport S, Bazinet R, Lee H (2007) Chronic NMDA administration to rats up-regulates frontal cortex cytosolic phospholipase A2 and its transcription factor, activator protein-2. J Neurochem 102:1918–1927

Ribeiro JA, Sebastiao AM, de Mendonca A (2003) Participation of adenosine receptors in neuroprotection. Drug News Perspect 16(2):80–86

Sanacora G et al (2004) Subtype-specific alterations of gamma-aminobutyric acid and glutamate in patients with major depression. Arch Gen Psychiatry 61(7):705–713

Sanacora G, Kendell SF, Levin Y, Simen AA, Fenton LR, Coric V, Krystal JH (2007) Preliminary evidence of riluzole efficacy in antidepressant-treated patients with residual depressive symptoms. Biol Psychiatry 61:822–825

Sanacora G, Zarate CA, Krystal JH, Manji HK (2008) Targeting the glutamatergic system to develop novel, improved therapeutics for mood disorders. Nat Rev Drug Discov 7:426–437

Sanz JM, Di Virgilio F (2000) Kinetics and mechanism of ATP-dependent IL-1 beta release from microglial cells. J Immunol 164(9):4893–4898

Scaini G, Rezin GT, Carvalho AF, Streck EL, Berk M, Quevedo J (2016) Mitochondrial dysfunction in bipolar disorder: evidence, pathophysiology and translational implications. Neurosci Biobehav Rev 68:694–713

Scaini G, Fries GR, Valvassori SS, Zeni CP, Zunta-Soares G, Berk M et al (2017) Perturbations in the apoptotic pathway and mitochondrial network dynamics in peripheral blood mononuclear cells from bipolar disorder patients. Transl Psychiatry 7(5):e1111

Scola G, Andreazza AC (2015) The role of neurotrophins in bipolar disorder. Prog Neuropsychopharmacol Biol Psychiatry 56:122–128

Shiah IS, Yatham LN (2000) Serotonin in mania and in the mechanism of action of mood stabilizers: a review of clinical studies. Bipolar Disord 2(2):77–92

Steckert AV, Valvassori SS, Moretti M, Dal-Pizzol F, Quevedo J (2010) Role of oxidative stress in the pathophysiology of bipolar disorder. Neurochem Res 35(9):1295–1301

van Leeuwen N, Bellingrath S, de Kloet ER, Zitman FG, DeRijk RH, Kudielka BM, Wüst S (2011) Human mineralocorticoid receptor (MR) gene haplotypes modulate MR expression and transactivation: implication for the stress response. Psychoneuroendocrinology 36(5):699–709

Wang JF, Shao L, Sun X, Young LT (2009) Increased oxidative stress in the anterior cingulate cortex of subjects with bipolar disorder and schizophrenia. Bipolar Disord 11:523–529

Watson S, Gallagher P, Ritchie JC, Ferrier IN, Young AH (2004) Hypothalamic-pituitary-adrenal axis function in patients with bipolar disorder. Br J Psychiatry 184:496–502

Wilkinson ST, Sanacora G (2019) A new generation of antidepressants: an update on the pharmaceutical pipeline for novel and rapid-acting therapeutics in mood disorders based on glutamate/GABA neurotransmitter systems. Drug Discov Today 24(2):606–615

Yoshimura R, Nakano Y, Hori A et al (2006) Effect of risperidone on plasma catecholamine metabolites and brain-derived neurotrophic factor in patients with bipolar disorders. Hum Psychopharmacol 21:433–438

Young LT, Warsh JJ, Kish SJ et al (1994) Reduced brain 5-HT and elevated NE turnover and metabolites in bipolar affective disorder. Biol Psychiatry 35:121–127

Zarate CA Jr, Quiroz JA, Singh JB, Denicoff KD, De Jesus G, Luckenbaugh DA, Charney DS, Manji HK (2005) An open-label trial of the glutamate-modulating agent riluzole in combination with lithium for the treatment of bipolar depression. Biol Psychiatry 57(4):430–432

Zarate CJ, Singh J, Manji H (2006) Cellular plasticity cascades: targets for the development of novel therapeutics for bipolar disorder. Biol Psychiatry 59:1006–1020

The Role of Stress in Bipolar Disorder

Eduardo H. L. Umeoka, Judith M. C. van Leeuwen, Christiaan H. Vinkers, and Marian Joëls

Contents

Abstract Stress is a major risk factor for bipolar disorder. Even though we do not completely understand how stress increases the risk for the onset and poorer course of bipolar disorder, knowledge of stress physiology is rapidly evolving. Following

E. H. L. Umeoka (✉)
Faculty of Medicine, University Center Unicerrado, Goiatuba, GO, Brazil

J. M. C. van Leeuwen
Donders Institute for Brain, Cognition, and Behaviour, Radboud University Medical Center, Nijmegen, The Netherlands

C. H. Vinkers
Amsterdam UMC, Vrije Universiteit Amsterdam, Department of Psychiatry, Amsterdam Neuroscience, Amsterdam, The Netherlands

Amsterdam UMC, Department of Anatomy and Neurosciences, Amsterdam Neuroscience, Amsterdam, The Netherlands

M. Joëls
Brain Center Rudolf Magnus, University Medical Center Utrecht, Utrecht, The Netherlands

University Medical Center Groningen, Groningen, The Netherlands

Curr Topics Behav Neurosci (2021) 48: 21–40
https://doi.org/10.1007/7854_2020_151
Published Online: 4 August 2020

stress, stress hormones – including (nor)adrenaline and corticosteroid – reach the brain and change neuronal function in a time-, region-, and receptor-dependent manner. Stress has direct consequences for a range of cognitive functions which are time-dependent. Directly after stress, emotional processing is increased at the cost of higher brain functions. In the aftermath of stress, the reverse is seen, i.e., increased executive function and contextualization of information. In bipolar disorder, basal corticosteroid levels (under non-stressed conditions) are generally found to be increased with blunted responses in response to experimental stress. Moreover, patients who have bipolar disorder generally show impaired brain function, including reward processing. There is some evidence for a causal role of (dysfunction of) the stress system in the etiology of bipolar disorder and their effects on brain system functionality. However, longitudinal studies investigating the functionality of the stress systems in conjunction with detailed information on the development and course of bipolar disorder are vital to understand in detail how stress increases the risk for bipolar disorder.

Keywords Blunted response · Cognition · Cortisol awakening response (CAR) · Hypercortisolemia · Hypothalamus-pituitary-adrenal (HPA) axis · Network · Trier social stress test (TSST)

1 Introduction

1.1 Activation of Hormonal Systems After Stress

Situations of perceived threat, i.e., stressors, which are subjectively experienced as "stress," activate a cascade of events eventually resulting in the release of multiple stress hormones (Herman 2018; see Fig. 1). Thus, directly after stress, activation of the sympathetic nervous system causes the release of adrenaline from cells in the adrenal medulla. Adrenaline allows the individual to quickly respond to a stressor, in part by increasing heart rate, blood circulation, and respiration. Via indirect pathways, adrenaline also increases the release of noradrenaline in the brain, contributing to a quick cognitive response to the stressful situation.

Slightly later, a second system is activated, i.e., the hypothalamus-pituitary-adrenal (HPA) axis. Information about the stressful situation is funneled through the paraventricular nucleus of the hypothalamus (PVN), inducing the release of corticotropin-releasing hormone (CRH) – and to a lesser degree of vasopressin – from the median eminence into portal vessels surrounding the anterior pituitary gland. Corticotrophic cells in the anterior pituitary respond to CRH and vasopressin by releasing adrenocorticotropin releasing hormone (ACTH) into the circulation. In the adrenal cortex, ACTH stimulates the production of cortisol (the predominant adrenocortical hormone in humans) or corticosterone (the main hormone in most

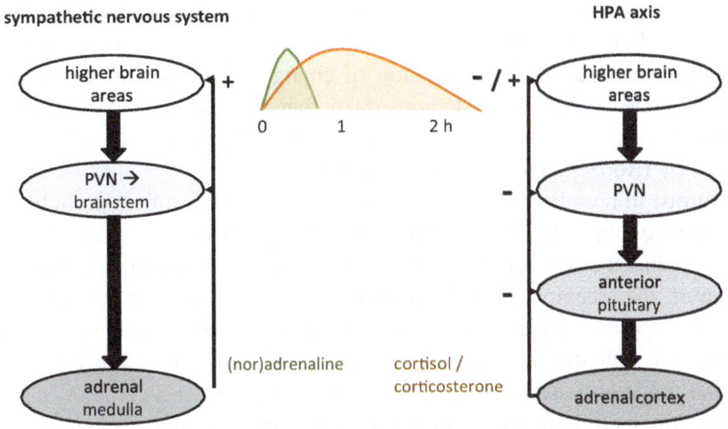

Fig. 1 Schematic representation of the hormonal systems activated after stress. Activation of the sympathetic nervous system causes the release of adrenaline. The end product of the hypothalamus-pituitary-adrenal (HPA) axis is cortisol (in humans) and corticosterone (in rodents). Adrenaline indirectly causes the release of noradrenaline in the brain. Corticosteroid hormones negatively feedback primarily on the hypothalamus and to a lesser degree on the anterior pituitary and higher brain areas. As a consequence of the hormone systems activated after stress, the brain is exposed to waves of hormones, i.e., a quick and short-lasting wave of (nor)adrenaline (in green) and a slower and longer-lasting wave of corticosteroids (in orange) which is normalized after approximately 2 h

rodents) which is then released into the bloodstream. Like adrenaline, cortisol and corticosterone affect numerous peripheral organs to promote and replenish resources that allow the individual to face the stressor. Corticosteroid hormones are lipophilic and therefore easily enter the brain where they reach virtually all cells. In the PVN – and to a lesser extent the anterior pituitary and higher brain regions – corticosteroids negatively impact cellular activity, thus turning down the release of CRH, ACTH, and eventually corticosteroid production. This occurs approximately 2 h after the onset of the stressful situation. Overall after stress, cells in the body, including the brain, are exposed to consecutive waves of stress hormones (Fig. 1). The corticosteroid wave occurs on top of circadian variations – which are determined by underlying ultradian pulses (Lightman et al. 2008), with a peak just before awakening and a nadir at the end of the active phase. The amount of corticosteroids reaching brain cells also depends on other factors, e.g., the expression of p-glycoproteins which determine the transport of particularly cortisol over the blood-brain barrier and the cellular plasma membrane (Pariante 2008).

1.2 *Stress Hormone Receptors*

Noradrenaline binds to G-protein-coupled receptors in the plasma membrane and, through this pathway, exerts its actions in seconds to minutes. Secondarily to these rapid actions, noradrenaline can also affect the transcriptional machinery and hence alter neuronal function over the course of hours. Although probably all adrenoceptor subtypes are involved in the mediation of changes in brain function after stress, pharmacological studies have demonstrated that particularly β-adrenoceptors are important for stress-induced effects on memory formation of adverse events (reviewed by (Roozendaal and McGaugh 2011)).

In contrast to noradrenaline – and peptides like CRH – corticosteroid hormones bind to intracellularly located receptors, which in their inactive state are bound to various proteins including heat shock protein 90. Upon binding of the hormone to the receptor, the receptor complex dissociates, and the hormone-receptor molecule translocates to the nuclear compartment. The activated receptor binds as a homodimer to palindromic recognition sites in the DNA of responsive genes and affects the transcriptional activity of that particular gene.

In addition, it has been shown that activated receptor monomers can bind to other transcriptional regulators and, in this indirect manner, change gene transcription. It has been shown that approximately 1–2% of all genes are potentially altered in their transcriptional activity when exposed to corticosteroids, which explains the pleiotropic action of the hormone. The transcriptional activity not only depends on the expression of receptors but is also determined by local expression of cofactors (Meijer 2002; Meijer et al. 2019). Due to these genomic signaling pathways, corticosteroid hormones generally exert actions that start with a delay of >1 h yet can last for hours to days. More recently, it has become evident that in some cases, corticosteroids can also evoke rapid effects, i.e., within minutes (Joëls et al. 2012). Presumably, these actions are mediated by receptors located in the vicinity of the plasma membrane and do not involve transcription and translation. Whether the receptor molecules mediating these rapid effects form a pool separate from the intracellular receptors has not been resolved to date.

Two types of corticosteroid receptors have been identified (for review see De Kloet et al. 2005). First, the high-affinity mineralocorticoid receptor (MR), which is identical to the receptor expressed in the kidney. Cells in the kidney express 11-β-hydroxysteroid dehydrogenase 2 (11βHSD2) which converts corticosterone and cortisol into their inactive 11-keto congeners (Seckl 2004). This allows the less prevalent adrenal hormone aldosterone to bind with high affinity to the MR, thus exerting its role in the maintenance of the mineral balance. Most cells in the brain, by contrast, express 11βHSD1 rather than 11βHSD2, which promotes the recovery of corticosterone and cortisol. Most brain MRs, therefore, bind corticosterone or cortisol instead of aldosterone; the affinity for corticosterone and cortisol is very high, causing the MR to be mostly in its active state, even with low (nadir) levels of corticosterone or cortisol. Expression of MRs is particularly high in all hippocampal subfields, in the lateral septum, in some motor nuclei in the brain stem, and, to a lesser degree, in amygdalar nuclei and neocortical layers. The second receptor type is the glucocorticoid receptor (GR), which is much more ubiquitously expressed, in neurons and glial cells. Neurons in the CA1 hippocampal region, the dentate gyrus, and classical feedback regions like the PVN contain high levels of GR. The GR has a tenfold lower affinity than MR for corticosterone and cortisol. In some cells that express both MR and GR, e.g., hippocampal CA1 neurons, basal levels of corticosteroid hormones – such as circulate under rest at the circadian nadir – result in substantial activation of MR, while GR is only partly activated. Upon stress, the remainder of the available GR will be activated. These cells "shuttle" between a situation of predominant MR activation and a situation where both receptor types are substantially activated.

2 Stress Hormone Actions on the Brain in Healthy Individuals

2.1 Cellular Effects of Stress Hormones on Brain Circuits

As pointed out above, noradrenaline is released in specific pathways in the brain and locally exerts primarily rapid actions through pre- and/or postsynaptically localized receptors. Via β-adrenoceptors, neurons are mostly quickly excited (reviewed in (Joëls et al. 2012)), but the exact effect after stress depends on the concentration of noradrenaline, the local expression of receptor subtypes, and other processes such as reuptake. CRH acts in a similar fashion, through G-protein-coupled receptors. The expression of CRH receptor subtypes (CRH-R1 and CRH-R2) in particular cell populations but also, e.g., the availability of CRH-binding proteins, determines the overall outcome (Joëls and Baram 2009).

In limbic neurons, such as in the CA1 hippocampal area, dentate gyrus, and basolateral amygdala, corticosterone was shown to rapidly increase spontaneous glutamatergic transmission, via a nongenomic MR-dependent pathway (Joëls et al. 2012). These rapid nongenomic effects of corticosteroids are state- and region-dependent. Thus, basolateral amygdala neurons of mice that were earlier stressed respond to corticosterone with a rapid nongenomic *suppression* of glutamate transmission, which involves GR. Similarly, parvocellular neurons in the PVN show a rapid nongenomic GR-dependent suppression of glutamatergic transmission (see Levy and Tasker 2012).

Rapid non-genomic signaling in hippocampal cells is complemented by slow-onset gene-dependent effects, which in all cases investigated involved the GR. These GR effects promote the signal-to-noise ratio, by enhancing specific glutamatergic signals while suppressing background activity of neurons. The latter has not only been described for hippocampal CA1 (and CA3) neurons but also principal neurons in the prefrontal cortex.

The waves of stress hormones to which neurons are exposed partly overlap in time, which could mean that cells are subject to concomitant actions of several stress hormones. In addition, consecutive waves may affect each other's action. A clear example is the effect of stress hormones on neurons in the basolateral amygdala (Karst and Joëls 2016). These neurons respond in vitro to a moderate to high concentration of the β-adrenoceptor agonist isoproterenol with a short-lived burst of glutamate-mediated excitatory activity, followed approximately 1 h later by suppression of glutamate signaling. However, if the wave of isoproterenol was followed 20 min later by a wave of corticosterone (at a high concentration), the secondary inhibitory phase did not occur; instead, a prolonged period of excitation was observed: The delayed brake on amygdalar activity was in this case "overruled" by the subsequent wave of corticosterone.

All in all, waves of stress hormones change brain function in a time-, region-, and receptor-dependent manner. To what extent local receptors are being activated depends on the type and severity of the stressor. Although it is not simple to translate this body of knowledge to an overall picture how stress affects entire brain circuits, it nevertheless did guide an extensive series of studies examining time-dependent effects of cortisol and/or stress on cognitive function in rodents and humans.

2.2 Neuronal Circuits and Cognitive Function

To test time-dependent effects of stress hormones on cognitive function, one can administer, e.g., yohimbine (which indirectly causes the release of noradrenaline in the brain), hydrocortisone (in humans), or corticosterone (in rodents), and test changes in cognition after various intervals. Preferably, the effect of stress (rather than exogenous hormone exposure) is studied. To examine rapid nongenomic and delayed gene-mediated effects, respectively, circuit activity and cognitive function were probed directly after stress exposure or >1 h later, allowing to distinguish between rapid nongenomic and delayed gene-mediated actions, respectively. To focus on the contribution of a particular receptor type, exogenous hormone administration or stress was in some cases combined with pretreatment with specific receptor antagonists or, in the case of rodents, with the use of brain-selective receptor knockout.

Functional neuroimaging and behavioral studies revealed that directly after stress, the activity of the salience network (SN), and particularly of the amygdala, is strongly increased, involving β-adrenoceptors (Hermans et al. 2011). Interestingly, stress also increases the connection of the amygdala nuclei with striatal areas, at the cost of pathways to higher brain areas such as the hippocampus (Vogel et al. 2016; Schwabe 2017). This connection with the striatum was shown to depend on MR function and to be required for optimal behavioral performance under stress. The rapid β-adrenergic and MR-dependent phase is important for alertness, vigilance, emotion, and rapid decisions necessary for the immediate survival of stressful situations.

These rapid actions are complemented by effects that appear with a delay of at least 1 h. At that time, amygdala activity is normalized (or even suppressed), while behavior involving the hippocampus or frontal cortex is facilitated compared to non-stressed controls. Rational decision-making, driven by the executive control network (ECN), is improved. In rodents, these later behavioral effects were found to depend on GR; this has not been tested specifically in humans to date. Behaviorally,

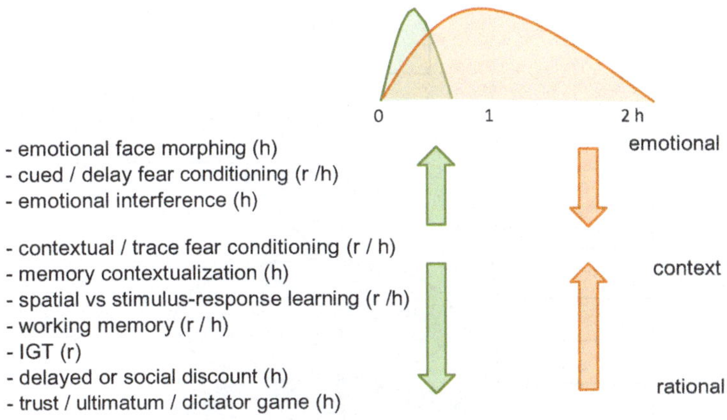

Fig. 2 Summary of behavioral observations in rodents (r) and human subjects (h) directly after stress/corticosteroid administration (rapid) and >1 h after stress/corticosteroid administration (delayed). The tests are arranged from those involving primarily amygdalar/striatal circuits (top), through hippocampal circuits (middle) to prefrontal circuits (bottom). Directly after stress monoamines (green) and corticosteroids acting primarily via MR promote emotional processing, at the cost of higher cognitive functions such as contextual memory formation or reward-based decision-making. At a longer interval >1 h after stress or corticosteroid administration (orange), the reverse is seen. *IGT* Iowa gambling task

this late phase is associated with stronger (than in control conditions) contextualization of information and more rational/less emotionally driven decision-making. Late, as opposed to early, effects of stress were also shown to promote altruistic behavior. In general, delayed actions of stress and corticosteroids promote cognitive processes that are beneficial for the future survival of the individual.

Figure 2 summarizes the time-dependent effects of stress and/or corticosteroid hormones on cognitive function. It is assumed that a good balance between both phases is important for optimal survival of individuals in the face of stress, i.e., one needs to react quickly for immediate survival but also needs storage, contextualization, and rationalization of stressful information to restore cognitive performance and to be well-prepared for similar challenging conditions in the future.

3 Changes in Stress Responsiveness in Bipolar Disorder

3.1 Imbalance in the Stress System: Importance of Genetic and (Early) Life History

The effect of stress hormones depends on many factors, including the age, sex, health, and medication of the individual. Also, the genetic background is of eminent importance. For instance, multiple single nucleotide polymorphisms (SNPs) have been reported for both MR and GR; some combinations are inherited as haplotypes

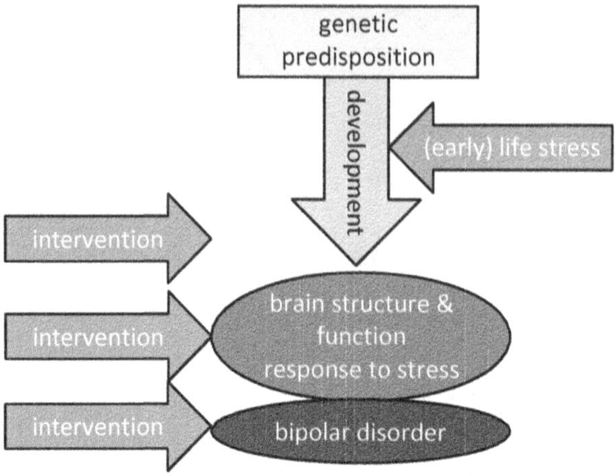

Fig. 3 Development of the brain and the stress system is determined to an important extent by an individual's genetic background. The effect of small genetic variances that predispose to a hyper- or hyporesponsive stress system may be amplified in the face of multiple life events, especially when these are experienced early in life when both the stress system and brain connections are still developing. This may have lasting consequences for brain structure and function as well as the response of individuals to stress (and the impact thereof on the brain). These changes may form an added risk factor for the precipitation of bipolar disorder in genetically vulnerable individuals. Based on these insights, intervention aimed at normalization of the stress system could be effective

(Koper et al. 2014; ter Heegde et al. 2015). Particular SNPs or haplotypes may be associated with a higher incidence of psychopathology. Given the role of corticosteroids in the negative feedback loop, it is to be expected that such genetic variants also affect the stress-induced release of cortisol and hence the exposure of the brain to waves of hormones after stress. For instance, a low functionality of the GR may cause insufficient feedback, resulting in a prolonged wave of cortisol after stress. Such elevated hormone levels, in turn, may downregulate receptor expression, which further exacerbates the insufficient feedback. This may eventually result in a vicious circle. A mild genetic predisposition to less (or more) functional GR or MR may become amplified when life events frequently challenge the stress system of the individual. Especially when these events take place during the sensitive period of development, this may not only have an impact on a stress system that has not yet stabilized but also on brain circuits that are still being shaped. Over (or under) exposure of the brain to corticosteroids during that time may have lasting consequences for the balance in the stress system and the way in which brain regions are interconnected and thus for their role in cognitive performance (see Fig. 3).

In agreement with this view, adverse conditions experienced early in life are generally reported to be a risk factor for the development of brain disorders, including psychiatric disorders such as bipolar disorder or schizophrenia (Turecki et al. 2014; Jawahar et al. 2015; Cancel et al. 2019).

In the following sections, we first summarize the current evidence for the altered function of the HPA axis in bipolar disorder and next highlight how the disorder affects time-dependent cognitive effects of stress. We compare the characteristics reported for bipolar disorder (BD) with those seen in relation to schizophrenia.

3.2 Changes in the HPA Axis in Bipolar Disorder Patients

A common finding in psychiatry is the disruption of the HPA axis in mood disorders, with most studies reporting hypercortisolemia in depressed patients (Gold et al. 1988; Gillespie and Nemeroff 2005). As for patients diagnosed with BD, high levels of basal cortisol were previously reported to be prevalent when the measurements were taken in the morning, before 10:00 AM (Girshkin et al. 2014). The difference between cortisol levels in BD patients compared to controls is even more pronounced when the cortisol awakening response (CAR) is determined (Belvederi Murri et al. 2016). However, basal cortisol levels were also found to be augmented in the afternoon and night hours, as recently demonstrated by a meta-analysis performed by Belvederi Murri et al. (2016) including 37 studies on basal cortisol levels in BD patients.

It would be an oversimplification to determine stress (cortisol levels) in BD patients regardless of the phase of the disease in which the samples were taken. Therefore, Belvedere Murri and colleagues also assessed effects sizes (Hedges's g) of basal cortisol levels (at various moments of the day) during the different phases of the illness. The subgroup analyses showed that in BD patients during the depressive episodes, hypercortisolemia were only evident in the studies which samples were continuously obtained throughout the day (Hedges's $g = 0.44$). The studies in which samples were obtained during the euthymic phase indicated significant high cortisol levels at wakening ($g = 0.59$), morning ($g = 0.41$), afternoon ($g = 0.31$), and 24-h continuous sampling ($g = 0.28$). When samples were obtained during the manic phase, significant hypercortisolemia was found at morning ($g = 0.66$), night ($g = 0.15$), and 24-h continuous sampling ($g = 0.64$). Additionally, meta-regression analyses pointed out that assessing basal cortisol levels during the manic phase was associated with higher effects sizes (Belvederi Murri et al. 2016).

Although the literature points to increased cortisol levels in BD patients under basal conditions, there might also be differences during real-life stress, but cortisol levels are difficult to determine and standardize under such conditions. Nevertheless, one study did assess cortisol level in response to negative life events and found no significant differences between BD and controls (Havermans et al. 2011). There are several studies suggesting that under a controlled stressful laboratory situation, such as the Trier social stress test (Kirschbaum et al. 1993), the cortisol response of BD patients is *blunted* compared to controls (Wieck et al. 2013; Houtepen et al. 2015). Houtepen and colleagues found that from 20 to 90 min after the TSST, BD patients presented a blunted cortisol response compared to healthy controls, whereas BD patients' unaffected siblings showed a cortisol response that was similar to the healthy controls.

As high levels of basal cortisol are observed in BD patients, this could point to impaired negative feedback control of the HPA axis possibly due to diminished GR sensitivity, as previously indicated in pharmacological studies (Schmider et al. 1995; Rush et al. 1996; Rybakowski and Twardowska 1999; Watson et al. 2004). The blunted response to the TSST does not seem to support this idea. Yet Houtepen et al. (2015) argued that the chronically elevated cortisol levels of BD patients might gradually impair the HPA axis' ability to respond to stress, eventually resulting in a blunted stress response. Unfortunately, studies that closely follow the functionality of the stress response and the development of BD are scarce.

Many BD patients are under medication. Houtepen et al. (Houtepen et al. 2015) showed that BP patients under antipsychotics treatment presented a flat cortisol response to the TSST compared to BD patients under non-antipsychotics (anticonvulsants, benzodiazepines, and antidepressants) treatment and healthy controls, suggesting a substantial influence of antipsychotics on the stress response measured under these circumstances. Interestingly, prior to the start of the TSST in this study, no antipsychotic effects were observed, suggesting that they might exert no effect on basal cortisol level. However, a meta-regression analysis indicated that the use antipsychotics is associated with a reduction of basal hypercortisolemia in BD patients compared to controls; no association was observed between basal cortisol in BD patients and antidepressants, lithium, or mood stabilizers (Belvederi Murri et al. 2016). The latter finding is in line with a study by Girshkin et al. (2014), who evaluated morning cortisol data of BD patients who were under treatment mostly with antidepressants and mood stabilizers (or medication-free) and found no significant differences compared to controls.

Antipsychotics are mostly prescribed for schizophrenia (SZ), another psychiatric condition in which the HPA axis is dysregulated (Brenner et al. 2009). Similar to what was found in BD patients, SZ patients also present high basal cortisol levels in early stages of the disease (Chaumette et al. 2016), high morning cortisol levels (Girshkin et al. 2014), and a blunted cortisol response to social stress (Ciufolini et al. 2014; Zorn et al. 2017). However, in contrast to BD patients, SZ patients show a reduced CAR compared to controls (Berger et al. 2016). Data on cortisol levels in BD and SZ patients are summarized in Table 1.

4 Changes in Cognitive Function in Bipolar Disorder Related to Stress

As described in Sect. 2, glucocorticoids modulate neuronal activity and network function and, when chronically elevated, also morphology, as was shown already some decades ago in preclinical experiments (Magariños and McEwen 1995; Magariños et al. 1996). Neurons in the medial prefrontal cortex are also sensitive

Table 1 Overview of cortisol levels in bipolar disorder and schizophrenic patients compared to controls under basal condition and in response to awakening social stress

Cortisol levels	BD vs controls	SZ vs controls	Type of study	Reference
Basal (overall)	Augmented[1]	Augmented[2]	[1, 2]Meta-analysis	[1]Belvederi Murri et al. (2016); [2]Chaumette et al. (2016)
Basal (morning)	Augmented[1]	Augmented[1]	[1]Meta-analysis	[1]Girshkin et al. (2014)
Basal (afternoon)	Augmented[1]	N/A	[1]Meta-analysis	[1]Belvederi Murri et al. (2016)
Basal (night)	Augmented	N/A	[1]Meta-analysis	[1]Belvederi Murri et al. (2016)
Cortisol awakening response (CAR)	Augmented[1]	Diminished[1]	[1]Meta-analysis	[1]Berger et al. (2016)
Stress-induced cortisol (TSST)	Blunted[1, 2]	Blunted[3, 4]	[1, 2]Original research [3, 4]Meta-analysis	[1]Wieck et al. (2013), [2]Houtepen et al. (2015), [3]Ciufolini et al. (2014), [4]Zorn et al. (2017)
Illness phase and medication effects on cortisol levels in BD patients	**BD vs controls**	**SZ vs controls**	**Type of study**	**Reference**
Depressive	Augmented[1] (only if 24 h measurements were taken)	N/A	[1]Meta-analysis	[1]Belvederi Murri et al. (2016)
Euthymia	Augmented[1]	N/A	[1]Meta-analysis	[1]Belvederi Murri et al. (2016)
Mania	Augmented[1]	N/A	[1]Meta-analysis	[1]Belvederi Murri et al. (2016)
Medication (antipsychotics)	Blunted TSST response[1], lower hypercortisolemia[2]	N/A	[1]Original Research [2]Meta-Analysis	[1]Houtepen et al. (2015), [2]Belvederi Murri et al. (2016)

Illness phases and medication effects on cortisol levels are also shown for bipolar patients. Most of the data summarized in this table were obtained from meta-analyses published in the last few years
Superscript numbers indicate from which studies data were obtained

to high glucocorticoid levels and, similar to CA3 neurons, show decreased dendritic complexity even after a mild stress paradigm (Brown et al. 2005). In contrast, pyramidal and stellate neurons in the amygdala, as well as pyramidal neurons in the orbitofrontal cortex, show increased dendritic arborization (Vyas et al. 2002; Mitra et al. 2005). Overexposure to glucocorticoids as generally seen in BD patients – at least under basal (non-stressed) conditions – might thus result in altered morphology and function of limbic structures such as the hippocampus and amygdala, areas that are important for (emotional) memory. In agreement, hippocampal atrophy has been found in BD patients, particularly the CA3 area of the hippocampus (Bertolino et al. 2003; Konradi et al. 2011; Mathew et al. 2014).

4.1 Time-Dependent Changes in Cognitive Processing Following Stress in BD Patients

Given the changes observed in BP patients regarding circulating cortisol levels and the potential consequences thereof for neuronal function and morphology, one might expect that neuronal circuits are altered as well as the cognitive processes for which these circuits are crucial. BD is indeed related to dysfunction of several brain networks, including the salience network, executive control network, the default mode network, and the reward circuitry (Bora et al. 2010; Mamah et al. 2013; Whitton et al. 2015; Goya-Maldonado et al. 2016; Schreiter et al. 2016; Gong et al. 2019; Karcher et al. 2019). We will first discuss the evidence for disturbed reward processing in BD patients, under basal and stress conditions.

Reward processing is affected during the acute and recovery phase of stress. During and directly after stress in healthy controls, reward-seeking is increased and anticipatory responses in the brain's reward circuitry amplified, including in the orbitofrontal cortex and ventral striatum (Kumar et al. 2014; Lewis et al. 2014); responses to reward consumption are reduced (Bogdan and Pizzagalli 2006; Porcelli et al. 2012; Berghorst et al. 2013; Kumar et al. 2014). By contrast, in the late aftermath of stress – i.e., during stress recovery – anticipatory reward responses are reduced (Montoya et al. 2014), whereas neural responses to reward outcomes are increased (van Leeuwen et al. 2019a, 2019b).

In BD patients, though, these acute and late effects of stress on the reward circuitry are absent (see Fig. 4). More specifically, reward-related responses were not decreased directly after stress (Berghorst et al. 2016) and were not increased in the aftermath of stress (van Leeuwen et al. 2019b). These findings suggest that neurobiological adjustments after stress are necessary for the maintenance of good mental health. Frequent exposure to stressors may ultimately lead to generally impaired reward processing. Indeed, the number of stressful life events was negatively associated with striatal responses after stress (van Leeuwen et al. 2019b). Reduced upregulation of the reward circuitry following stress was also observed in siblings of schizophrenia patients (van Leeuwen et al. 2019a), suggesting diagnosis spanning alterations in the reward circuitry in relation to stress in psychiatric disorders.

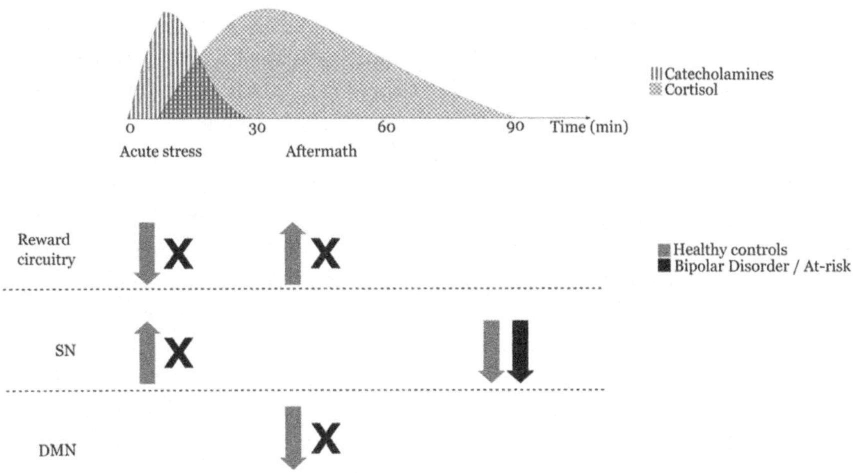

Fig. 4 Model illustrating the time-dependent effects of stress on brain functioning in healthy controls and individuals at risk for/suffering from bipolar disorder. The studies summarized in this review implicate that resilience to stress most likely requires active regulation of neurocognitive processes aimed to effectively cope with and recover from environmental stressors, rather than the absence of a reaction to such stressors. *SN* salience network, *DMN* default mode network

4.2 Network Function in BP Patients and Individuals at Risk for Psychopathology

Connectivity within and between the salience network (SN), executive control network (ECN), and default mode network (DMN) that also play a role in response to stress has consistently been shown to be affected in bipolar disorder, even in the absence of stress (Öngür et al. 2010; Chai et al. 2011; Gong et al. 2019; Sha et al. 2019).

Unfortunately, there are currently no published studies that looked at the effects of stress on these networks in BD patients. However, studies are focusing on the effects of stress in individuals with an *increased familiar risk* for BD. Studies have revealed etiological relationships between schizophrenia and BD and found that family background is the largest risk factor for both disorders (Merikangas et al. 2007). Indeed, relatives of schizophrenia patients have a higher chance to develop BD compared to the general population (Cheng et al. 2017). Given the large heritability of both schizophrenia and BD and their interaction with environmental factors, investigating the neural responses to stress in family members of schizophrenia patients could be a valuable paradigm to investigate the gene-environment interactions in bipolar disorder. Here we summarize previous findings on the effects of stress on these networks in vulnerable individuals.

First, the results on the salience network (SN). The activation of the sympathetic nervous system and the release of (nor)epinephrine during and directly after stress increase connectivity within the SN in healthy individuals, leading to increased threat detection and thereby aiding active coping (van Marle et al. 2009; Oei et al. 2012; Hermans et al. 2014; van Oort et al. 2017). In contrast, SN functional connectivity in siblings of schizophrenia patients is lower during acute stress, indicating impaired detection of relevant stimuli and inadequate response selection under stressful circumstances (van Leeuwen et al. 2018). In addition, salivary alpha-amylase level, an indirect marker of adrenergic activity (Van Stegeren et al. 2006), was higher in at-risk individuals exposed to a placebo test than in healthy controls, which did not further increase in response to stress, suggesting higher chronic levels of norepinephrine. The (late) aftermath of stress is characterized by downregulation of SN activity (Hermans et al. 2014), which did not differ between controls and siblings (van Leeuwen et al. 2018). Overall, these findings suggest reduced responsiveness to changes in the environment, possibly caused by a reduced dynamic range in the sympathetic/noradrenergic response. Evidence that BD patients also display exaggerated adrenergic signaling at trend level (van Leeuwen et al. 2019b) could point to a potentially reduced SN activation after stress as well.

Second, the default mode network (DMN). Previous studies showed that acute stress temporarily increases activity in the DMN, increasing interference from internal emotional states (Qin et al. 2009); deactivation was observed in the aftermath of stress (Van Leeuwen et al. 2018). Several studies have already demonstrated that the inability to downregulate the DMN in the absence of stress is associated with poor clinical outcome in BP and schizophrenia and with rumination in depression (Grimm et al. 2009; Pomarol-Clotet et al. 2012; Bartova et al. 2015; Wang et al. 2017). Moreover, there is one case report in the literature that shows normalization of DMN activity after successful treatment in a BD patient (Landin-Romero et al. 2013), suggesting a role for DMN dysregulation in the course of the disorder. In individuals at risk for schizophrenia, it was indeed found that DMN activity does not decrease in the aftermath of stress (Van Leeuwen et al. 2018). These findings indicate that good mental health requires a dynamic shift away from the DMN in the aftermath of stress. In vulnerable individuals, sustained activity within the DMN may result in increased rumination following stress and maybe a precipitating factor in the development of BD.

Finally, the executive control network (ECN). Only a few studies found altered ECN functional connectivity in BD. Regarding the situation after stress, ECN functional connectivity is reduced during and directly after stress but increased in the (late) aftermath of stress, the latter presumably through genomic actions of cortisol (Arnsten 2009; Qin et al. 2009; Henckens et al. 2011; Hermans et al. 2014). To date, no functional MRI studies have examined the effects of stress on ECN connectivity in BD. We previously observed an increase in functional connectivity between the ECN and the cerebellum in healthy controls but also in individuals at risk for schizophrenia in the aftermath of stress (van Leeuwen et al. 2018). The role of the ECN and its relation to stress in BD require further investigation.

5 Concluding Remarks

In this overview, we highlighted that stress causes consecutive yet overlapping waves of hormones to reach the brain. These hormones change neuronal function in a time-, region-, and receptor-dependent manner. While it is difficult to predict how this affects entire circuits and thereby cognitive processing, an extensive series of studies point to an overarching picture: During and directly after stress, the salience network and default mode network are increased in activity, at the cost of networks involved in contextualization. In the aftermath of stress (starting at least 30 min after stress onset), the earlier activation of these networks is dampened, while networks involved in executive control and contextualization are enhanced in their activity. Both phases are necessary for (quick) correct appraisal of the situation at hand and (later) rational decision-making and context-related storage of information for future use.

If the stress-induced release of hormones is altered, as appears to be the case in BD, this in itself will already change any functional process resulting from hormonal actions. The effects that have been described in terms of cortisol release are not entirely consistent, though. Basal levels generally seem to be increased, yet there may be diminished stress-induced cortisol release, at least under laboratory stress conditions. How BD patients respond to real-life stress is still unresolved. Moreover, cortisol levels are not the only factor determining the effect of stress on brain function. Other factors of relevance comprise, e.g., the extent to which cortisol gets into the brain, what happens to receptor expression, and potential effects downstream of the receptors. In addition, stress causes the release of multiple hormones that potentially interact with each other, and many of these hormones have not (yet) been determined in relation to BD.

Importantly, how HPA function alters along the course of the disease is largely unknown. Possibly, hyperactivity of the system occurs early on in the disorder, which could slowly evolve into hypoactivity in the face of prolonged overexposure of the brain and body to cortisol. Precise measurement of both the stress system and the development of BD is necessary to address this issue. This might also give insight whether HPA disturbances are the cause or consequence of the disorder. There is some evidence for a causal role. Longitudinal studies revealed that in BD patients relapse of depressive and (hypo)manic episodes is preceded by a higher number of stressful life events, compared to euthymic periods (absence of depressive or (hypo)manic symptoms) (Lex et al. 2017). Moreover, patients reported increased emotional reactivity to daily life stressors compared to controls (Myin-Germeys et al. 2003; Havermans et al. 2010). These findings suggest that an impaired initial response to stress and/or a poor recovery from stress may play a role in the development and clinical course of bipolar disorder.

Related to the disease and/or alterations in the stress system, brain circuits may be altered in BD patients. We supplied some evidence for this notion in Sect. 4, although there is clearly a paucity in information, particularly with regard to studies investigating time-dependent effects of stress on cognitive processing in BD patients versus controls. While most human stress studies focus on the acute phase of stress, it is becoming clear that it is important to include all phases of the stress response, particularly the recovery period which has not received much attention. We observed quite striking differences between controls and both at-risk individuals and patients in the early aftermath of stress. In general, healthy controls displayed stress-dependent changes in neuronal activity during cognitive tasks (e.g., emotional information or reward processing), while this was diminished in at-risk individuals and BD patients. We propose that resilience to stress most likely requires active and dynamic regulation of neurocognitive processes aimed to effectively cope with and recover from environmental stressors, rather than the absence of a reaction to such stressors. These findings could provide a starting point for more research elucidating the mechanisms of stress vulnerability in BD, to identify who is at risk, and make the first steps toward preventive rather than reactive therapeutic interventions.

References

Arnsten AFT (2009) Stress signalling pathways that impair prefrontal cortex structure and function. Nat Rev Neurosci 10:410–422

Bartova L, Meyer BM, Diers K et al (2015) Reduced default mode network suppression during a working memory task in remitted major depression. J Psychiatr Res 64:9–18

Belvederi Murri M, Prestia D, Mondelli V et al (2016) The HPA axis in bipolar disorder: systematic review and meta-analysis. Psychoneuroendocrinology 63:327–342

Berger M, Kraeuter AK, Romanik D et al (2016) Cortisol awakening response in patients with psychosis: systematic review and meta-analysis. Neurosci Biobehav Rev 68:157–166

Berghorst LH, Bogdan R, Frank MJ, Pizzagalli DA (2013) Acute stress selectively reduces reward sensitivity. Front Hum Neurosci 7:133

Berghorst LH, Kumar P, Greve DN et al (2016) Stress and reward processing in bipolar disorder: a functional magnetic resonance imaging study. Bipolar Disord 18:602–611

Bertolino A, Frye M, Callicott JH et al (2003) Neuronal pathology in the hippocampal area of patients with bipolar disorder: a study with proton magnetic resonance spectroscopic imaging. Biol Psychiatry 53:906–913

Bogdan R, Pizzagalli DA (2006) Acute stress reduces reward responsiveness: implications for depression. Biol Psychiatry 60:1147–1154

Bora E, Fornito A, Yücel M, Pantelis C (2010) Voxelwise meta-analysis of gray matter abnormalities in bipolar disorder. Biol Psychiatry 67:1097–1105

Brenner K, Liu A, Laplante DP et al (2009) Cortisol response to a psychosocial stressor in schizophrenia: blunted, delayed, or normal? Psychoneuroendocrinology 34:859–868

Brown SM, Henning S, Wellman CL (2005) Mild, short-term stress alters dendritic morphology in rat medial prefrontal cortex. Cereb Cortex 15:1714–1722

Cancel A, Dallel S, Zine A et al (2019) Understanding the link between childhood trauma and schizophrenia: a systematic review of neuroimaging studies. Neurosci Biobehav Rev 107:492–504

Chai XJ, Whitfield-Gabrieli S, Shinn AK et al (2011) Abnormal medial prefrontal cortex resting-state connectivity in bipolar disorder and schizophrenia. Neuropsychopharmacology 36:2009–2017

Chaumette B, Kebir O, Mam-Lam-Fook C et al (2016) Salivary cortisol in early psychosis: new findings and meta-analysis. Psychoneuroendocrinology 63:262–270

Cheng C-M, Chang W-H, Chen M-H et al (2017) Co-aggregation of major psychiatric disorders in individuals with first-degree relatives with schizophrenia: a nationwide population-based study. Mol Psychiatry 23:1756–1763

Ciufolini S, Dazzan P, Kempton MJ et al (2014) HPA axis response to social stress is attenuated in schizophrenia but normal in depression: evidence from a meta-analysis of existing studies. Neurosci Biobehav Rev 47:359–368

De Kloet ER, Joëls M, Holsboer F (2005) Stress and the brain: from adaptation to disease. Nat Rev Neurosci 6:463–475

Gillespie CF, Nemeroff CB (2005) Hypercortisolemia and depression. Psychosom Med 67:S1 26–S1 28

Girshkin L, Matheson SL, Shepherd AM, Green MJ (2014) Morning cortisol levels in schizophrenia and bipolar disorder: a meta-analysis. Psychoneuroendocrinology 49:187–206

Gold PW, Goodwin FK, Chrousos GP (1988) Clinical and biochemical manifestations of depression. N Engl J Med 319:413–420

Gong JY, Chen G, Jia Y et al (2019) Disrupted functional connectivity within the default mode network and salience network in unmedicated bipolar II disorder. Prog Neuro Psychopharmacol Biol Psychiatry 88:11–18

Goya-Maldonado R, Brodmann K, Keil M et al (2016) Differentiating unipolar and bipolar depression by alterations in large-scale brain networks. Hum Brain Mapp 37:808–818

Grimm S, Boesiger P, Beck J et al (2009) Altered negative BOLD responses in the default-mode network during emotion processing in depressed subjects. Neuropsychopharmacology 34:932–943

Havermans R, Nicolson NA, Berkhof J, deVries MW (2010) Mood reactivity to daily events in patients with remitted bipolar disorder. Psychiatry Res 179:47–52

Havermans R, Nicolson NA, Berkhof J, deVries MW (2011) Patterns of salivary cortisol secretion and responses to daily events in patients with remitted bipolar disorder. Psychoneuroendocrinology 36:258–265

Henckens MJAG, van Wingen GA, Joëls M, Fernández G (2011) Time-dependent corticosteroid modulation of prefrontal working memory processing. Proc Natl Acad Sci U S A 108:5801–5806

Herman JP (2018) Regulation of Hypothalamo-pituitary-adrenocortical responses to stressors by the nucleus of the solitary tract/dorsal vagal complex. Cell Mol Neurobiol 38:25–35

Hermans EJ, Van Marle HJF, Ossewaarde L et al (2011) Stress-related noradrenergic activity prompts large-scale neural network reconfiguration. Science 334(80):1151–1153

Hermans EJ, Henckens MJAG, Joëls M, Fernández G (2014) Dynamic adaptation of large-scale brain networks in response to acute stressors. Trends Neurosci 37:304–314

Houtepen LC, Boks MPM, Kahn RS et al (2015) Antipsychotic use is associated with a blunted cortisol stress response: a study in euthymic bipolar disorder patients and their unaffected siblings. Eur Neuropsychopharmacol 25:77–84

Jawahar MC, Murgatroyd C, Harrison EL, Baune BT (2015) Epigenetic alterations following early postnatal stress: a review on novel aetiological mechanisms of common psychiatric disorders. Clin Epigenetics 7:122

Joëls M, Baram TZ (2009) The neuro-symphony of stress. Nat Rev Neurosci 10:459–466

Joëls M, Sarabdjitsingh RA, Karst H (2012) Unraveling the time domains of corticosteroid hormone influences on brain activity: rapid, slow, and chronic modes. Pharmacol Rev 64:901–938

Karcher NR, Rogers BP, Woodward ND (2019) Functional connectivity of the striatum in schizophrenia and psychotic bipolar disorder. Biol Psychiatry Cogn Neurosci Neuroimaging 4:956–965

Karst H, Joëls M (2016) Severe stress hormone conditions cause an extended window of excitability in the mouse basolateral amygdala. Neuropharmacology 110:175–180

Kirschbaum C, Pirke KM, Hellhammer DH (1993) The "Trier social stress test" – a tool for investigating psychobiological stress responses in a laboratory setting. Neuropsychobiology 28:76–81

Konradi C, Zimmerman EI, Yang CK et al (2011) Hippocampal interneurons in bipolar disorder. Arch Gen Psychiatry 68:340–350
Koper JW, van Rossum EFC, van den Akker ELT (2014) Glucocorticoid receptor polymorphisms and haplotypes and their expression in health and disease. Steroids 92:62–73
Kumar P, Berghorst LH, Nickerson LD et al (2014) Differential effects of acute stress on anticipatory and consummatory phases of reward processing. Neuroscience 266:1–12
Landin-Romero R, Novo P, Vicens V et al (2013) EMDR therapy modulates the default mode network in a subsyndromal, traumatized bipolar patient. Neuropsychobiology 67:181–184
Levy BH, Tasker JG (2012) Synaptic regulation of the hypothalamic-pituitary-adrenal axis and its modulation by glucocorticoids and stress. Front Cell Neurosci 6:24
Lewis AH, Porcelli AJ, Delgado MR (2014) The effects of acute stress exposure on striatal activity during Pavlovian conditioning with monetary gains and losses. Front Behav Neurosci 8:179
Lex C, Bäzner E, Meyer TD (2017) Does stress play a significant role in bipolar disorder? A meta-analysis. J Affect Disord 208:298–308
Lightman SL, Wiles CC, Atkinson HC et al (2008) The significance of glucocorticoid pulsatility. Eur J Pharmacol 583:255–262
Magariños AM, McEwen BS (1995) Stress-induced atrophy of apical dendrites of hippocampal CA3c neurons: comparison of stressors. Neuroscience 69:83–88
Magariños AM, McEwen BS, Flügge G, Fuchs E (1996) Chronic psychosocial stress causes apical dendritic atrophy of hippocampal CA3 pyramidal neurons in subordinate tree shrews. J Neurosci 16:3534–3540
Mamah D, Barch DM, Repovš G (2013) Resting state functional connectivity of five neural networks in bipolar disorder and schizophrenia. J Affect Disord 150:601–609
Mathew I, Gardin TM, Tandon N et al (2014) Medial temporal lobe structures and hippocampal subfields in psychotic disorders: findings from the bipolar-schizophrenia network on intermediate phenotypes (B-SNIP) study. JAMA Psychiat 71:769–777
Meijer OC (2002) Coregulator proteins and corticosteroid action in the brain. J Neuroendocrinol 14:499–505
Meijer OC, Buurstede JC, Schaaf MJM (2019) Corticosteroid receptors in the brain: transcriptional mechanisms for specificity and context-dependent effects. Cell Mol Neurobiol 39:539–549
Merikangas KR, Akiskal HS, Angst J et al (2007) Lifetime and 12-month prevalence of bipolar spectrum disorder in the national comorbidity survey replication. Arch Gen Psychiatry 64:543–552
Mitra R, Jadhav S, McEwen BS et al (2005) Stress duration modulates the spatiotemporal patterns of spine formation in the basolateral amygdala. Proc Natl Acad Sci U S A 102:9371–9376
Montoya ER, Bos PA, Terburg D et al (2014) Cortisol administration induces global down-regulation of the brain's reward circuitry. Psychoneuroendocrinology 47:31–42
Myin-Germeys I, Peeters F, Havermans R et al (2003) Emotional reactivity to daily life stress in psychosis and affective disorder: an experience sampling study. Acta Psychiatr Scand 107:124–131
Oei NYL, Veer IM, Wolf OT et al (2012) Stress shifts brain activation towards ventral "affective" areas during emotional distraction. Soc Cogn Affect Neurosci 7:403–412
Öngür D, Lundy M, Greenhouse I et al (2010) Default mode network abnormalities in bipolar disorder and schizophrenia. Psychiatry Res Neuroimaging 183:59–68
Pariante CM (2008) The role of multi-drug resistance p-glycoprotein in glucocorticoid function: studies in animals and relevance in humans. Eur J Pharmacol 583:263–271
Pomarol-Clotet E, Moro N, Sarró S et al (2012) Failure of de-activation in the medial frontal cortex in mania: evidence for default mode network dysfunction in the disorder. World J Biol Psychiatry 13:616–626
Porcelli AJ, Lewis AH, Delgado MR (2012) Acute stress influences neural circuits of reward processing. Front Neurosci 6:157
Qin S, Hermans EJ, van Marle HJF et al (2009) Acute psychological stress reduces working memory-related activity in the dorsolateral prefrontal cortex. Biol Psychiatry 66:25–32

Roozendaal B, McGaugh JL (2011) Memory modulation. Behav Neurosci 125:797–824
Rush AJ, Giles DE, Schiesser MA et al (1996) The dexamethasone suppression test in patients with mood disorders. J Clin Psychiatry 57:470–484
Rybakowski JK, Twardowska K (1999) The dexamethasone/corticotropin-releasing hormone test in depression in bipolar and unipolar affective illness. J Psychiatr Res 33:363–370
Schmider J, Lammers CH, Gotthardt U et al (1995) Combined dexamethasone/corticotropin-releasing hormone test in acute and remitted manic patients, in acute depression, and in normal controls: I. Biol Psychiatry 38:797–802
Schreiter S, Spengler S, Willert A et al (2016) Neural alterations of fronto-striatal circuitry during reward anticipation in euthymic bipolar disorder. Psychol Med 46:3187–3198
Schwabe L (2017) Memory under stress: from single systems to network changes. Eur J Neurosci 45:478–489
Seckl JR (2004) 11β-hydroxysteroid dehydrogenases: changing glucocorticoid action. Curr Opin Pharmacol 4:597–602
Sha Z, Wager TD, Mechelli A, He Y (2019) Common dysfunction of large-scale neurocognitive networks across psychiatric disorders. Biol Psychiatry 85:379–388
ter Heegde F, De Rijk RH, Vinkers CH (2015) The brain mineralocorticoid receptor and stress resilience. Psychoneuroendocrinology 52:92–110
Turecki G, Ota VK, Belangero SI et al (2014) Early life adversity, genomic plasticity, and psychopathology. Lancet Psychiatry 1:461–466
van Leeuwen JMC, Vink M, Fernández G, Hermans EJ, Joëls M, Kahn RS, Vinkers CH (2018) At-risk individuals display altered brain activity following stress. Neuropsychopharmacology. 43(9):1954–1960
van Leeuwen JMC, Vink M, Joëls M et al (2019a) Increased responses of the reward circuitry to positive task feedback following acute stress in healthy controls but not in siblings of schizophrenia patients. NeuroImage 184:547–554
van Leeuwen JMC, Vink M, Joëls M et al (2019b) Reward-related striatal responses following stress in healthy individuals and patients with bipolar disorder. Biol Psychiatry Cogn Neurosci Neuroimaging 4:966–974
van Marle HJF, Hermans EJ, Qin S, Fernández G (2009) From specificity to sensitivity: how acute stress affects amygdala processing of biologically salient stimuli. Biol Psychiatry 66:649–655
van Oort J, Tendolkar I, Hermans EJ et al (2017) How the brain connects in response to acute stress: a review at the human brain systems level. Neurosci Biobehav Rev 83:281–297
Van Stegeren A, Rohleder N, Everaerd W, Wolf OT (2006) Salivary alpha amylase as marker for adrenergic activity during stress: effect of betablockade. Psychoneuroendocrinology 31:137–141
Vogel S, Fernández G, Joëls M, Schwabe L (2016) Cognitive adaptation under stress: a case for the mineralocorticoid receptor. Trends Cogn Sci 20:192–203
Vyas A, Mitra R, Shankaranarayana Rao BS, Chattarji S (2002) Chronic stress induces contrasting patterns of dendritic remodeling in hippocampal and amygdaloid neurons. J Neurosci 22:6810–6818
Wang Y, Tang W, Fan X et al (2017) Resting-state functional connectivity changes within the default mode network and the salience network after antipsychotic treatment in early-phase schizophrenia. Neuropsychiatr Dis Treat 13:397–406
Watson S, Gallagher P, Ritchie JC et al (2004) Hypothalamic-pituitary-adrenal axis function in patients with bipolar disorder. Br J Psychiatry 184:496–502
Whitton AE, Treadway MT, Pizzagalli DA (2015) Reward processing dysfunction in major depression, bipolar disorder and schizophrenia. Curr Opin Psychiatry 28:7–12
Wieck A, Grassi-Oliveira R, do Prado CH et al (2013) Differential neuroendocrine and immune responses to acute psychosocial stress in women with type 1 bipolar disorder. Brain Behav Immun 34:47–55
Zorn JV, Schür RR, Boks MP et al (2017) Cortisol stress reactivity across psychiatric disorders: a systematic review and meta-analysis. Psychoneuroendocrinology 77:25–36

The Role of Genetics in Bipolar Disorder

Chiara Fabbri

Contents

Abstract Bipolar disorder (BP) is a highly heritable disease, with heritability estimated between 60 and 85% by twin studies. The underlying genetic architecture was poorly understood for years since the available technology was limited to the candidate gene approach that did not allow to explore the contribution of multiple loci throughout the genome. BP is a complex disorder, which pathogenesis is influenced by a number of genetic variants, each with small effect size, and environmental exposures. Genome-wide association studies (GWAS) provided meaningful insights into the genetics of BP, including replicated genetic variants, and allowed the development of novel multi-marker methods for gene/pathway analysis and for estimating the genetic overlap between BP and other traits. However, the existing GWAS had also relevant limitations. Notably insufficient statistical power and lack of consideration of rare variants, which may be responsible for the relatively low heritability explained (~20% in the largest GWAS) compared to twin studies.

C. Fabbri (✉)
Social, Genetic and Developmental Psychiatry Centre, Institute of Psychiatry, Psychology and Neuroscience, King's College London, London, UK
e-mail: chiara.fabbri@kcl.ac.uk

Curr Topics Behav Neurosci (2021) 48: 41–60
https://doi.org/10.1007/7854_2020_153
Published Online: 7 August 2020

The availability of data from large biobanks and automated phenotyping from electronic health records or digital phenotyping represent key steps for providing samples with adequate power for genetic analysis. Next-generation sequencing is becoming more and more feasible in terms of costs, leading to the rapid growth in the number of samples with whole-genome or whole-exome sequence data. These recent and unprecedented resources are of key importance for a more comprehensive understanding of the specific genetic factors involved in BP and their mechanistic action in determining disease onset and prognosis.

Keywords Bipolar disorder · Gene · Genetics · Genome-wide association study · Heritability · Polygenic risk score · Polymorphism

1 Introduction: Why Genetics Matters in the Susceptibility to Bipolar Disorder?

Each human being is different from another one in many different ways, and it is such a common experience that we probably do not spend much time thinking about the underlying reasons. However, that diversity has fascinating biological bases explaining its reason for being there: nature has favored flexibility and change in order to make life adaptable to the environment. The way we respond to environmental stimuli is dependent from our genes, or, in other words, we all start our lives with a certain genetic makeup that defines our potential in terms of personality, intelligence, body shape, disease risk, and so on, and then the environment acts on this potential and leads to different outcomes in each one of us, at least for most of the traits. The percentage of contribution coming from genes is indeed variable, depending on the condition (trait or phenotype) that we consider. Some traits have complete penetrance, meaning that 100% of the subjects carrying a certain genetic variant or polymorphism exhibit the trait, independently from environmental exposures. For example, a specific polymorphism in the fibroblast growth factor receptor 3 (*FGFR3*) gene always results in achondroplasia, a disorder characterized by dwarfism. However, most of the traits have incomplete penetrance and variable expressivity, meaning that not all the subjects carrying certain genetic variant (s) manifest the trait, and in those who manifest it, there are different degrees of expressivity. These phenomena are caused not only by interactions with the environment but also by interactions with modifier genes (Lobo 2008). Traits showing incomplete penetrance and variable expressivity are indeed typically influenced by a number of genetic variants, and they are defined as complex or polygenic. Examples of this type of trait are intelligence, height, weight, and psychiatric disorders (Plomin and Deary 2015). Intelligence, height, and weight are normally distributed and can be quite easily interpreted as a result of the adaptation to the requests of different environments, but it is more difficult to interpret psychiatric disorders under an evolutionary perspective. However, psychiatric disorders can also be seen in a continuum spectrum, where the pathological manifestation is at one extreme of the distribution, and there are different degrees of subthreshold manifestations. Under this perspective, depressive and hypomanic subthreshold symptoms may confer an advantage in situations that require learning from negative experiences and creativity/high productivity, respectively. This mechanism maintains more or less constant the frequency of genetic variants conferring risk for potentially lethal diseases, such as sickle cell anemia and cystic fibrosis, wherein heterozygous individuals (one copy of the risk allele) may benefit from increased resistance to malaria and tuberculosis, respectively. However, homozygous individuals (two copies of the risk allele) develop a life-threatening disease. In the case of psychiatric disorders, of course, the genetic factors implicated are more complex; as discussed above they are not monogenic diseases, but the underlying theory explaining the maintenance of the risk variants in the population is comparable (Keller 2018). If there is not an individual genetic variant involved in the risk of bipolar disorder (BP), then how many variants or genes are involved? There is no definitive answer to this question, and different methodological approaches were applied, depending on the working hypothesis, as discussed in Sect. 3. However, as introductory information, it is useful to know that two individuals differ for five million genetic variants on average, the greatest number of which are represented by single nucleotide polymorphisms (SNPs, i.e. substitutions of a single DNA base pair). Copy number variations (CNVs, i.e. insertions or deletions of a certain number of base pairs) are less numerous than SNPs, but they are responsible for a higher percentage of inter-individual variability in terms of number of base pairs (The 1000 Genomes Project Consortium, 2015). SNPs are responsible for the lowest part of the variability between two individuals in terms of number of DNA base pairs, but they are more frequent in the general population (i.e., the same SNP is often found in more than 1% of the population). At the same time, the other types of genetic variants are rarer and somehow more unique of a relatively restricted group of individuals. For this reason, the largest part of existing studies was focused on investigating the role of SNPs in the susceptibility to BP. It is indeed much easier to design genetic studies looking at variants which are known to be represented in a population with a certain frequency instead of genotyping known rare variants which may not be seen at all in the studied sample or search for unknown rare variants by DNA sequencing. Whole-genome or whole-exome sequencing is becoming more and more feasible in the last years in terms of costs, but it is still marginally represented in the existing literature. The starting point for developing all the available methodological approaches to the study of the genetic basis of BP was however the same: the evidence of a high genetic component of this disorder found by family, adoption, and twin studies, as discussed in the next section.

2 Bipolar Disorder Is Heritable: Twin, Adoption, and Family Studies

The observation that some disorders recur within the same families for generations goes back in ancient times, starting from the Old Testament passages which say that God "punishes the children and their children for the sins of the fathers to the third and fourth generation" (Exodus 34:7). BP is one of those diseases which recurs among generations of the same family since it shows a high heritability as demonstrated by a number of studies published since the 1960s. Heritability estimates the degree of variation in a trait that is due to genetic variation between individuals, and it can vary between zero (no genetic component at all) and one (disorders with complete penetrance, for which the presence of a genotype is necessary and sufficient to manifest the trait). Traits with a heritability of one are often autosomal dominant, such as achondroplasia that was taken as an example in the "Introduction" section, meaning that the inheritance of one copy of the risk variant from one of the parents is associated with the manifestation of the trait. Thus, these traits are often seen in many generations within the same family. BP is not a monogenic trait, and it has incomplete penetrance. However, it has quite high heritability that was estimated to be between 60 and 85% by twin studies (Smoller and Finn 2003). Twin studies estimate the heritability of a trait by comparing the concordance of the trait between dizygotic twins (DZ, who share on average 50% of their genetic variation) and monozygotic twins (MZ, who share 100% of their genetic variation), providing a natural condition optimal for performing genetic studies. Assuming that shared environmental influences on MZ twins are not different from environmental influences on DZ twins, significantly higher concordance rates in MZ twins reflect the effect of genes. Other types of studies which contributed to estimating the genetic contribution to BP were adoption studies and family studies. Adoption studies compare the rate of a disorder in biological family members to those in adoptive family members in order to distinguish the genetic component from the environmental one. These studies are logistically difficult to conduct, and their availability is limited. However, they confirmed that there is a significantly higher rate of affective illness in the biological parents (31%) than in the adoptive parents (12%) of bipolar probands (Smoller and Finn 2003). Finally, family studies investigate if a disorder aggregates in families by comparing the prevalence of the disorder among first-degree relatives of affected probands (cases) to the prevalence in the population or among relatives of unaffected probands (controls). A relevant limitation of these studies consists in the fact that they cannot distinguish if a condition aggregating in families is caused by genetic or environmental factors or a combination of the two; however, they are still useful to estimate if there is an increased risk in relatives of cases compared to controls and to quantify this risk. According to the results of family studies, the recurrence risk of bipolar disorder for first-degree relatives of bipolar probands is 8.7%. In comparison, the risk for unipolar depression is 14.1%, indicating that positive family history for BP increases the risk of both bipolar and unipolar disorders, while interestingly family history of unipolar depression does not seem to increase the risk of BP (Smoller and Finn 2003). The estimated absolute risks correspond to a recurrence risk ratio for BP and major depressive disorder (MDD) in relatives of bipolar probands of around 4 and 2, respectively. Several studies have also demonstrated that clinical features might be associated with greater familiarity with BP. The most replicated are represented by early onset of the disease, presence of psychotic symptoms, and good lithium response. Early-onset BP may represent a more severe subtype with stronger genetic loading, as demonstrated by a greater familial risk of mood disorders in relatives of probands with early-onset BP, and it could also represent a distinct subtype of BP, characterized by a greater neurodevelopmental component compared to non-early-onset BP. Patients with early-onset BP and psychotic symptoms have more frequently cognitive impairment and neurological signs, reduced frontal gray matter at the time of their first psychotic episode, and greater brain changes than healthy controls, in a pattern similar to early-onset schizophrenia cases (Arango et al. 2014). The prepubertal onset of psychopathology was associated with a poorer response to lithium, and lithium responsiveness was reported to aggregate in families, supporting the hypothesis of a genetic component (Smoller and Finn 2003).

3 How Many Genes Modulate the Risk of Bipolar Disorder? Linkage Studies, Candidate Gene Studies, and Genome-Wide Association Studies

Starting from the evidence emerged from twin studies, the first experimental approach used to identify the specific genetic regions responsible for the heritability of BP was represented by linkage studies. These studies use information from family members who are affected and unaffected with the disorder and examine which genetic regions are coinherited with the disease within the family. More than 40 linkage scans for BP, including three meta-analyses, have been published, implicating many areas of the genome but with little consistency between studies. The strongest evidence was found for linkage on chromosomes 6q for BP type I and 8q for BP type I and type II; however, the genes responsible for these linkage signals have not been identified (Barnett and Smoller 2009). The main reason for these inconsistent and inconclusive findings is that BP has a polygenic architecture, while linkage studies work well when the genetic risk is conferred by a relatively small number of genes, such as for cystic fibrosis. Candidate gene association studies suffered from the same limitation. In this case, specific alleles in genes with a hypothesized link with the pathogenesis of BP are investigated in terms of possible different distribution between cases affected with BP and healthy controls. The most part of the investigated candidate genes plays a central role in the activity of the dopaminergic, serotoninergic, and glutamatergic pathways or the modulation of circadian rhythms. Examples of genes which have been associated with the risk of BP in independent samples or meta-analyses include disrupted in schizophrenia

1 (*DISC1*), D-amino acid oxidase activator (*DAOA*, or *G72*), the dopamine transporter (*SLC6A3*) and the serotonin transporter (*SLC6A4*), brain-derived neurotrophic factor (*BDNF*), the NMDA glutamate receptor subunit 2B (*GRIN2B*), the kainate class ionotropic glutamate receptor gene *GRIK4*, and neuregulin 1 (*NRG1*) (Barnett and Smoller 2009). However, none of these has been established as a BP susceptibility gene. Candidate gene association studies showed several relevant limitations: (1) earlier studies reported more often positive and larger effect associations than later studies, a phenomenon which is also referred to as "winner's curse"; (2) most of the studies were performed on underpowered sample sizes and without applying appropriate multiple testing correction; and (3) many genes are involved in the risk of BP, part of which has likely no known connection with BP and cannot be investigated using the candidate gene method. Given these limitations, genome-wide association studies (GWAS) became the most common approach to the study of the genetics of complex disorders such as BP in the last 10 years. GWAS are based on a technology called microarray, which can genotype hundreds of thousands or millions of genetic variants cost-effectively, by multiplexed and parallel processing. GWAS include common genetic variants in a population, typically those variants having a frequency above 1%. As noted in the "Introduction" section, these common variants are usually represented by SNPs, which can be studied in an easier way than rare variants (in terms of costs, time, and sample size needed). GWAS have been successful for identifying susceptibility alleles for a broad range of common complex disorders including diabetes, cardiovascular disease, prostate and breast cancer, and many others. The early GWAS of BP did not identify any loci achieving genome-wide significance, that is, a p threshold $<5 \times 10^{-8}$, which takes into account multiple testing correction. The top variants emerging from these early GWAS included *DGKH*, coding for diacylglycerol kinase eta, an enzyme involved in modulating the effects of mood stabilizers, and *CACNA1C*, coding for the alpha-1C subunit of the L-type voltage-gated calcium channel, which was then confirmed by later studies (Barnett and Smoller 2009). The main issue of GWAS is represented by the need for large samples to have adequate power to detect small effect sizes at the genome-wide significance threshold. Recent studies indeed showed that most of the genome-wide significant loci have an effect size measured as odds ratio (OR) $\leq$ 1.1 (Stahl et al. 2019). In order to have an adequate power (i.e., $\geq$80%) for detecting this magnitude of effect size for a common variant with the frequency of 10% in a population, a sample of about one million subjects was estimated to be needed (Visscher et al. 2017), while currently, the largest GWAS of BP included 20,352 cases and 31,358 controls (Stahl et al. 2019). This study identified 30 independent loci associated with BP at the genome-wide significance threshold that is expected to be a much lower number compared to those that would be identified in a sample size providing adequate power. The significant loci contain genes encoding ion channels and transporters (e.g., *CACNA1C*, *SCN2A* (sodium voltage-gated channel alpha subunit 2), *SLC4A1* (solute carrier family 4 member 1)), neurotransmitter receptors (e.g., *GRIN2A*), and synaptic components (e.g., *ANK3* (ankyrin 3) and *RIMS1* (regulating synaptic membrane exocytosis 1)) (Stahl et al. 2019). These processes are important in neuronal hyperexcitability, which is a pathogenetic

mechanism implicated in BP using pluripotent stem cell-derived neurons from patients, and by the fact that the reduction of neuronal hyperexcitability is one of the mechanisms of action of mood stabilizers such as lithium (Mertens et al. 2015). Other significant loci which were reported in earlier GWAS were within the *TRANK1* (tetratricopeptide repeat and ankyrin repeat containing 1), *NCAN* (neurocan), and *TENM4* (teneurin transmembrane protein 4) genes. In the same study, the integration of the GWAS results with information on SNPs modulating gene expression implicated *GLT8D1* (glycosyltransferase 8 domain containing 1), which is involved in proliferation and differentiation of neural stem cells. Pathway analyses reveal genetic evidence for insulin secretion and endocannabinoid signaling. Top genes in these pathways included calcium and potassium channel subunits, MAP kinase, and GABA-A receptor subunit genes. The SNP-based heritability estimated by this study was 17–23% depending from the considered population prevalence of BP (between 0.5 and 2%), and heritability estimates were higher for BP type I than BP type II (25% vs 11%) (Stahl et al. 2019). These heritability estimates are much lower than the heritability estimated by twin studies, because they are based on common genetic variants only. An overview of the significant findings of GWAS is provided in Table 1.

4 Genetic Overlap Between Bipolar Disorder and Other Brain Disorders: Disorder-Specific or General Genetic Influences?

New methods recently developed to study the genetics of complex traits estimate the genetic correlation between different traits and the predictive ability of the genetics of one trait on another one. One of the most used approaches is linkage disequilibrium score regression (LDSR) because it can be performed using GWAS summary statistics (no need for raw genotype data). LDSR was initially developed to differentiate between GWAS *p*-values inflation caused by polygenicity and by confounding effects, such as population stratification. Other applications were developed; in particular, the method can be used to estimate genetic correlations between different traits. For providing these estimations, LDSR uses the test statistics of the SNPs from a GWAS (summary statistics) and the degree of linkage disequilibrium (LD) between SNPs. LD refers to the nonrandom association of alleles at two or more loci in a population, or, in other words, to the fact that two or more alleles are found together more often than expected by chance, and this phenomenon is influenced by different factors (e.g., rate of genetic recombination, mutation rate). The genetic variants associated with BP are expected to show a high degree of LD among each other (LD score), because the trait is polygenic, and for each associated SNPs, a number of nearby SNPs is expected to show association signals as well, inflating the GWAS statistics. In contrast, inflation caused by other factors would not be associated with LD score. Thus, the regression of the SNP test

Table 1 An overview of the results of genome-wide association studies (GWAS) of bipolar disorder

Year	Ethnicity	N of cases/N of controls	N of significant signals[a]	Nearest gene(s)	Reference
2007	Caucasian (UK)	1,868/2,938	0	/	Wellcome Trust Case Control Consortium (2007)
2008	Caucasian (USA/Germany)	1,233/1,439	1	*DGKH*	Baum et al. (2008a)
2008	Caucasian (USA/UK)	1,461/2,008	0	/	Sklar et al. (2008)
2008	Caucasian (USA/UK/Ireland)	4,387/6,209	1	*ANK3*	Ferreira et al. (2008)
2008	Caucasian (USA/UK/Germany)	3,101/4,377	0	/	Baum et al. (2008b)
2009	Caucasian (USA/UK/Canada)	3,683/14,507	0	/	Scott et al. (2009)
2009	Caucasian (USA)	1,001/1,033	0	/	Smith et al. (2009)
	African-American (USA)	345/670	0	/	
2011	Caucasian (USA/-Australia/Europe)	2,411/3,613	1	*NCAN*	Cichon et al. (2011)
2011	Caucasian (USA/-Canada/Europe)	7,481/9,250	2	*CACNA1C*, *TENM4*, *ANK3*, *SYNE1*	Psychiatric GWAS Consortium Bipolar Disorder Working Group (2011)
2013	Caucasian (USA/-Canada/Europe)	8,699/12,163	4	*CACNA1C*, *TENM4*, *RHEBL1*, *DHH*, *TRPC4AP*, *GSS*, *MYH7B*	Green et al. (2013)
2013	Caucasian (USA/Europe)	6,773/9,915	6	*ANK3*, *TRANK1*, *LMAN2L*, *PTGFR*	Chen et al. (2013)
	Asian (Taiwan)	1,000/1,000	0		
2014	Caucasian (USA/Europe)	9,747/14,278	18	*ANK3*, *TENM4*, *TRANK1*, *ADCY2*, *MIR2113*, *POU3F2*	Mühleisen et al. (2014)
2016	Caucasian (USA/Europe)	9,784/30,471	6	*TRANK1*, *DDN*, *MAD1L1*, *ELAVL2*, *ERBB2*, *MIR2113*, *POU3F2*	Hou et al. (2016)

(continued)

Table 1 (continued)

Year	Ethnicity	N of cases/N of controls	N of significant signals[a]	Nearest gene(s)	Reference
2017	Caucasian (USA/Europe)	13,902/19,279	8	*TRANK1*, *LMAN2L*, *FER1L5*, *CNNM4*, *SYNE1*, *ADD3*, *XPNPEP1*, *CACNA1C*	Charney et al. (2017)
2018	Japanese	2,964/61,887	7	*FADS1*, *FADS2*, *FADS3*	Ikeda et al. (2018)
	Japanese + Caucasian	10,445/71,137	7	*MAD1L1*, *TRANK1*, *TENM4*, *SYNE1*, *MLL2*, *FADS2*, *NFIX*	
2019	Caucasian (USA/Europe)	20,352/31,358	30	*LMAN2L*, *TRANK1*, *ANK3*, *CACNA1C*, *NCAN*, *GRIN2A*, *TENM4*, and other 28 genes	Stahl et al. (2019)

Earlier studies were strongly underpowered to detect variants with small effect on disease risk
[a]Genome-wide significant ($p < 5 \times 10^{-8}$) independent variants

statistics from a GWAS against the LD score provides an estimation of the polygenicity of a trait and was used to confirm the polygenic architecture of psychiatric disorders (Bulik-Sullivan et al. 2015a). The same approach can be adapted to estimate the genetic correlation between two traits (Bulik-Sullivan et al. 2015b), and it has been largely applied in the field of neuropsychiatry.

Using LDSR, BP was demonstrated to be genetically correlated with other psychiatric disorders, particularly with schizophrenia (SCZ) and secondly with MDD and obsessive-compulsive disorder, while less strongly with anorexia nervosa and autism spectrum disorders (Stahl et al. 2019; Brainstorm Consortium et al. 2018). BP type I was estimated to be more strongly genetically correlated with SCZ, while BP type II was more strongly genetically correlated with MDD. The genetic correlation between BP and SCZ was estimated to be 0.70, supporting a widely shared genetic predisposition. In a large GWAS of SCZ and BP, 114 independent genetic loci were found associated with both disorders, and genetic pathways implicated in the common genetic susceptibility were modulators of neuron projection development, synaptic plasticity, and neurogenesis. On the other hand, response to potassium ion was reported to be the only pathway associated with SCZ only and not BP (Bipolar Disorder and Schizophrenia Working Group of the Psychiatric Genomics Consortium 2018).

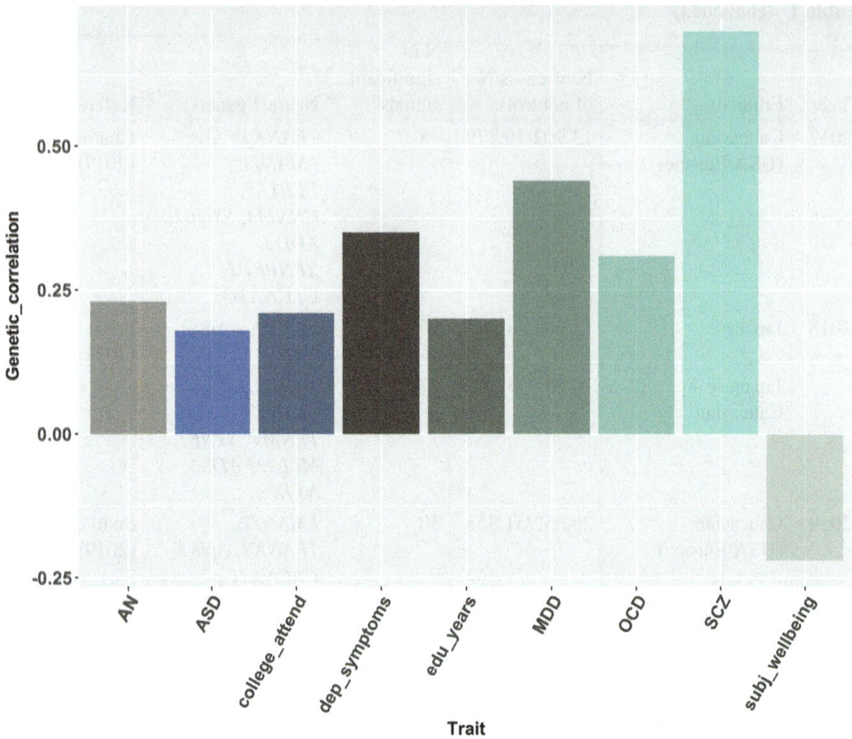

Fig. 1 Barplot representing the genetic correlations between bipolar disorder and other psychiatric and nonpsychiatric traits. Genetic correlations were estimated using linkage disequilibrium score regression (LDSR), and references to the corresponding literature are reported in Sect. 4. *AN* anorexia nervosa, *ASD* autism spectrum disorder, *college_attend* college attendance, *dep_symptoms* depressive symptoms, *edu_years* years of education, *MDD* major depressive disorder, *OCD* obsessive-compulsive disorder, *SCZ* schizophrenia, *subj_wellbeing* subjective well-being

Moving beyond psychiatric disorders, positive genetic correlations were reported for BP and education years, as well as college attendance, but not with either adult or childhood IQ, suggesting that the role of BP genetics in educational attainment may be independent of general intelligence (Stahl et al. 2019). This finding is consistent with evidence obtained in clinical samples, suggesting that IQ in BP may be affected by the onset and progression of the disease, while in the pre-onset phase, prodromal symptoms like elevated energy may contribute to good educational attainment (Vreeker et al. 2016). BP does not seem to be genetically correlated with a number of neurological diseases, such as Alzheimer's disease, Parkinson's disease, multiple sclerosis, and ischemic stroke, despite a nonsignificant trend of positive genetic correlation was reported with the last (Brainstorm Consortium et al. 2018). An overview of the genetic correlations of BP with other disorders is reported in Fig. 1.

Another approach used to estimate the genetic overlap between two traits is represented by polygenic risk scores (PRS). This method, contrary to LDSR, requires raw genotypes, though some recent methods can be applied to summary statistics. PRS are used to estimate if the genetics of one trait (according to the GWAS results obtained in a base sample) is able to predict the same trait or a different trait in an independent sample (target sample) and the strength of prediction. PRS are calculated in the target sample as the sum of the risk alleles associated with the trait weighted by their effect size calculated in the base sample. This approach contributed to a better understanding of the genetic overlap between BP and transdiagnostic clinical dimensions. Among 24 clinical dimensions or characteristics, a positive correlation was reported between BP PRS and manic symptoms in SCZ, between SCZ PRS and psychosis in BP cases, and between BP+SCZ PRS and psychotic features in BP and negative symptoms in SCZ (Bipolar Disorder and Schizophrenia Working Group of the Psychiatric Genomics Consortium 2018). Overall, these findings support the hypothesis that the main clinical dimension associated with the shared genetic liability between BP and SCZ is represented by psychosis. Consistently, across a number of disorders, including BP, SCZ PRS was shown to predict the risk of psychotic symptoms, despite the phenotypic variance explained was very small (4.4%) (Calafato et al. 2018).

5 The Role of Rare Genetic Variants

Large (>100 kb), rare copy number variants (CNVs) have been shown to confer risk for SCZ and other neurodevelopmental disorders, while the available evidence suggests that this type of variant is less strongly involved in the risk of BP. The strongest evidence for association with BP was obtained for duplications at 16p11.2, which were found in 0.13% of BP cases compared with 0.03% of healthy controls. Duplications at *ATF7IP2*, upstream of *GRIN2A*, a gene associated with BP by GWAS, and at the gene *CGNL1* (cingulin-like 1) were also reported as risk factors. The burden of very large (>500 kb) CNVs in BP was shown to be not different from controls, contrary to what was observed in SCZ (Green et al. 2016).

Whole-exome sequencing in families including affected individuals led to the identification of rare variants segregating with BP in the G protein-coupled receptor (GPCR) family genes, including the corticotropin-releasing hormone receptor 2 (*CRHR2*) gene and the metabotropic glutamate receptor 1 (*GRM1*). GPCR are involved in the transduction of cellular signals, in response to stimuli such as neurotransmitters and neuropeptides. Receptors that activate G proteins include serotonin, glutamatergic, and dopaminergic receptors (e.g., *HTR1B*, *GRM1*, *GRM4*, and *DRD5*). Through the activation of G proteins, these receptors modulate two major signaling pathways, cAMP and phosphatidylinositol signaling pathways, both of which are associated with the pathophysiology of BP and the mechanism of action of drugs commonly prescribed in BP (Cruceanu et al. 2018).

Rare damaging variants were hypothesized to play a role in severe BP cases comorbid with other neuropsychiatric disorders, particularly those having a neurodevelopmental component, such as attention deficit hyperactivity disorder, seizure disorders, and learning disabilities. In such cases, rare variants damaging the protein structure or function were identified in affected individuals and not in controls, particularly in genes coding for proteins with GTPase-activating function which are among the targets of lithium (Rao et al. 2017). CNVs in the neurexin 1 (*NRXN1*) gene have been associated with BP comorbid with intellectual disability and have replicated associations with cognitive ability, autism, and SCZ. Neurexins are crucial for the regulation of neurotransmission and the formation of synaptic contacts. Thus, they are hypothesized to play a role in neurodevelopment (Viñas-Jornet et al. 2014, 2018).

6 Gene × Environment Studies

Apart from genetic risk factors, a number of environmental exposures including urbanicity, stressful life events, early life stress, and substance abuse may underlie the development of BP. Environmental risk factors are hypothesized to interact with the individual's genetic predisposition to BP, and they may partly explain the incomplete penetrance of the disorder. Gene x environment (G × E) studies aim to understand how different genotypes affect the risk of developing a disorder depending on the exposure to environmental factors. Most of the G × E studies were focused on variants in candidate genes rather than multiple variants across the genome (GWAS), and BP was under-investigated compared to other psychiatric disorders, leading to inconclusive results so far. Most studies considered the impact of childhood trauma in interaction with the rs6265 SNP of the brain-derived neurotrophic factor (*BDNF*) gene or SNPs in genes regulating the monoaminergic neurotransmitter system (the 5-HTTLPR variant of the *SLC6A4* gene and the rs4680 SNP in the catechol-O-methyltransferase (*COMT*) gene). However, none of these findings was convincingly replicated, and the effect of potential confounders was implicated as a possible reason (Misiak et al. 2018).

More extensive G × E research has been conducted on the risk of developing a psychotic disorder, including but not limited to a psychotic affective disorder. A strong candidate gene which has modulating effect on the risk of psychosis in cannabis users and has been replicated is serine/threonine kinase 1 (*AKT1*), a gene encoding a serine/threonine kinase involved in the transduction of signal following cannabinoid receptor activation. Another group of studies used individual's family history and environmental exposures to derive a proxy of G × E interaction. Individuals with a family history of psychotic illness appear to be particularly sensitive to the effects of multiple environmental risk factors, including cannabis use and urban upbringing. However, positive family history can also be affected by environmental factors and not only genetics; furthermore, genetic variants that increase sensitivity to the environment may be distinct from genetic variants that directly increase the risk of illness. The demonstration corroborated this last hypothesis that a PRS of overall sensitivity to the environment (both positive and negative) predicted both the effects of parenting on emotional problems and response to psychological treatment among children with anxiety disorders. However, it did not directly predict psychopathology (Zwicker et al. 2018). Further research is needed to understand how genetic variants modulate psychopathology in interaction with the environment, particularly in BP, which has been under-investigated so far.

7 Nongenetic Mechanisms Contributing to the Regulation of Gene Expression: Epigenetics

Gene expression is modulated not only by genetic variants affecting gene transcription or transduction in proteins but also through fluid mechanisms that can respond to internal or external stimuli, in a way to adapt gene expression profiles to the environment, namely, epigenetics. Epigenetics may partly explain the mechanism by which environmental factors modulate the risk of BP. Examples of epigenetic mechanisms of gene expression regulation are DNA methylation and histone modification. DNA methylation consists in the addition of a methyl (CH3) group to the carbon-5 of cytosines in the DNA sequence, generating a modified nucleotide called 5-methylcytosine (5mC). DNA methylation occurs throughout the entire genome at cytosines of CpG dinucleotides, which are frequently enriched around promoter regions, in so-called CpG islands. These regions are responsible for DNA methylation-dependent control of gene promoter activity, and higher methylation is typically associated with gene repression. Mechanistically, it represses the activation of a promoter by either sterically inhibiting the binding of transcription factors or by actively recruiting repressor proteins such as histone deacetylases (HDAC), which will ultimately lead to the formation of heterochromatin (a highly condensed form of DNA which is not accessible by transcription factors). DNA methylation patterns are heritable, possibly explaining part of the heritability of BP. However, they are also modulated by environmental exposure through the activity of a specific group of enzymes called DNA methyltransferases (DNMTs) and a number of translocation (TET) family of proteins, responsible for DNA methylation and demethylation, respectively.

The published studies were heterogeneous in terms of the applied methodology to evaluate DNA methylation (type of assay), genes which were tested (global genome methylation, methylation in selected candidate genes, or genome-wide methylation), and type of tissue used (blood in most studies, but in some cases postmortem brain tissues). DNA methylation is tissue-specific, but high concordance between blood and brain was reported, suggesting the utility of peripheral blood as a proxy of brain tissues. Studies examining global genome methylation in BP compared to healthy controls did not find consistent findings. Moreover, the interpretation of alternations in global methylation levels would not be easy under the pathogenetic point of view. The most investigated candidate genes were in line with those reported by studies looking at genetic variation, namely, *BDNF*; monoaminergic genes, particularly *SCL6A4*; the serotonin receptors 1A, 2A, and 3A (*HTR1A*, *HTR2A*, and *HTR3A*); and the membrane-bound catechol-O-methyltransferase (*COMT*). In addition, methylation in the dystrobrevin-binding protein 1 (*DTNBP1*) gene was implicated in BP, which product is involved in the modulation of glutamatergic neurotransmission by influencing exocytotic glutamate release. Methylation candidate gene association studies had sparse and often not replicated findings, mainly because they do not capture the polygenicity of the disease, as discussed in Sect. 3. Genome-wide methylation studies did not provide encouraging findings either since they did not identify replicated findings. They used different arrays and different preprocessing and analysis methods and considered different tissues, making them poorly comparable. However, an interesting finding was that differentially methylated regions between BP and controls depend from the brain region considered. For example, in the frontal cortex (particularly in Brodmann's area 9) and anterior cingulate, only a few differentially methylated regions overlapped with promoters, whereas a greater proportion occurred in introns and intergenic regions (Fries et al. 2016). These noncoding regions may contribute to gene expression regulation since they were found to overlap with long intergenic noncoding RNAs (lincRNAs) and microRNAs. The function of these noncoding RNAs is only partially known, but they are hypothesized to regulate gene expression by interacting with transcription factors and by regulating posttranscriptional processing, such as mRNA transport, translation, and degradation (Ransohoff et al. 2018).

The role of epigenetics in the pathogenesis of BP remains poorly understood so far, in terms of genetic regions/genes showing abnormal methylation, differences between tissues, and causal mechanisms responsible for the different methylation patterns seen in BP compared to controls. A possible mechanism that may partially be responsible for alterations in methylation in BP is represented by overexpression of DNMT1 found by several independent studies in Brodmann's area 9, which was associated with the downregulation of specific GABAergic and glutamatergic genes (Fries et al. 2016). Under this perspective, DNMT1 may be the target of future new treatments.

8 Current and Future Lines of Research

From the previous sections of this chapter, it is quite evident that a strong genetic basis of BP has been established, but the specific genetic variants or regions, as well as the mechanistic process through which they lead to the disease, are still largely unknown. However, a number of new and promising research resources have emerged in the last years.

The recent improvements in genotyping/sequencing technologies but also in the available storing and computational resources allowed the creation and growth of large biobanks in a number of countries, such as the UK Biobank in the UK and All of Us in the USA (UK Biobank 2019) (National Institute of Health (NIH) 2019). These initiatives collected a wide range of health-related measures and biomarkers in hundreds of thousand subjects from the general population, in order to study the individual (genetic and nongenetic) factors associated with the development of diseases, but also with well-being, lifestyles, and other traits of interest such as aging. Thus, not only disease risk factors can be identified, but it is possible also to study those variables modulating disease onset and prognosis and, in this way, develop preventive strategies. These unprecedented resources have been available in the last few years and are in an expansion phase, in terms of number of subjects but also phenotypic information, genetic data, and other biomarkers (e.g., brain imaging). For example, whole-exome sequencing is expected to be completed in all 500,000 subjects included in UK Biobank at the end of 2020, providing the complete DNA sequence of the coding regions of the genome in a sample of unprecedented size and richness of phenotypic characterization. Digital phenotyping has been increasingly used to collect phenotype data from personal digital devices, for example, for remotely monitoring psychiatric symptoms over time and collecting activity data (Insel 2018). Another very powerful resource for extracting phenotypic information is represented by electronic health records (EHR), which typically include both structured information (e.g., diagnostic codes) and narrative notes from which the information of interest can be automatically extracted using natural language processing. Automated definitions of BP using three rule-based classifiers based on information extracted from EHR were demonstrated to have genetic correlations of 0.66, 0.74, and 1 (the most stringent definition) with a traditional interview-based diagnosis of BP, confirming that EHR can be used for automated definition of diagnosis (Chen et al. 2018).

Data from different biobanks are more and more often included in collaborative research efforts, thanks to international consortia, such as the Psychiatric Genomics Consortium (PGC), that made possible the largest GWAS of BP so far (Stahl et al. 2019), and it is actively committed to the recruitment of more samples and the development of new analysis methods. The last point has been object of particular interest since the analysis approaches used in psychiatric genetics had suffered from important limitations for many years. A poor understanding of the genetic architecture of BP and other psychiatric disorders, alongside with limited technological resources, had restricted the existing research to the study of candidate genes and individual variant effects for years. The drop in the costs of genotyping and next-generation sequencing has played a fundamental role in the shift from individual genes and individual variants to the study of the combined effect of a number of variants across the genome. Methods such as LDSR and PRS are examples of analysis approaches reflecting the polygenicity of complex traits, being able to incorporate the effects of all the SNPs with evidence of association with a trait. Future improvements of polygenic analysis methods are going to (a) better estimate the effects of individual variants contributing to a trait, (b) take into account interaction effects, and (c) include also rare genetic variants, which analysis has been challenging.

Another recent interest in psychiatric genetics has been the identification of causality between traits. Genetics can indeed contribute to understanding the causal relationship between traits that are commonly seen together, but where there is no clear knowledge indicating which one is causing the other or if they have independent causes. Mendelian randomization (MR) allows to test this kind of hypothesis on observational data, using genetic variants randomly allocated at conception as instrumental variables for a modifiable exposure (or trait) of interest. MR is analogous to an RCT where instead of the allocation of participants to different treatment groups, individuals are randomized by nature to carry or not carry genetic variants that may modify the risk of exposure. MR uses genetic variants associated with exposure (hypothetical causal trait) as instruments to infer causality on an outcome. For a genetic variant to be a valid instrument, three assumptions must be satisfied: (1) the genetic variant is strongly associated with the exposure of interest (at the genome-wide level); (2) the genetic variant is not associated with any confounder of the exposure-outcome association; and (3) the genetic variant is only associated with the outcome through the exposure (Pagoni et al. 2019). Interestingly, MR was used to assess the causal relationship between BP and a number of cardiometabolic traits, being increased rates of cardiovascular diseases a major concern in patients with BP. Using BP as the exposure, no causal relationship with cardiometabolic traits was identified, suggesting that cardiometabolic abnormalities in these patients may be mostly secondary (e.g., to lifestyles and medications) and not related to the underlying pathophysiology of BP (So et al. 2019). These results further corroborate the importance of preventing and monitoring environmental risk factors of cardiovascular disease in BP.

9 Conclusion

Notable progress has been achieved toward a better understanding of the genetics of BP in the last years, thanks to technological advances and development of new analysis methods. The genetic pathways mediating the disease, as well as the genetic overlap of BP with other disorders or traits, have been partially elucidated. However, the genetic contribution to BP and its interactions with the environment are still not completely understood, and the available scientific findings have no produced clinical applications so far.

The use of PRS to predict the risk of psychiatric disorders is currently one of the most promising future applications, and it has attracted the interest of public and private enterprises. The PRS for a trait can be converted into a standardized score that follows a normal distribution, with higher PRS corresponding to higher risk, in a way that could be used to determine an individual's risk of the corresponding trait based on his/her position on this distribution. Individuals falling above a certain threshold could be informed of their risk and benefit from preventive strategies. It is unclear how extreme a score would have to be to achieve clinical relevance. However, it was speculated that a PRS in the top 1–5% of the population would warrant feedback (Lewis and Vassos 2017) (Fig. 2). PRS could be used as a screening tool to identify individuals at risk and allow an evidence-based allocation of health-care resources (e.g., clinical monitoring, education about the environmental risk factors associated with the disease). PRSs are currently able to explain between 1 and 15% of the variation between cases and controls (Visscher et al.

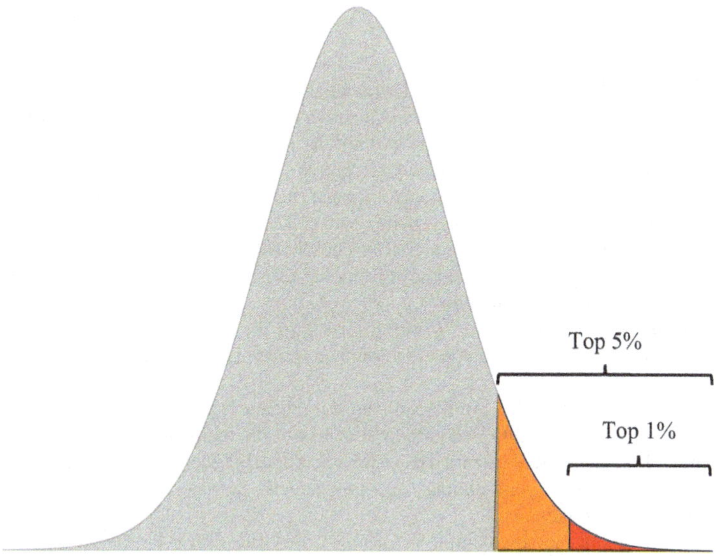

Fig. 2 Example of bipolar disorder polygenic risk score (PRS) distribution. Subjects in the highest 1–5% of the PRS distribution are those who might warrant feedback and might benefit from preventive interventions. This plot is meant to serve as a conceptual example and does not reflect the distribution of real data

2017); however in an individual at the top end of the risk distribution (top 1–5%), the risk could be significantly increased compared to the general population. Apart from the need for further data for assessing the clinical relevance of PRS in psychiatry, other issues should be addressed in order to develop clinical applications of psychiatric genetics. The most relevant of them are represented by ethical concerns, education of physicians, and availability of tools for standard interpretation of genetic results and standardized corresponding clinical interventions. Potential ethical concerns include misinterpretation of findings due to insufficient education and information, stigma and discrimination, commercial use of PRS having unclear clinical relevance as direct-to-consumer product, and PRS use for embryo selection (eugenics). Another issue that GWAS have only recently started to address is the inclusion of people of non-European ancestry, in order to improve our knowledge about the genetic variants that may be specifically observed only in some ethnic groups and relevant to disease risk (Sullivan et al. 2018).

Hopefully, the valuable resources available to researchers within biobanks and consortia will contribute in solving the still unanswered questions and to the translation of genetic findings in clinical applications with demonstrated beneficial effects to the individual and the society. Appropriate regulations about the consented use of genetic information should be developed in order to avoid discrimination and nonethical uses.

References

Arango C, Fraguas D, Parellada M (2014) Differential neurodevelopmental trajectories in patients with early-onset bipolar and schizophrenia disorders. Schizophr Bull 40(Suppl 2):S138–S146

Barnett JH, Smoller JW (2009) The genetics of bipolar disorder. Neuroscience 164(1):331–343

Baum AE, Akula N, Cabanero M, Cardona I, Corona W, Klemens B et al (2008a) A genome-wide association study implicates diacylglycerol kinase eta (DGKH) and several other genes in the etiology of bipolar disorder. Mol Psychiatry 13(2):197–207

Baum AE, Hamshere M, Green E, Cichon S, Rietschel M, Noethen MM et al (2008b) Meta-analysis of two genome-wide association studies of bipolar disorder reveals important points of agreement. Mol Psychiatry 13(5):466–467

Bipolar Disorder and Schizophrenia Working Group of the Psychiatric Genomics Consortium (2018) Genomic dissection of bipolar disorder and schizophrenia, including 28 subphenotypes. Cell 173(7):1705–1715.e16

Brainstorm Consortium, Anttila V, Bulik-Sullivan B, Finucane HK, Walters RK, Bras J et al (2018) Analysis of shared heritability in common disorders of the brain. Science 360(6395):eaap8757

Bulik-Sullivan BK, Loh P-R, Finucane HK, Ripke S, Yang J, Patterson N et al (2015a) LD score regression distinguishes confounding from polygenicity in genome-wide association studies. Nat Genet 47(3):291–295

Bulik-Sullivan B, Finucane HK, Anttila V, Gusev A, Day FR, Loh P-R et al (2015b) An atlas of genetic correlations across human diseases and traits. Nat Genet 47(11):1236–1241

Calafato MS, Thygesen JH, Ranlund S, Zartaloudi E, Cahn W, Crespo-Facorro B et al (2018) Use of schizophrenia and bipolar disorder polygenic risk scores to identify psychotic disorders. Br J Psychiatry 213(3):535–541

Charney AW, Ruderfer DM, Stahl EA, Moran JL, Chambert K, Belliveau RA et al (2017) Evidence for genetic heterogeneity between clinical subtypes of bipolar disorder. Transl Psychiatry 7(1): e993

Chen DT, Jiang X, Akula N, Shugart YY, Wendland JR, Steele CJM et al (2013) Genome-wide association study meta-analysis of European and Asian-ancestry samples identifies three novel loci associated with bipolar disorder. Mol Psychiatry 18(2):195–205

Chen C-Y, Lee PH, Castro VM, Minnier J, Charney AW, Stahl EA et al (2018) Genetic validation of bipolar disorder identified by automated phenotyping using electronic health records. Transl Psychiatry 8(1):86

Cichon S, Mühleisen TW, Degenhardt FA, Mattheisen M, Miró X, Strohmaier J et al (2011) Genome-wide association study identifies genetic variation in neurocan as a susceptibility factor for bipolar disorder. Am J Hum Genet 88(3):372–381

Cruceanu C, Schmouth J-F, Torres-Platas SG, Lopez JP, Ambalavanan A, Darcq E et al (2018) Rare susceptibility variants for bipolar disorder suggest a role for G protein-coupled receptors. Mol Psychiatry 23(10):2050–2056

Ferreira MAR, O'Donovan MC, Meng YA, Jones IR, Ruderfer DM, Jones L et al (2008) Collaborative genome-wide association analysis supports a role for ANK3 and CACNA1C in bipolar disorder. Nat Genet 40(9):1056–1058

Fries GR, Li Q, McAlpin B, Rein T, Walss-Bass C, Soares JC et al (2016) The role of DNA methylation in the pathophysiology and treatment of bipolar disorder. Neurosci Biobehav Rev 68:474–488

Green EK, Hamshere M, Forty L, Gordon-Smith K, Fraser C, Russell E et al (2013) Replication of bipolar disorder susceptibility alleles and identification of two novel genome-wide significant associations in a new bipolar disorder case-control sample. Mol Psychiatry 18(12):1302–1307

Green EK, Rees E, Walters JTR, Smith K-G, Forty L, Grozeva D et al (2016) Copy number variation in bipolar disorder. Mol Psychiatry 21(1):89–93

Hou L, Bergen SE, Akula N, Song J, Hultman CM, Landén M et al (2016) Genome-wide association study of 40,000 individuals identifies two novel loci associated with bipolar disorder. Hum Mol Genet 25(15):3383–3394

Ikeda M, Takahashi A, Kamatani Y, Okahisa Y, Kunugi H, Mori N et al (2018) A genome-wide association study identifies two novel susceptibility loci and trans population polygenicity associated with bipolar disorder. Mol Psychiatry 23(3):639–647

Insel TR (2018) Digital phenotyping: a global tool for psychiatry. World Psychiatry 17(3):276–277

Keller MC (2018) Evolutionary perspectives on genetic and environmental risk factors for psychiatric disorders. Annu Rev Clin Psychol 14:471–493

Lewis CM, Vassos E (2017) Prospects for using risk scores in polygenic medicine. Genome Med 9 (1):96

Lobo I (2008) Three individuals carry the same disease-causing mutation; two suffer from the disease but exhibit different symptoms, while the third is completely unaffected. Why? Nat Educ 1(1):64

Mertens J, Wang Q-W, Kim Y, Yu DX, Pham S, Yang B et al (2015) Differential responses to lithium in hyperexcitable neurons from patients with bipolar disorder. Nature 527(7576):95–99

Misiak B, Stramecki F, Gawęda Ł, Prochwicz K, Sąsiadek MM, Moustafa AA et al (2018) Interactions between variation in candidate genes and environmental factors in the etiology of schizophrenia and bipolar disorder: a systematic review. Mol Neurobiol 55(6):5075–5100

Mühleisen TW, Leber M, Schulze TG, Strohmaier J, Degenhardt F, Treutlein J et al (2014) Genome-wide association study reveals two new risk loci for bipolar disorder. Nat Commun 5:3339

National Institute of Health (NIH) (2019) All of us. https://www.joinallofus.org/en/biobank. Accessed 21 Nov 2019

Pagoni P, Dimou NL, Murphy N, Stergiakouli E (2019) Using Mendelian randomisation to assess causality in observational studies. Evid Based Ment Health 22(2):67–71

Plomin R, Deary IJ (2015) Genetics and intelligence differences: five special findings. Mol Psychiatry 20(1):98–108

Psychiatric GWAS Consortium Bipolar Disorder Working Group (2011) Large-scale genome-wide association analysis of bipolar disorder identifies a new susceptibility locus near ODZ4. Nat Genet 43(10):977–983

Ransohoff JD, Wei Y, Khavari PA (2018) The functions and unique features of long intergenic non-coding RNA. Nat Rev Mol Cell Biol 19(3):143–157

Rao AR, Yourshaw M, Christensen B, Nelson SF, Kerner B (2017) Rare deleterious mutations are associated with disease in bipolar disorder families. Mol Psychiatry 22(7):1009–1014

Scott LJ, Muglia P, Kong XQ, Guan W, Flickinger M, Upmanyu R et al (2009) Genome-wide association and meta-analysis of bipolar disorder in individuals of European ancestry. Proc Natl Acad Sci U S A 106(18):7501–7506

Sklar P, Smoller JW, Fan J, MAR F, Perlis RH, Chambert K et al (2008) Whole-genome association study of bipolar disorder. Mol Psychiatry 13(6):558–569

Smith EN, Bloss CS, Badner JA, Barrett T, Belmonte PL, Berrettini W et al (2009) Genome-wide association study of bipolar disorder in European American and African American individuals. Mol Psychiatry 14(8):755–763

Smoller JW, Finn CT (2003) Family, twin, and adoption studies of bipolar disorder. Am J Med Genet C Semin Med Genet 123C(1):48–58

So H-C, Chau K-L, Ao F-K, Mo C-H, Sham P-C (2019) Exploring shared genetic bases and causal relationships of schizophrenia and bipolar disorder with 28 cardiovascular and metabolic traits. Psychol Med 49(8):1286–1298

Stahl EA, Breen G, Forstner AJ, McQuillin A, Ripke S, Trubetskoy V et al (2019) Genome-wide association study identifies 30 loci associated with bipolar disorder. Nat Genet 51(5):793–803

Sullivan PF, Agrawal A, Bulik CM, Andreassen OA, Børglum AD, Breen G et al (2018) Psychiatric genomics: an update and an agenda. Am J Psychiatry 175(1):15–27

The 1000 Genomes Project Consortium (2015) A global reference for human genetic variation. Nature 526(7571): 68–74

UK Biobank (2019) UK Biobank: improving the health of future generations. https://www.ukbiobank.ac.uk. Accessed 21 Nov 2019

Viñas-Jornet M, Esteba-Castillo S, Gabau E, Ribas-Vidal N, Baena N, San J et al (2014) A common cognitive, psychiatric, and dysmorphic phenotype in carriers of NRXN1 deletion. Mol Genet Genomic Med 2(6):512–521

Viñas-Jornet M, Esteba-Castillo S, Baena N, Ribas-Vidal N, Ruiz A, Torrents-Rodas D et al (2018) High incidence of copy number variants in adults with intellectual disability and co-morbid psychiatric disorders. Behav Genet 48(4):323–336

Visscher PM, Wray NR, Zhang Q, Sklar P, McCarthy MI, Brown MA et al (2017) 10 years of GWAS discovery: biology, function, and translation. Am J Hum Genet 101(1):5–22

Vreeker A, Boks MPM, Abramovic L, Verkooijen S, van Bergen AH, Hillegers MHJ et al (2016) High educational performance is a distinctive feature of bipolar disorder: a study on cognition in bipolar disorder, schizophrenia patients, relatives and controls. Psychol Med 46(4):807–818

Wellcome Trust Case Control Consortium (2007) Genome-wide association study of 14,000 cases of seven common diseases and 3,000 shared controls. Nature 447(7145):661–678

Zwicker A, Denovan-Wright EM, Uher R (2018) Gene-environment interplay in the etiology of psychosis. Psychol Med 48(12):1925–1936

Targeting Mitochondrial Dysfunction for Bipolar Disorder

Maya Kuperberg, Sophie L. A. Greenebaum, and Andrew A. Nierenberg

Contents

Abstract People with bipolar disorder (BD) all too often have suboptimal long-term outcomes with existing treatment options. They experience relapsing episodes of depression and mania and also have interepisodic mood and anxiety symptoms. We need to have a better understanding of the pathophysiology of BD if we are to make progress in improving these outcomes. This chapter will focus on the critical role of mitochondria in human functioning, oxidative stress, and the biological

M. Kuperberg
Beer Yaakov-Ness Ziona Mental Health Center, Tel Aviv University, Tel-Aviv, Israel

S. L. A. Greenebaum (✉)
Department of Psychiatry, Massachusetts General Hospital, Boston, MA, USA
e-mail: sgreenebaum@mgh.harvard.edu

A. A. Nierenberg
Department of Psychiatry, Massachusetts General Hospital, Boston, MA, USA

Harvard Medical School, Boston, MA, USA

Curr Topics Behav Neurosci (2021) 48: 61–100
https://doi.org/10.1007/7854_2020_152
Published Online: 13 February 2021

mechanisms of mitochondria in BD. Additionally, this chapter will present the evidence that, at least for some people, BD is a product of mitochondrial dysregulation. We review the modulators of mitochondria, the connection between current BD medication treatments and mitochondria, and additional medications that have theoretical potential to treat BD.

Keywords Bipolar disorder · Medication · Mitochondria · Treatment

Patients with bipolar disorder (BD) frequently experience recurrent episodes of depression and mania, along with interepisodic mood and anxiety symptoms. Despite effective treatments, even with the best guideline-concordant pharmacotherapy, relapse rates remain high, and interepisodic symptoms persist. Better treatments are needed. Advances in the neurobiological understanding of BD, coupled with the growing trend of repurposing medications for new indications, have led researchers to study interventions which target mitochondria to treat BD. This chapter aims to provide an introduction to the biological mechanisms underlying mitochondrial function and discuss the evidence for mitochondrial dysregulation in BD. We will then take a look at mitochondrial modulators and review their effects or theoretical potential in treating patients suffering from BD.

1 The Mitochondria

Mitochondria are intracellular organelles that have co-existed symbiotically with eukaryotic cells for nearly 2 billion years, serving many different functions within the cells. Most notably, mitochondria produce essential cellular energy in the form of adenosine triphosphate (ATP), regulate calcium homeostasis, and are involved in cellular death pathways (Cuperfain et al. 2018). We will now explore each of these functions and their relationship with bipolar disorder (BD).

1.1 Mitochondria as an Energy Source

The primary source of energy for the cell comes from glucose. Glucose can be metabolized in three distinct steps, the first occurring in the cell cytoplasm and the latter two in the mitochondria. The first step is *glycolysis*, a process that with a net energetic yield of only two molecules of ATP per molecule of glucose, but this process also provides other molecules essential for further pathways: pyruvate and nicotinamide adenine dinucleotide hydride (NADH). The second step of glucose metabolism is the *citric acid cycle* (*Krebs cycle*), which utilizes pyruvate (obtained either as a by-product of glycolysis from the first glycolysis step or other metabolic sources) with a net yield of one ATP molecule, again providing essential molecules for the final step, NADH and flavin adenine dinucleotide (FADH2), which are passed on to the last phase of cellular respiration. The third and final step is oxidative phosphorylation (OXPHOS), in which the proteins of the electron transport chain (ETC) remove electrons from the substrates created in the previous two steps and shuttle them along with different complexes. This transport of electrons induces pumping of protons from the mitochondrial matrix to the mitochondrial intermembrane space (Bonora et al. 2012), creating the electrochemical gradient that drives the enzymatic conversion of ADP into ATP in the fifth and final complex of the ETC aptly named *ATP synthase* (Cuperfain et al. 2018). All in all, by going all the way through the OXPHOS pathway, one molecule of glucose will generate between 36 and 38 molecules of ATP, making it by and large the most efficient pathway for energy production (Li et al. 2015).

All cells require energy in order to function, but the human brain has an extremely high energetic demand, consuming about 20% of oxygen (at rest) despite weighing only 2% of body weight (Clarke 1999). Neurons require large amounts of ATP to maintain functions such as neurotransmission, neuronal plasticity, protein synthesis, osmolarity, and cell division (Jardim et al. 2018). In order to meet these demands, the mitochondria move to where they are needed, using microtubule networks, to provide energy in regions of intense energy demands, e.g., around synapses (where ATP is used for synaptic transmission) or in areas of neurite outgrowth (Srivastava et al. 2018; McCann and Ross 2018). As one of the major consumers of energy, the human brain may be especially sensitive to dysregulations in mitochondrial functioning, with the implication that even mild disruptions in energy supplies – so mild that they might not present systematically as a mitochondrial disease – could still have neuropsychiatric manifestations (Peters et al. 2004).

As in most other tissues, the brain's primary source of energy is the mitochondria, producing the vast majority of ATP by OXPHOS. While it has previously been thought that only 1% of brain ATP is provided by glycolysis – that first step in glucose metabolism detailed above (Erecinska and Silver 1989) – glycolysis is more central in neurons, providing between 8 and 15% of the brain's ATP (Blazey et al. 2018). Even though glycolysis is a much less efficient route for ATP production than OXPHOS, generating about 17 times less ATP per glucose molecule, it is 100 times faster (Liberti and Locasale 2016). This mean glycolysis can serve as a rapid source of energy under conditions of limited oxygen supply when impairments occur in the OXPHOS pathway for any reason or simply when energy demands exceed supply. The process of glycolysis leads to an accumulation of pyruvate and lactate, which can serve as detectable biomarkers for this pathway's activity (Dogan et al. 2018). Lactate itself is now known to be more than strictly a waste product of glycolysis, with functions as an intermediary in metabolic processes and even as quick accessible fuel (Gladden 2004). While there is no doubt that glucose is the preferred source of energy during both rest and activation, there is an ongoing debate regarding lactate's role in the brain's energy metabolism – some speculate lactate may be preferred when neurons fire at a high rate (Baltan 2015), while others see lactate as an "opportunistic" glucose-sparing substrate (Dienel 2012). Either way, changes in concentrations of lactate might represent mitochondrial and metabolic dysfunction that shift the cell toward reliance on less efficient sources of energy (Lin et al. 2003).

Cells keep their ATP levels tightly controlled within a narrow range, but the ATP molecules themselves cannot be easily stored, due to their low stability in water and the high rate of ATP-dependent processes that would quickly utilize and deplete its levels (Bonora et al. 2012). When ATP levels rise beyond the energetic demand of the cell, excess energy needs to be stored away, and this can be done by attaching ATP's high-energy phosphate to creatine, generating one phosphocreatine (PCr) and one ADP (Guimaraes-Ferreira 2014). This reaction is catalyzed by the enzymes Creatine Kinase (CK), a bidirectional enzyme capable of both transferring a phosphate group from PCr to ADP (forming ATP and creatine) in times of energetic demand and transferring a phosphate from ADP to creatine (forming PCr and ADP) in times of energy surplus. Interestingly, since acute cellular activity depletes ATP, in order to maintain its tightly controlled levels, the cell needs to rapidly break down PCr and reproduce ATP – an action that leads to a measurable drop in PCr. With that mechanism in place, we can measure PCr levels as indicative of the cellular energy storage status (Allen 2012; Du et al. 2018; Sahlin and Harris 2011) – with high levels of PCr reflecting good or excess energy production and low levels of PCr reflecting insufficient energy production. It has even been suggested that chronically depleted PCr might be indicative of cellular hypometabolism due to mitochondrial dysfunction that caused chronic insufficient ATP supply (Modica-Napolitano and Renshaw 2004).

1.2 Other Functions of Mitochondria

As stated before, while the mitochondria are essential for cellular energy metabolism, they take an integral part in many more cellular functions. Our review of these functions is just the tip of the iceberg, as in-depth details of the molecular mechanisms are beyond the scope of this chapter. Interested readers are highly encouraged to delve into the finer details in other sources.

Calcium Homeostasis Calcium (Ca2+) are ions essential for the physiology of the organism, involved in many cellular processes and functions including metabolism, secretion, gene expression, cell survival, and cell death. The mitochondria and endoplasmic reticulum (ER) are two organelles serving as major reservoirs of intracellular calcium (Srivastava et al. 2018), by taking up calcium ions, releasing them back to the cytoplasm, and buffering intracellular calcium concentration to avoid high cytosolic levels that could be toxic to the cell (de Sousa et al. 2014). Calcium ions are not passive players in these interactions, and they can, in turn, affect mitochondrial functioning: not only can they affect mitochondrial membrane depolarization, but well-balanced calcium concentrations also are beneficial for mitochondrial functioning. They can contribute to faster activity of the respiratory chain enzymes and eventual higher ATP output. Calcium ions can also modulate the clearance of reactive oxygen species (ROS) by increasing antioxidant defenses, helping to sustain the increased metabolic rate (Zhang et al. 2016). Despite all these beneficial effects, it turns out excessive calcium can cause problems: an overload of mitochondrial calcium can cascade and culminate in disturbed mitochondrial functioning, increased ROS production, and even cell death with apoptotic or necrotic mechanisms (Javadov et al. 2018). Calcium concentrations, like that of many other cellular components, need to be balanced in order to guarantee maximum gains at minimal costs.

In addition to the effects of calcium on the mitochondria, calcium also affects other biological systems relevant to BD and psychiatric illness. In the brain, calcium is essential for the proper effects of neurotransmission: the release of neurotransmitters can result in a rapid (but transient) rise in calcium levels in the post-synaptic neuron, leading to changes in neuron excitability and membrane structure for both the short- and long-term. Outside of the specific synaptic context, the influx of calcium to the post-synaptic neuron initiates a cascade of signaling events that affect cellular gene expression for processes such as dendritic growth, synapse development, and neuronal plasticity (Greer and Greenberg 2008). Lastly, calcium levels have been shown to have effects on the activity of the circadian clock and circadian rhythms. A construct was shown to be involved in the pathophysiology of BD (McCarthy et al. 2016).

Apoptosis Mitochondria have a major role in apoptosis, or programmed cellular death, via both its intrinsic and extrinsic pathways (de Sousa et al. 2014), for example, due to triggers such as excessively high cytosolic levels of calcium, excessive activity of free radicals, and other perturbations in the balance of pro- and anti-apoptotic factors. Apoptosis is not necessarily a deleterious process, but a physiological part of normal cell turnover and brain development, even in adults. For example, apoptosis can lead to the selective destruction of synapses (Lee et al. 2018; Flippo and Strack 2017) – which at appropriate levels could be related to beneficiary neuroplasticity and pruning, but when utilized inappropriately could result in an extensive synaptic loss (Baranov et al. 2019; de Sousa et al. 2014).

In the context of mitochondria's role, and like any other balanced biological processes, cells have developed defense mechanisms to prevent excessive cellular death by inappropriate apoptotic mechanisms brought about by aberrant mitochondria. By the process of mitochondrial autophagy (mitophagy), cells segregate their damaged components and divide them by mitochondrial fission, degrading the dysfunctional mitochondrion and retaining healthy ones. Modest mitophagy can be compensated for by the cells' mitochondrial reserve and mitochondrial biogenesis to maintain energy production. Excessive mitophagy, however, or lack of compensatory mitochondrial biogenesis, will result in cell death (Kubli and Gustafsson 2012). A dysregulation in the process of mitophagy could lead to accumulation of damaged mitochondria, resulting in decreased energy production, increased oxidative stress (explained in a subsequent section), and decreased mitochondrial calcium buffering capacity – conditions that are especially harmful to postmitotic cells, such as neurons in the brain (Scaini et al. 2019).

2 Reactive Oxygen Species and Oxidative Stress

Mitochondria are a major source of reactive oxygen species (ROS, and to a lesser degree of reactive nitrogen species, RNS). ROS are highly reactive oxygen-containing free radical molecules that continuously form in all aerobic organisms. Free radicals are highly reactive due to their chemical structure, containing a single unpaired electron in their outermost shell of electrons, which makes them able to interact with and damage cellular components such as proteins, lipids, and nucleic acids (Zhang et al. 2016), leading to cell injury and death. For example, in the case of the mitochondria, ROS activity could damage mitochondrial DNA, proteins of the respiratory chain, or lipids of the mitochondrial membrane lipids. This resulting dysfunction could increase membrane permeability and disrupt the calcium homeostasis (Guo et al. 2013), affecting the antioxidant activity and ROS production (Zhang et al. 2016). In this manner, excessive or unopposed ROS activity in the mitochondria could set off a chain of events resulting in cellular damage.

During normal aerobic cellular respiration, a small percentage of the electrons passing through the electron transfer chain (ETC) in the mitochondria can escape, and this leakage can lead to the formation of ROS (Srivastava et al. 2018). This is generally a normal and balanced process, but under pathological conditions – such as diminished or heightened activity of the ETC complexes – mitochondrial ROS formation can increase further, beyond the limits of normal and balanced concentrations (Raha and Robinson 2000). In addition to the electron leakage that happens during oxidative phosphorylation (OXPHOS), many cellular processes (both physiological and pathological) as well as environmental exposures can also lead to generation of reactive species. These include – but are not limited to – immune system activation, tissue ischemia, mental stress, aging, synthesis of prostaglandins, phagocytosis, cytochrome P450 activity, exposure to pollutants, or ionizing radiations and more (Pizzino et al. 2017).

While ROS have potentially harmful effects, they also have important functions in cellular functioning. These reactive species can be used as signaling and modulating molecules, involved in major biological pathways such as adaptation to hypoxia, cell differentiation, and phagocytosis (Sena and Chandel 2012; Pizzino et al. 2017; Droge 2002). To counterbalance the toxic potential, tight regulation is required so that ROS could participate in physiological cell signaling without causing structural damage or cellular death. Cells have developed antioxidant defenses and ROS scavenging capacities, allowing moderate levels of ROS for their beneficial roles while preventing their accumulation to dangerous levels. Antioxidant defenses include endogenous molecules such as glutathione (GSH), catalase (CAT), superoxide dismutase (SOD), coenzyme Q10, melatonin or uric acid, and exogenous antioxidants such as vitamin C, vitamin E, zinc, and drugs, for example, acetylcysteine, a precursor of GSH (Liguori et al. 2018). GSH is the most important antioxidant molecule, and the mitochondria contain about 10–15% of the total GSH in the cell (Matschke et al. 2019). It should be noted, of course, that medicine and biology are rarely unidirectional – the oxidative balance is a complex process, with many antioxidants also having pro-oxidant abilities under certain conditions or concentrations (Matschke et al. 2019).

If the ROS-homeostasis is perturbed, due to either increased production or reduced antioxidant defenses, cells might reach a state termed *oxidative stress*, meaning the balance between pro-oxidant and antioxidant mechanisms has shifted in favor of the pro-oxidants (Pizzino et al. 2017). When oxidative stress occurs, cells are exposed to oxidative damage of essential macromolecules including DNA, lipids, proteins, and carbohydrates. The consequences of damage caused by free radicals can be extremely deleterious and lead to many ailments, both acute and chronic, including cancer, cardiovascular and neurodegenerative disorders, as well as aging processes (Pizzino et al. 2017; Pham-Huy et al. 2008). In particular, the mitochondrial DNA (mtDNA) that has no protective histones and lacks repair mechanisms present for nuclear DNA is particularly vulnerable to damage caused by such oxidative modifications (McCann and Ross 2018). Researchers have sought whether increasing the body's antioxidant defenses could slow or even stop the progression of many of these diseases, with generally conflicting results (Liguori et al. 2018).

The brain is especially vulnerable to shifts in the ROS balance and oxidative stress for several reasons: first, because of its high metabolic rate and high lipid content, it is easier to achieve high concentrations of reactive species in the brain, naturally shifting the balance in favor of the pro-oxidants (Massaad and Klann 2011). Second, neurons have lower levels of antioxidant defenses than other cells in the nervous system, naturally making the shift toward the pro-oxidant factors easier (Steckert et al. 2010; Matschke et al. 2019). Third, since neuronal cells are mostly postmitotic, they accumulate damage throughout life (Matschke et al. 2019). Studies of the nervous system have shown that psychiatric patients endure excessive levels of oxidative stress (Srivastava et al. 2018), and this is a possible mechanism by which neuropsychiatric illnesses result in pathophysiologic changes – and a possible mechanism that can be targeted by medications.

3 Mitochondria in Bipolar Disorder

Bipolar disorder (BD) is a neuropsychiatric disorder with a cyclical nature, alternating between periods of euthymia, mania, and depression. The precise pathophysiological pathways responsible for BD remain elusive even after many years of research, likely due to multifactorial etiology involving many different molecular disturbances, arising from mutations in many small genetic risk areas and accumulation of environmental stressors (Szczepankiewicz 2013).

Originally, BD was thought to reflect an adrenergic-cholinergic imbalance, with higher adrenaline activity in mania and higher acetylcholine activity in depression. This hypothesis has since been updated, describing mania as a state of higher *catecholaminergic* status (i.e., both dopaminergic and adrenergic) relative to cholinergic status and vice versa in depression. Medication studies generally supported this hypothesis: drugs that increase acetylcholine in the central nervous system (CNS) increase depressive symptoms, and drugs that increase adrenaline and dopamine can induce or exacerbate manic symptoms (van Enkhuizen et al. 2015). The efficacy of selective serotonin reuptake inhibitors (SSRI) in the treatment of depression and their propensity to induce a switch to a manic state in patients with BD have led to a more general "monoamine dysregulation" theory of BD that includes not only catecholamines but also serotonin. This hypothesis is backed by the effects of serotonergic agents but also by molecular imaging studies, and associated genetic studies pointed at genes that influence monoamine systems, including transporters and catalyzers (Sigitova et al. 2017). However, decades of studies and many similar drug developments later, there is disagreement regarding the precise mechanisms causing the monoamine imbalances. Nowadays, it is apparent that additional neurotransmission systems participate in the pathophysiology of BD other than the monoamines, including glutamatergic, GABAergic, and even opioid dysregulation (Blacker et al. 2017; Lutz and Kieffer 2013).

Despite the many drugs that have gained FDA approval for BD, current treatment options remain unsatisfactory, and many patients continue to experience intra-episodic symptoms or even full-blown episodes that are resistant to treatment, particularly of the depressive polarity (Geddes and Miklowitz 2013). In the search for greater pathophysiological understanding and effective treatment modalities, research has been inspecting the role of mitochondrial dysfunction and oxidative damage in BD. This idea has been proposed by Kato and Kato in 2000 and is based on relevant study findings, the increased likelihood of maternal inheritance in generational transmission of BD, abnormal findings in the mitochondrial DNA (mtDNA) of patients with BD, and comorbidity of affective disorder with mitochondrial disorders (Kato and Kato 2000; Kato et al. 2018). For example, the manic phase in BD seems to be particularly prone to increases in energy production and oxidative stress, with manic patients showing increased mitochondrial respiration and ATP utilization (Weber et al. 2013). One proposed mechanism for these observations is that an initial increase in oxidative stress at the beginning of a manic episode leads to increased mitochondrial functioning, with consequent increases both in ROS production and in the activity of mitigating factors that are attempting to control the potentially toxic state of oxidative stress. By increasing antioxidant defenses, mitochondrial activity (and ROS production) can be elevated without yet triggering apoptotic mechanisms. Eventually, however, the defensive pathways become overwhelmed, mitochondrial functioning begins to deteriorate, and cellular damage occurs. Nearing the end of the manic episode, oxidative stress levels decrease, potentially provoking a transition from mania to euthymia or depression (Morris et al. 2017). This is just one proposed mechanism, and not all evidence is consistent with it.

While behaviorally, BD can easily be described as a disorder of energy levels – an abundance in mania versus a deficit in depression – attempts at direct translations from the phenomenological level to the molecular level are scantily successful, and evidence is accumulating that aberrations in bioenergetic mechanisms exist in *all* phases of the illness. In addition to the specific processes and pathways, patients with

BD demonstrated abnormalities in the structure and distribution of mitochondria within the cell, both in the brain and peripheral cells (Cataldo et al. 2010). Results from different studies are often inconsistent, limited by a small number of patients, different stages of disease, and different pharmacotherapies. It seems that while the mitochondrial function is overall somewhat disrupted, the specific pattern of disruption is inconclusive. Below we will explore potential molecular mechanisms by which mitochondrial dysfunction could be associated with BD.

3.1 Possible Mechanisms of Mitochondrial Dysfunction in Bipolar Disorder

3.1.1 A Shift from OXPHOS to Glycolysis

Studies demonstrate a shift from OXPHOS to glycolysis in at least several brain areas, if not globally, as evident by decreased phosphocreatine (PCr) and increased lactate in magnetic resonance spectroscopy (MRS) studies (Dudley et al. 2015, 2016; Dogan et al. 2018; Nierenberg et al. 2013). As a quick reminder, glycolysis is the fastest but least efficient pathway from glucose to ATP, while OXPHOS takes (relatively) much more time but yields many more ATP molecules per glucose. PCr serves as the cell's energy storage with lower PCr levels implying energy stores are low, either due to decreased production or increased utilization of ATP. Cellular conditions of high energetic demand, or a dysfunction in the OXPHOS pathways, could lead to increased reliance on glycolysis for ATP supply with the associated lactate accumulation and reduced energy production (Callaly et al. 2015). A 2018 meta-analysis reported that overall, and despite some contradictory findings, lactate levels were indeed elevated in the brains of patients with BD compared to healthy controls. That treatment with a mood stabilizer was somewhat able to restore brain lactate to levels comparable to healthy control (Kuang et al. 2018). These findings suggest that patients with BD indeed shift away from OXPHOS toward the largely inefficient process of glycolysis as an energy source, leading to reduced total energy output of the mitochondria.

Electron Transport Chain (ETC) Interestingly, it has been shown that under normal conditions, the expression of ETC genes was the same in patients with BD as in healthy controls, but during stress conditions of glucose deprivation where upregulation of ETC genes is expected and was indeed observed in healthy controls, patients with BD displayed *reduced* expression for the entire ETC. This pattern is consistent with the notion of mitochondrial inability to adapt to energetic stress in BD (Naydenov et al. 2007). Many studies detected disturbances in the activity and regulation of complexes of the ETC, but the direction of change is inconsistent. Most likely, these changes are dependent on the polarity of the current mood episode and on the specific brain and region assessed and possibly could also be affected by the type of psychotropic medication the patients received before assessment (Holper et al. 2019).

- *ETC Complex I*: Complex I is the primary site of ROS production, and its inhibition is linked to decreased energy production (Callaly et al. 2015) and increased oxidative stress (Onukwufor et al. 2019). Compared to healthy controls, a 2010 postmortem study found decreased activity and levels of complex I in the prefrontal cortex (PFC) of patients with BD. No differences were detected between patients who were treated with antipsychotic medications, known to inhibit complex I activity and those who were not (Andreazza et al. 2010). More recently, 2018 meta-analysis found complex I subunits expression to be generally *reduced*, but certain units' expression increased (Holper et al. 2019), while a 2019 study found *increased* complex I activity in medicated patients with bipolar depression. It has been speculated that this increased activity is a compensatory response to decreases in activity of other complexes (Zverova et al. 2019).
- *ETC Complex II*: During a bipolar depressive episode, activity of complex II decreased when compared to euthymic patients, and the activity was also negatively correlated with scores on a depression scale (Valvassori et al. 2018; Zverova et al. 2019) – that is, the lower the activity of complex II, the more severe their depression.
- *ETC Complex IV*: During a bipolar depressive episode, levels of complex IV were decreased (Zverova et al. 2019).

3.1.2 Creatine Kinase

CK is the enzyme responsible for the reversible transfer of a high-energy phosphate group between ATP and creatine, creating PCr stores when supply exceeds demand or rapidly generating ATP in times of energetic need (Yuksel et al. 2015; MacDonald et al. 2006). During periods of tissue activation, the predicted response observed in healthy individuals is a reduction in PCr levels with no change in ATP levels, reflecting the cell's ability to regenerate ATP from PCr – and the importance of constant ATP levels. In patients with BD who were undergoing a visual stimulation test, ATP levels decreased, while PCr levels did not change, implying no generation of ATP despite the need. Suspecting a specific defect in the CK reaction, a recent study found a significant reduction in the reaction rate constant of the CK enzymes in patients with BD, compared to healthy controls. This implies that patients with BD have abnormalities in their ability to generate ATP during times of stress, despite normal levels at rest – and this might be related to the patients' sensitivity to conditions of increased stress (Du et al. 2018).

3.1.3 Calcium

Impaired regulation of calcium signaling is a highly reproducible abnormality in BD. Peripheral blood cells of patients with BD show elevated calcium, associated with dysfunction in both endoplasmic reticulum (ER) and mitochondrial activity (Machado-Vieira et al. 2011). Dysregulated calcium homeostasis has also been associated with excessive ROS production (Fonseca et al. 2015). Interestingly, studies of patients with the mitochondrial disease chronic progressive external ophthalmoplegia (CPEO) showed that 16–21% of them were diagnosed as having BD, a rate much higher than the population reported prevalence of 1%. Mice with mutants in the gene responsible for CPEO, polymerase gamma (POLG, which is a mitochondrial DNA polymerase), exhibited depressive and manic-like symptoms, were affected by treatment with lithium, and also showed altered calcium signaling (Kato 2019).

Calcium signaling is affected by the *inositol signaling pathway* – inositol being an essential substrate for signaling molecules, involved in multiple cellular events. Of particular relevance, this pathway can lead to increased calcium release in the brain, and when inositol levels are deficient, the calcium response mediated by this pathway will be attenuated (Harwood 2005). Studies have shown that optimal mitochondrial functioning is required for common pathways of inositol-triggered calcium release, tying together inositol and mitochondrial functioning (Wilson et al. 2019). The "inositol depletion hypothesis" of BD claims that the periodic switching between manic and depressive episodes results from neuronal activity driven by an altered inositol signaling and calcium signaling (Berridge 2014). Higher inositol signals were detected during the manic phase and lower levels during the depressive phase (Yu and Greenberg 2016). Lithium, a commonly used and useful drug in the treatment of BD, reduces the supply of inositol by inhibiting inositol-monophosphatase (IMPase) – a key enzyme in the production of inositol – implying lithium can prevent abnormal increases in calcium as part of its therapeutic actions (Kato 2019).

3.1.4 Increased Oxidative Stress

Many studies have shown increased levels of lipid and protein peroxidation, reflecting oxidative damage, in patients with BD compared to healthy controls (Akarsu et al. 2018; Brown et al. 2014). Whether these findings reflect a state marker of acute episodes or a trait marker of BD, and precisely how they result from an overproduction of free radicals or decreased antioxidant defenses, are points of disagreement – with different studies reporting contradictory results (Valvassori et al. 2018; Siwek et al. 2016; Kim et al. 2017). For example, some studies find levels of the important antioxidant GSH to be reduced (Rosa et al. 2014; Tsai and Huang 2015), others find GSH levels increased (Ngamchuea et al. 2018), and others are not able to detect any significant differences from healthy controls at all (Lagopoulos et al. 2013; Soeiro-de-Souza et al. 2016). These are but a few examples and many other factors are at play in the ROS balance.

3.1.5 Neurotransmitters

One interesting focus of study combines oxidative stress, mitochondrial functioning, and neurotransmitter hypothesis of BD pathophysiology. This merge of fields could mean that the findings are links of the same chain, each affecting the others, and not separate pathophysiologic pathways.

Dopamine is a catecholamine neurotransmitter with many functions in the brain, including essential roles in the modulation of behavior and cognition and also serving as a precursor for (nor)epinephrine production. Dopamine is central in various neural pathways related to motivation, reward, sleep, mood, attention, and learning (Juarez Olguin et al. 2016). Studies have found that in patients with BD, the density of dopamine receptors is increased during psychotic mania, and manic patients also demonstrate hyperactivity in dopaminergic circuits of the reward system. Studies are less consistent regarding dopamine and its receptors during times of euthymia and depression (Ashok et al. 2017). Regarding mitochondrial and brain functioning, dopamine has complex actions. On the one hand, elevated levels can lead to mitochondrial dysfunction by inhibiting mitochondrial motility (Chen et al. 2008), inhibiting subunits of the ETC (Czerniczyniec et al. 2007) and increasing oxidative stress with possible eventual cell death (Monzani et al. 2019). On the other hand, dopamine also has neuroprotective properties, working in synergism with uric acid to repair DNA damage and protecting neurons against glutamatergic excitotoxicity (Vaarmann et al. 2013) – which is thought to be a possible process in BD.

Glutamate is the main excitatory neurotransmitter in the brain, primarily binding to *N*-methyl-D-aspartate (NMDA) receptors and affecting calcium channels. As described before, calcium is involved in many cellular pathways, and the mitochondria and the ER tightly regulate its levels. Activation of glutamate receptors induces calcium influx into the cytosol of neurons, where the mitochondria buffer the ions. Increased levels of glutamate may lead to calcium overload – also known as *glutamate excitotoxicity* – a process that can result in overproduction of ROS and eventual cell death. Abnormal increases in glutamate have been reported in patients with BD (Bustillo et al. 2019), and several mood stabilizers are capable of regulating its levels (Soeiro-de-Souza et al. 2018a). A possible mechanism is that a genetic mutation in mtDNA of patients with BD leads to abnormalities in the mitochondria's ability to buffer calcium, resulting in a decreased capacity to handle glutamate excitotoxicity without resorting to apoptosis (Callaly et al. 2015).

3.1.6 Brain-Derived Neurotrophic Factor (BDNF)

BDNF is a type of neurotrophic factor, which are molecules essential for neuronal proliferation and differentiation during brain development, and that also play critical roles in plasticity, survival, and connectivity in the adult brain. BDNF also has various neuroprotective effects – mitigating glutamate excitotoxicity, anti-apoptotic, and antioxidant activities, at times as a response to mitochondrial dysfunction (Chen et al. 2017; Markham et al. 2014). Impairment in BDNF signaling have been found in a wide range of neurologic and psychiatric disorders, including BD: patients during acute episodes, whether manic or depressive, had significantly lower levels of BDNF compared to healthy controls, while during euthymia no differences were detected (de Oliveira et al. 2009). The decreased levels of BDNF may make patients with BD less resilient to small impairments in mitochondrial functioning.

3.1.7 NAA

N-Acetylaspartate (NAA) is a metabolite highly concentrated in neurons, with roles in myelination and energy metabolism in mitochondria (likely through the generation and use of acetyl-CoA). Levels of NAA reflect the health of neurons and their mitochondria and act as reliable markers for neuronal energy impairment or dysfunction (Moffett et al. 2013). NAA can be measured using proton magnetic resonance spectroscopy (^{1}H-MRS) studies, but results are difficult to interpret as NAA levels are sensitive to pharmacological and behavioral treatments, the current mood state, and illness duration. In light of this, and as seems to be the theme when attempting to study a topic as diverse as "mitochondrial functioning" – different studies often end up with conflicting results (Soeiro-de-Souza et al. 2018b). Overall, studies do indicate abnormalities in energy balance, evidenced by differences in NAA levels between patients with BD and healthy controls.

3.1.8 Bcl-2

The anti-apoptotic protein B cell lymphoma protein-2 (Bcl-2) prevents cellular apoptosis by preventing the release of cytochrome-c in the mitochondria. Increased Bcl-2 activity is known to correspond to an increase in OXPHOS and ATP production (Manfredi et al. 2003), and elevated levels of Bcl-2 can increase the mitochondria's capacity for calcium uptake and resistance to calcium influx, preventing impairments due to incapacitating calcium overload (Murphy et al. 1996). In patients with BD, levels of Bcl-2 were found to be lower in the prefrontal cortex (Kim et al. 2010), while serum levels during a manic episode were negatively correlated with the severity of manic symptoms (Chen et al. 2015) – that is, lower levels of Bcl-2 were correlated with more severe mania. It should also be noted that BDNF exerts part of his anti-apoptotic effects via the Bcl-2 protein (Chen et al. 2017), and as noted before, its levels were found to be decreased in patients with BD.

3.2 Mitochondrial Genes

The vast majority of genes involved in mitochondrial function are located in the cell nucleus (about 1,200), but 37 genes are encoded directly by mitochondrial DNA (mtDNA) located within the organelle itself (Chinnery and Hudson 2013). mtDNA does not undergo recombination during meiosis like nuclear (autosomal) DNA, is strictly maternally inherited through the ovum, and lacks the protective repair mechanisms nuclear DNA has – leading to about 10 times the frequency of polymorphisms (Howell et al. 2003). The close physical proximity of mtDNA to ROS generated by the mitochondrion itself might be another factor in its high mutation rate (Pei and Wallace 2018). Mitochondria can increase the numbers of mtDNA copies when needed by using the polymerase POLG, and alterations in whole blood mtDNA copy number are considered indicative of mitochondrial dysfunction (Yamaki et al. 2018). Since POLG is a protein prone to suffering from oxidative damage, it has been postulated that oxidative stress could be causing POLG downregulation in acute phases of BD, resulting in the observed decreases in mtDNA copy numbers during those episodes (Wang et al. 2018). A recent study has shown that in patients with BD, the levels of mtDNA copy numbers are significantly lower during acute episodes (of either mania or depression) and at least in mania are inversely correlated to the number of previous mood episodes – that is, the more previous episodes a patient has had, the lower their mtDNA copy numbers. Studies on the topic are less consistent when assessing periods of euthymia, with some finding a reduction in mtDNA copy numbers even then, some limiting their findings to BD type I or advanced disease only, and others still finding no significant differences when comparing patients with BD to healthy controls (Kim et al. 2019; Yamaki et al. 2018). It should be noted that certain antipsychotic agents by themselves are associated with decreasing the mtDNA copy number (Kumar et al. 2018), and this might be at the root of some of these contradictory findings on the matter. In addition to these examples, many studies on patients with BD have found loci of interest in areas related to mitochondrial functioning. Since this chapter's aim is to focus on potential treatments for BD, we will not explore this topic further.

4 How Conventional Drugs for Bipolar Disorder Relate to Mitochondrial Functioning

Current available treatments for bipolar disorder (BD) address different pathways and, generally, have multiple molecular targets. Many of these drugs have effects on the mitochondria or related pathways, with more significant effects observed in patients who have shown a clinical response to the drug in question. While these pathways are not necessarily the drugs' sole mechanism of action in BD – they are more pieces of the puzzle.

Electron Transport Chain (ETC) Lithium seems to overall increase complex I expression and activity, albeit inconsistently (de Sousa et al. 2014, 2015). Lithium-responsive patients showed differential expression of mitochondrial-related genes involved in the ETC and OXPHOS, compared to lithium-unresponsive patients

(Stacey et al. 2018). Another mood stabilizer, valproate, also has effects on the ETC and is able to reverse methamphetamine-induced inhibition of ETC complexes in the brains of rodents (Valvassori et al. 2010).

Antioxidant Properties Both lithium and valproate increase the levels of the most important antioxidant in the brain, GSH (Nascimento et al. 2015; Chiu et al. 2013), and the mood stabilizer lamotrigine appears to have antioxidant properties as well (Ozkul et al. 2014; Kim et al. 2007). Patients with BD who were treated with lithium had lower levels of lipid peroxidation markers during mood episodes, with the effect more pronounced in patients who also demonstrated a clinical response to lithium (Data-Franco et al. 2017). Patients with BD suffer from more oxidative stress, especially during acute episodes, and so it is possible that reducing oxidative stress is one of the mechanisms by which these drugs exert their clinical effects.

Anti-apoptotic Properties Lithium, valproate, and electro-convulsive treatment (ECT) all increase the expression of the anti-apoptotic protein Bcl-2, which has the ability to inhibit mitochondrially mediated apoptosis, consequentially increasing the mitochondria's resistance to toxic effects in the environment (Orrenius 2004). It should be noted though that these results are not always reproduced, and studies on the matter are plagued by methodological difficulties (Odeya et al. 2018).

Hyperexcitability Pluripotent stem-cell (iPSC) technology allows researchers to take fibroblasts and differentiate them into neurons. When using fibroblasts obtained from patients with BD type I, these neurons then showed hyperexcitability, smaller mitochondria size, and enhanced mitochondrial function, with the authors suggesting these mitochondrial properties lead to the excessive neuronal activity. Of relevance, in neurons derived from patients who showed a clinical response to lithium, applying lithium to the cells had profound effects on hyperexcitability and mitochondria size. In contrast, in cells derived from lithium-nonresponders, the drug did not induce any obvious changes. Further analysis showed that when applying lithium to cells derived from the lithium-responsive patients, a change was observed in the expression of 560 genes, compared to merely 40 genes in the lithium-nonresponsive group (Mertens et al. 2015; Stern et al. 2018).

Calcium Mood stabilizers can modulate calcium channels and increase levels of brain-derived neurotrophic factor (BDNF), a neurotrophic factor with the ability to mitigate glutamate excitotoxicity (Callaly et al. 2015; Data-Franco et al. 2017; Kato 2019). In addition, and as discussed previously, lithium is an inhibitor of IMPase decreasing levels of inositol and attenuating the intra-cellular calcium response (Harwood 2005). It is possible that reducing calcium overload is one of the pathways targeted by these drugs.

Lactate Lithium and valproic acid have shown to normalize elevated lactate levels in patients with BD (Kuang et al. 2018), while quetiapine – an antipsychotic drug – is able to decrease lactate levels in manic patients, again with a more prominent decrease in patients who have shown a clinical response to treatment (Kim et al. 2007). These findings are likely not causal, as increased lactate levels could be indicative of a larger problem these drugs are targeting, but it is yet again tying together treatments for BD with mitochondrial dysfunction.

5 Mitochondrial Potential Treatments

After conceptualizing bipolar disorder (BD) as a disorder of mitochondrial and energetic dysregulation, we can now explore mitochondrial treatment modalities. Research has mainly focused on attempting to *balance* mitochondrial functioning – for example, enhancing it in depression and diminishing during mania or increasing antioxidant defenses to prevent damage from ROS due to insufficient antioxidant activity or increased cellular respiration (Nierenberg et al. 2013). One thing to keep in mind is that mitochondrial modulators may take a long time to bring about clinical effects, certainly longer than antipsychotics or mood stabilizers, and are hence often explored as adjuvant therapeutic options and not necessarily monotherapy (Berk et al. 2008).

5.1 *Likely Beneficial*

5.1.1 PPAR Agonists

The PPARs (peroxisome proliferator-activated receptor) are a family of three receptors in the cell's nuclear, with the isomers α, δ, and γ. These receptors function as transcription regulators, regulating the expression of genes related to energy homeostasis, oxidative stress, metabolism, cell differentiation, inflammation, and importantly – mitochondrial biogenesis. In the brain, they also regulate genes related to excitatory neurotransmission and myelination (Grings et al. 2017; Nierenberg et al. 2013). Traditionally, drugs targeting the PPARs were used in general medicine to treat hypertriglyceridemia and type II diabetes, but their wide array of effects makes them potential therapeutic targets to ameliorate many other disorders, including mitochondrial dysfunction. Of specific relevance to BD, the gene for a coactivator of PPAR, PGC-1α, has been weakly linked to clinical response to lithium (Geoffroy et al. 2016).

A PPAR agonist may target one of the receptor isomers selectively or several at the same time. The selective PPAR-γ agonists and antidiabetic drugs troglitazone, rosiglitazone, and pioglitazone have shown mostly positive effects in the treatment of depression, either alone or as add-on therapy. Alas, troglitazone has been withdrawn from the market due to hepatotoxicity, and rosiglitazone use is strictly limited by the FDA (and suspended by the European Medicine Agency) due to increased cardiovascular risk. This leaves mainly pioglitazone, but its use is also severely limited by side effects such as weight gain, congestive heart failure, edema, and bone fractures (Colle et al. 2017).

Another promising member of the PPAR-agonists family is bezafibrate, a pan-agonist that targets all isomers of PPAR – α, δ, and γ. Bezafibrate is used in general medicine as a common treatment for hypertriglyceridemia, can increase mitochondrial biogenesis, and also has antioxidant and anti-inflammatory effects (Grings et al. 2017). It is known to have favorable safety and side effect profile, and a clinical study attempting to repurpose this drug for bipolar depression is currently underway (Bezafibrate Treatment for Bipolar Depression: A Proof of Concept Study 2015).

5.1.2 Minocycline

Minocycline is a safe and well-tolerated tetracycline antibiotic. It was also found to have multiple interlocking mechanisms that converge on neuroprotective properties relevant to the putative mitochondrial pathophysiology of BD – with effects as an antioxidant, anti-apoptotic, and glutamate neurotransmission modulator. Minocycline is thought to exert its neuroprotective effects by reducing oxidative damage, scavenging ROS and normalizing levels of GSH and markers of oxidative stress, and mitigating glutamate excitotoxicity. It can also modulate apoptotic signaling pathways and increase the expression of the anti-apoptotic gene Bcl-2, affecting the mitochondrial pathway of apoptosis (Zheng et al. 2019; Shultz and Zhong 2017).

Minocycline has been shown to induce psychotropic effects in schizophrenia, improving symptoms and gray matter volume (Robertson et al. 2019). Its effects on mood disorders have not been as consistent, with some studies showing beneficial effects and others failing to detect differences from placebo (Zheng et al. 2019). Two studies are currently underway, assessing the efficacy of minocycline as add-on treatment for bipolar depression (A Pilot Study Investigating the Efficacy of Minocycline and N-Acetyl Cysteine for Bipolar Depression 2016; Minocycline for Bipolar Depression 2012).

5.1.3 N-Acetyl-Cysteine (NAC)

NAC has several mechanisms of action that make it relevant to psychiatric disorders. It has well-known anti-inflammatory, antioxidant, and glutamatergic modulating effects, including protection from glutamate excitotoxicity (Samuni et al. 2013; Naziroglu et al. 2013), and rodent models for neurodegenerative diseases have shown it is able to restore mitochondrial respiration and complex activity (Pereira et al. 2018). Among its many effects, NAC provides the rate-limiting cysteine for GSH production, and supplementation of NAC can increase GSH synthesis (Berk et al. 2008). By increasing GSH, NAC is both protecting the cell from oxidative damage and augmenting its mitochondrial respiratory capacity – as it is now able to withstand more of its own noxious by-products. Its effects on the glutamatergic system are of particular interest, as this system has been correlated with increased impulsivity (Ende et al. 2016; Pattij and Vanderschuren 2008), a trait often associated with BD. It is this effect that might be behind NAC's beneficial effects in the treatment of pathological gambling, a disorder characterized by deficiencies in impulse-control (Grant et al. 2014).

To date (Pereira et al. 2018), three trials have assessed NAC as an adjunctive treatment for bipolar depression, one study in 2008 and two in 2019. Two of the three have shown a similar pattern of results – improvement in depressive symptoms, but one that is only evident only after a long duration of treatment (between 16 and 20 weeks) (Berk et al. 2008; Bauer et al. 2018). The third study did not find any significant differences between the placebo and NAC at any time point. However, it did note that the group who received a combination of NAC and other mitochondrial agents (but not NAC alone) has shown improvement 20 weeks after discontinuation of the study drugs – suggesting either the delayed onset of effects or improvement upon withdrawal of the medications (Berk et al. 2019). This study also mentioned an increase in manic symptoms at week 4 for some participants. While this finding did not survive correction for multiple testing, it still brings to light the possibility of inducing a manic switch by driving mitochondrial biogenesis – even without improvement in depressive symptoms. All in all, it seems that NAC might indeed possess antidepressant effects that only become manifest after continued long-term treatment.

5.1.4 Co-enzyme Q10

Co-Q10, also known as ubiquinone, plays many roles in the cell: It is a vital cofactor in the mitochondrial ETC, assisting in the shuttling of electrons between subunits and helping to establish the proton gradient needed for ATP production (Neergheen et al. 2017). Co-Q10 also has antioxidant and anti-inflammatory properties, is involved in DNA replication and repair, supports membrane stabilization, mitigates glutamate-induced excitotoxicity, serves as an essential cofactor of uncoupling proteins, and can regulate gene expression and programmed cell death (Mantle and Hargreaves 2019; Alcazar-Fabra et al. 2018; Kumari et al. 2016). Most of the body's CoQ10 requirements are derived from endogenous synthesis that declines with age naturally (Morris et al. 2013). Low Co-Q10 levels are associated with various pathological conditions, such as hypertension, diabetes, cardiovascular diseases, and importantly neurological and psychiatric disorders (Morris et al. 2013; Maes et al. 2009). In general medicine, administration of supplemental Co-Q10 has shown, if inconsistently, to reduce inflammatory mediators (Fan et al. 2017); to have beneficial effects in medical conditions such as heart failure, atherosclerosis, hypertension, hyperlipidemia, and diabetes (Garrido-Maraver et al. 2014); and even to reduce symptoms in neurodegenerative disorders such as Parkinson's and multiple sclerosis (Sanoobar et al. 2015; Shults et al. 2002).

Regarding BD, two recent studies examining adjuvant Co-Q10 have shown it can reduce symptoms of depression (Forester et al. 2015; Mehrpooya et al. 2018), with one of the studies noting the effect might be delayed and not prominent until at least 8 weeks of use.

5.1.5 Melatonin

Melatonin is a hormone synthesized by many tissues, known mostly for its regulatory effects on rhythmic processes such as circadian rhythm, growth hormone stimulation, and insulin secretion. Evidence has shown melatonin, and its metabolites have antioxidant properties, acting as direct scavengers of ROS, stimulators of GSH production, and increasers of mRNA expression of antioxidant genes. In the mitochondria, melatonin can increase the activity and expression of ETC proteins, increase mitochondrial biogenesis, increase mitochondrial membrane fluidity, and prevent changes to the mitochondrial permeability transition pore (MPTP), protecting the organelle from calcium overload. It has additional effects on mitochondrial gene expression, by preventing degradation of mitochondrial DNA (mtDNA) and modulating the expression of genes encoded by mtDNA (Reiter et al. 2017; Acuna-Castroviejo et al. 2007; Hardeland 2017; Jou 2011). The net result of all these processes is an increase in OXPHOS and decreases in ATP depletion and cell death.

Patients with BD were found to have decreased serum levels of melatonin, with possible variations dependent on mood state (Novakova et al. 2015). In addition, about half of patients with BD show circadian rhythm disturbances that are associated with a higher risk of BD relapse (Kishi et al. 2019). Studies researching the effects of treatment with melatonin receptor agonists have shown mixed results. A 2019 meta-analysis found ramelteon to be well-tolerated and clinically superior to placebo in the prevention of a depressive relapse, with no effects on scales of mania, sleep, quality of life, or other causes for relapse such as a manic or mixed episode. While this result is promising – especially in light of melatonin's ability to improve cardiometabolic outcomes for patients receiving antipsychotic treatment – it is limited by the small number of studies and tendency for conflicting results in the scientific literature (Kishi et al. 2019).

5.2 Theoretically Beneficial, but No Studies Have Been Published

5.2.1 Ebselen

Ebselen is a seleno-organic compound – that is, a compound containing carbon-to-selenium chemical bonds. It has an antioxidant and ROS scavenging activity similar to that of glutathione (GSH) and has been shown to protect neurons from damage caused by ischemic and glutamate excitotoxicity (Jia et al. 2018; Slusarczyk et al. 2019). Ebselen also works as an inhibitor of inositol monophosphatase (IMP), making it an IMPase that can effectively lower inositol levels in the brain and prevent abnormal, potentially toxic increases in calcium levels. As discussed, inhibition of IMP is also one of the key effects of lithium when used at clinically effective doses (Forlenza et al. 2014). This shared mechanism of action, combined with studies showing ebselen can alter emotional processing and impulsivity in humans and reduce symptoms in animal models of mania and depression, makes it a candidate for the treatment for BD, perhaps as a "lithium-mimetic" with good safety and tolerability (Masaki et al. 2016). A clinical trial is currently underway to assess whether Ebselen can reduce symptoms of (hypo)mania in patients with BD (Ebselen as an add-on Treatment in Hypo/Mania 2017).

5.2.2 Mangosteen

Garcinia mangostana Linn, commonly known as mangosteen, is a tropical fruit with antioxidant, anti-inflammatory, and anti-apoptotic effects. Studies suggest mangosteen might have specific effects on the GSH system, with the ability to increase the levels of this protective antioxidant.

Rodent models have shown mangosteen pericarp to have antidepressant – and antipsychotic-like effects, specifically affecting the serotonergic system, which is implicated in the pathophysiology of both conditions. Human studies have shown some efficacy of mangosteen as an anti-inflammatory and antioxidant agent. However, most trials included mangosteen pericarp in combination with other bioactive compounds – and by doing so have limited our ability to conclude specifically on mangosteen's effects. Only one study directly assessed mangosteen's utility in mental health in a randomized controlled study (RCT), finding it was able to improve scores of psychotic and depressive symptoms in patients with schizophrenia or schizoaffective disorder (Ashton et al. 2019). As mangosteen pericarp has a good safety profile and is generally well-tolerated, these preliminary studies justify further assessment of its effect on mental health patients – all the while keeping in mind potential pharmacological interactions, especially due to serotonergic activity.

5.2.3 Ketogenic Diet

A ketogenic diet is a high-fat, very low-carbohydrate diet that forces the body to rely on ketone bodies as a source of energy instead of dietary glucose. Briefly, by depriving the body of carbohydrates that can be converted into glucose, the liver turns instead to convert fat into fatty acids and ketone bodies, who then pass into the brain and replace glucose as an energy source (Brietzke et al. 2018). In regard to mitochondrial functioning, the ketogenic diet has been shown to induce an increase in mitochondrial biogenesis proteins, OXPHOS and subsequent ATP production, increase levels of the antioxidant GSH, reduce ROS production, and induce epigenetic changes in genes related to the mitochondria (Campbell and Campbell 2019). Animal models have shown the ketogenic diet to have antidepressant-like effects similar to those of antidepressant drugs, and several human case reports have been published detailing patients with treatment-resistant mood disorders who have responded to a ketogenic diet (Kovacs et al. 2019). The ketogenic diet is a particularly interesting option because of patients' own reports of improvements in their symptoms when adhering to it (Campbell and Campbell 2019), and even though this treatment is not without its side effects – notably gastrointestinal disturbances, dyslipidemia, and renal calculi – for patients who are able to comply with the diet, it might yet prove to be a beneficial solution.

5.2.4 Resveratrol

Resveratrol is a polyphenol naturally found in grapes and berries and has been considered for many years to be part of the mechanism for the "French paradox" – a term coined in 1992 to describe the low incidence of coronary heart diseases in France, despite a diet high in saturated fats. Originally, it was proposed that moderate red wine consumption – and hence resveratrol consumption – explained this unexpected finding, although it is now known that its levels in red wine are likely not high enough to account for the paradox fully. Regardless, resveratrol has many health and longevity promoting effects, with antioxidant, anti-inflammatory, anti-apoptotic, and even anti-carcinogenic effects (Jardim et al. 2018). Resveratrol has the ability to modulate the central nervous system, increase levels of monoamines, and act as a neuroprotective agent and has been studies in animal models as a potential aid for improving sleep quality, anxiety, and depression (Moore et al. 2018). Regarding mitochondrial functioning, resveratrol can improve mitochondrial function, mitochondrial biogenesis, and oxidative stress (Jardim et al. 2018). Several research groups have shown resveratrol's antidepressant-like effects in animal models, and a study in humans is currently underway to assess its efficacy in the treatment of depression (Efficacy of Resveratrol in Depression 2017). No studies have been done or registered involving patients with BD, but the postulated mechanism of action implies potential beneficial effects.

Pterostilbene A naturally occurring, dimethylated analog of resveratrol, with higher bioavailability. It likewise has antioxidant and neuroprotective effects and in rodent studies was shown to reverse the deleterious effects of aging on cognitive performance (Lange and Li 2018) and improve depression-like behaviors (Yang et al. 2019), but no human studies have been done or registered for mood disorders yet.

5.2.5 Taurine

A free amino acid with antioxidant and neuromodulator functions, able to protect the neuron against glutamate-induced neurotoxicity, reduce oxidative stress, and maintain mitochondrial function (Jakaria et al. 2019). In animals, it has shown an antidepressant-like effect (Wu et al. 2017), but no studies have been published specifically on mood disorders as of now. One study of its effects as an antimanic agent has been completed, but its results have not been published so far (Bezafibrate Treatment for Bipolar Depression: A Proof of Concept Study 2015).

5.3 Unlikely to Be Beneficial

5.3.1 Alpha-Lipoic Acid (ALA)

ALA, also known as thioctic acid, is a naturally occurring and dietarily obtained substance. It can be synthesized in the mitochondria, where it functions as a coenzyme for the formation of pyruvate dehydrogenase and α-ketoglutarate, components of the Krebs cycle. Its ability to increase the activity of the Krebs cycle leads to a reduction in glycolysis and lactate levels (Gomes and Negrato 2014). ALA also has antioxidant and anti-inflammatory effects, with studies demonstrating beneficial effects on metabolic syndrome and related diagnosis, coronary vascular diseases, and even cancer (Haghighatdoost and Hariri 2019; de Sousa et al. 2019). The specific mechanisms by which it exerts these effects are not fully understood. In the central nervous system, ALA affects the levels of the neurotransmitters norepinephrine, dopamine, and acetylcholine and blocks the dopamine D2 receptor. Studies have shown it can improve symptom severity and side effects for patients who have schizophrenia, prevent the progression of Alzheimer's disease, and improve outcomes for stroke survivors (de Sousa et al. 2019).

Regarding mood disorders, rodent models have shown beneficial effects of ALA as an antidepressant and antimanic drug, but only one study assessed ALA in patients with bipolar depression and did not reach a statistically significant result (Brennan et al. 2013). However, due to the positive effects of ALA in other psychiatric and neurological disorders, additional well-designed clinical trials are warranted before reaching a verdict on its efficiency for BD.

5.3.2 Pyrimidines

Uridine and its prodrug triacetyluridine (TAU) are pyrimidine nucleosides of RNA, with roles in glutamatergic transmission, metabolism of a cerebral phospholipid, mitochondrial functioning, and catecholamine synthesis – processes that have been linked to the pathophysiology of BD (Pereira et al. 2018). Cytidine is also a pyrimidine nucleoside of RNA that can be metabolized into uridine when administered orally, as cytidine-diphosphocholine (CDP-choline) (Wurtman et al. 2000).

Uridine While some preliminary studies have shown promising results in bipolar depression, including an open-label study on adolescent patients (Kondo et al. 2011), a recent study on adult patients with bipolar depression was unable to statistically distinguish the effects of uridine from those of placebo (Herlihy 2011). No other studies have been published since.

Triacetyluridine (TAU) A uridine prodrug, that is, our body can metabolize it into free uridine. One study assessing the efficacy of TAU in bipolar depression found it is effective as adjuvant treatment, and it seems its effects were greatest for people with worse depression severity at baseline. In addition to improvements in clinical measures of depression, patients who responded to treatment with TAU demonstrated greater increases in pH levels compared to nonresponders (Jensen et al. 2008).

Cytidine Cytidine has shown antidepressant-like effects in rodents and was reported to improve symptoms of depression in humans. A 2009 study of adjuvant cytidine in the treatment of bipolar depression has shown cytidine results in earlier improvement in depressive symptoms and in a greater reduction in glutamate/glutamine levels in the brain – pointing at a potential mechanism for its clinical effect (Yoon et al. 2009). A more recent study, on the combination of cytidine with omega-3 fatty acids as add-on treatment for BD, was not able to prove clinical benefits greater than placebo (Murphy et al. 2012).

5.4 Potential Risk of a Manic Switch

5.4.1 ALC (Acetyl-L-Carnitine)

ALC is an acetylated version of L-carnitine, rendering it better absorbed and more able to cross the BBB. L-Carnitine is a compound that can be endogenously generated or obtained through diet – as is the case for about 75% of carnitine in humans. Carnitines transport fatty acids into mitochondria, where they can be used as a source of energy. While the brain prefers glucose as its primary energy source, under metabolically compromised conditions, fatty acids become pivotal energy substrates. Carnitines can also directly affect OXPHOS by upregulating gene expression and protein activity of mitochondrial respiratory structures, enhancing mtDNA transcription, and stabilizing mitochondrial mRNA and membrane integrity against lipid peroxidation. As we age and our muscle mass is reduced, the plasma concentrations of carnitine decline as well.

ALC has been shown to upregulate genes related to proteins with antioxidant, anti-apoptotic, and neuroprotective properties. In the central nervous system, ALC assists in acetylcholine synthesis, enhances dopamine release, prevents loss of D1 dopamine receptor activity with age, counters glutamate-induced excitotoxicity, and increases GABA levels (Traina 2016; Pereira et al. 2018). It should be noted, however, that high levels of ACL are not without their cost – they can lead to lowered antioxidant status in the liver (possibly as a result of increased OXPHOS), meaning that any treatment with supplemental ALC should be combined with antioxidants, for example, ALA or NAC mentioned before (Hagen et al. 2002).

Studies in rats have shown it to improve cognitive functioning and lifespan, reversing the age-associated decline of mitochondrial functions. Rodent models were also able to show antidepressant-like effects. However, the only RCT performed so far in humans reported no effects for the combined administration of ALC and ALA, nor did it detect changes in PCr levels reflecting energy stores – a finding that was previously reported in case reports of depressed geriatric patients treated with ALC (Brennan et al. 2013). In addition, there have been two case reports of patients with BD who developed psychotic and manic episodes during treatment with ALC, suggesting caution when considering its clinical use in this population.

5.4.2 Creatine Monohydrate (CM)

Creatine is a nonessential dietary component that can be found in protein-rich foods such as meat, milk, and nuts and can also be synthesized endogenously in the body. In addition to its antioxidant properties, creatine is the precursor of PCr that holds an integral role in energy metabolism as a reservoir of inorganic phosphate, buffering energy concentrations in tissues with significant and fluctuating energy demands – such as the brain (Pereira et al. 2018; Allen 2012). Oral consumption of creatine monohydrate (as a dietary supplement) increases both brain creatine and PCr concentrations (Lyoo et al. 2003) and was therefore studied as an adjunctive treatment for depression. In the treatment of patients with MDD, this approach has generally shown promising results, but several patients with BD experienced a (hypo)manic switch during CM treatment (Toniolo et al. 2018; Roitman et al. 2007). Mood switching is particularly intriguing in this context, as we have previously noted patients with BD might have a specific deficit in turning the energy-storing PCr into available energy in the form of ATP (Du et al. 2018).

5.4.3 SAMe (S-Adenosyl-Methionine)

SAMe is a naturally occurring biological component of all living cells, formed from the combination of methionine and ATP. In addition to its critical role as a methyl donor, SAMe is also a precursor molecule for GSH, one of the body's most potent antioxidants. In the CNS, SAMe is critical in the synthesis and regulation of monoamines, making it possibly relevant in the pathophysiology – and treatment – of mood disorders (Pereira et al. 2018). As far back as 1989, trials demonstrated its efficacy in the treatment of depressive symptoms, but the risk of a manic switch in BD population was not negligible (Carney et al. 1989; Abeysundera and Gill 2018). SAMe augmentation of modern-day antidepressants and even as monotherapy has been shown to be beneficial in patients with MDD (Papakostas et al. 2010; Sarris et al. 2014), but two recent trials have not been able to replicate these results (Sarris et al. 2018, 2019). In addition to strong placebo effects, gender differences in reaction to SAMe may be responsible for some of the negative studies, with

SAMe having a greater effect in males than females (Sarris et al. 2015). Due to the high risk of manic switching, SAMe is contraindicated for patients with BD.

5.5 Vitamins

5.5.1 Vitamin A

Vitamin A is an essential molecule for many physiological processes in the body, including vision, immunity, and gene transcription and as a co-factor for redox activation (Hammerling 2016). Intoxication with vitamin A can lead to increased oxidative stress and dangerous side effects, such as cognitive decline, depression, and suicidality (de Oliveira 2015). It has been shown that the brains of persons suffering from mood disorders or schizophrenia have abnormalities in pathways related to vitamin A signaling (Haybaeck et al. 2015), reflecting the increased tone of retinoic acid (a derivative of vitamin A) and marking this as a possible pathway in the pathophysiology and eventual treatment of BD. No studies have been conducted so far on vitamin A-related drugs in the treatment of mood disorders.

5.5.2 Vitamin C

Known as ascorbic acid, vitamin C is an antioxidant acting both directly and as a co-substrate for many important cellular oxidation and reduction processes (Pehlivan 2017). Vitamin C has been linked to improvements in anxiety and mood (Pullar et al. 2018; Kocot et al. 2017), but studies could not prove statistically significant effects on mood (Sahraian et al. 2015). Few studies on BD population from the 1960s and 1980s have shown positive results, but have not been replicated since.

5.5.3 Vitamin D

Vitamin D is involved in calcium and phosphate metabolism but has many more roles in physiological pathways – including immune modulation, antioxidant and anti-inflammatory effects, and monoamine metabolism (Jamilian et al. 2019; Sabir et al. 2018). Vitamin D receptors (VDR) can even translocate into the mitochondria and affect their function (Ricca et al. 2018). Vitamin D-related disturbances have been linked to depressive and manic symptoms (Altunsoy et al. 2018), but when taking several meta-analyses into account, there are no conclusive results whether vitamin D supplementation is useful in the treatment of depressive symptoms (Gowda et al. 2015; Vellekkatt and Menon 2019). Despite the fact that many patients with mood disorders are vitamin D deficient (Cuomo et al. 2019; Petrov et al. 2018), supplementation has not been found beneficial for the treatment of bipolar depression (Marsh et al. 2017), and only one small study reported beneficiary effects on mania ratings (Sikoglu et al. 2015).

5.5.4 Vitamin E

Also called tocopherol, vitamin E is a fat-soluble antioxidant exclusively obtained from the diet, whose ability for ROS scavenging and lipid peroxidation reduction can lead to stabilization cellular and mitochondrial membranes (Rizvi et al. 2014). Vitamin E seems to be more potent when combined with vitamin C or Co-Q10 (Kontush and Schekatolina 2004; Dhitavat et al. 2005), and animal models show some improved neurological and mitochondrial function in aging mice. One study on older human adults was completed in 2018, seeking to assess the effectiveness of ascorbic acid (vitamin C) and tocopherol (vitamin E) in the treatment of depression (Effectiveness of Ascorbic Acid and Tocopherol for Depression in Elderly 2016), but its results have not yet been published.

5.5.5 Vitamins B

B1: Thiamine Thiamine is an an essential cofactor for the Krebs cycle. A 2016 study on patients with major depressive disorder (MDD) showed that while vitamin B1 supplementation did not improve depressive symptoms any more than placebo, it was able to bring about the positive change earlier (Ghaleiha et al. 2016). No studies have been done on BD population.

B3: Niacin Nicotinic acid, nicotinamide (also called *niacinamide*), and nicotinamide riboside, collectively termed vitamin B3, are precursor molecules for NAD+ and NADP+ in all body tissues – meaning they are indirectly involved in hundreds of enzymatic reactions related to OXPHOS, glycolysis, and lipid oxidation (Depeint et al. 2006; Conze et al. 2019). Each precursor has unique effects: nicotinic acid can lower "bad" lipids such as LDL or triglycerides and elevate "good" lipids like HDL and is taken as a treatment for dyslipidemia with a notable side effect of flushing (Boden et al. 2014). In contrast, niacinamide does not affect blood lipids, nor does it cause flushing, but it affects other systems – for example, it can elevate homocysteine levels and thereby increase the risk of vascular disease. Nicotinamide riboside also does not induce flushing and only elevates homocysteine to a lesser degree. It has been shown in rodent models to prevent adverse outcomes in various systems such as diet-induced weight gain, neuropathy, hearing loss, heart failure, irradiation damage, and central brain injury (Conze et al. 2019). Mouse models have shown supplementation of mothers with nicotinamide riboside could lead to stronger, less anxious, and more cognitively able offsprings (Ear et al. 2019).

Regarding mood, nicotinic acid is used as an active placebo in psychedelic studies due to its ability to induce flushing without altering the psychological state (Ross et al. 2016; Grob et al. 2011), although a case report of a manic episode during treatment with it for dyslipidemia was reported in a 54-year-old man with no history of mental illness (Loebl and Raskin 2013). Regarding niacinamide, animal models for depression have shown it to have similar effects on fluoxetine (Rex et al. 2004), but no human studies have been published showing similar efficacy. Niacinamide riboside seems to be the most promising of the three, and a study assessing its effects on mood has been recently completed, with the results of yet unpublished (A Study by ChromaDex to Assess the Effects of TRU NIAGEN on Cognitive Function, Mood and Sleep in Older Adults 2018).

Vitamin B6 Vitamin B6 again refers to a group of chemically similar compounds, with the active form serving as a coenzyme in many biological processes including preventing ROS generation and lipid peroxidation (Kannan and Jain 2004). Most studies on vitamin B6, especially on its own and not in addition to other supplementals, failed to detect effects on mood. In addition, vitamin B6 supplementation is not without risks, and patients were reported to suffer from adverse drug reactions that resemble Wernicke's encephalopathy, thankfully a reaction that was able to be reversed by vitamin B1 (thiamine) supplementation (Pereira et al. 2018).

B9: Folate Folic acid plays an essential role in mitochondrial energy production (Depeint et al. 2006), and deficiencies have been associated with depression, BD, and cognitive dysfunction (Baek et al. 2013). According to a recent review, several studies from the 1980s found associations between folate and affective morbidity, and more recently a couple of studies have shown supplementation with folate (or its metabolites) to have an effect on both manic and depressive symptoms. However, other studies found no discernible effect on patients who were not specifically folate-deficient (Pereira et al. 2018), and folic acid has the potential to negate the therapeutic effects of lamotrigine, a popular treatment for bipolar depression (Simon et al. 2018). It also appears that folic acid supplementation does not reduce the incidence of a mood disorder, albeit it may be able to delay the onset of the first mood episode or mitigate its severity (Sharpley et al. 2014; Okereke et al. 2015).

B12: Cobalamin Vitamin B12 is a cofactor for methionine, DNA, and myelin synthesis and is necessary for the maintenance of neuronal integrity and regulating neurotransmitters. Despite the fact that vitamin B12 deficiency is a known cause for mood deficits (Issac et al. 2015), cobalamin supplementation has not been proven to improve depressive symptoms (Pereira et al. 2018; Almeida et al. 2015).

Vitamins B2 (riboflavin), *B5* (pantothenic acid), and *B7* (biotin) are all important for biological mitochondrial pathways, and their deficiencies might be related to neurological deficits, but they have not been studied as specific treatments for depression or BD.

6 Summary

In this chapter, we have described how the central nervous system is particularly vulnerable to dysregulations of mitochondrial function due to its high energetic demands. Convergent data implicated subtle mitochondrial dysfunction as an important component of the pathophysiology of bipolar disorder (BD) and led to the exploration of alternative therapeutic targets. While some negative results have been published, targeting mitochondrial pathways seems to have a measurable effect on mood, evidenced not only by alleviation of depressive symptoms but also by induction of manic switches for certain compounds. When discussing mitochondrial treatments, combinations of several mitochondrial agents could be of interest as their actions can be synergistic or complementary to the prevention of side effects. For example, increasing energy production comes at the cost of increased free radical production and might justify a "mitochondrial cocktail" that includes an energy-enhancing agent with an antioxidant agent. Many studies have been conducted, and more are underway attempting to elucidate the role of mitochondrial agents in the treatment of BD. While effect sizes at the moment do not support any single drug as possible monotherapy or "the new lithium," it is important to keep in mind that BD is not one disease caused by one pathway, but has heterogenic presentations and heterogenic underlying pathways. The mitochondrial hypothesis and mitochondrial modulators provide viable alternatives to current treatments and a path toward drug discovery to achieve better outcomes for people with BD.

	ATP production	Mitochondrial biogenesis	Oxidative stress	Clinical evidence in mood disorder
Bezafibrate (Ioannou et al. 2010; Huang et al. 2017)	Increase	Increase	Decrease	Evidence in depression. Studies on BD underway
Minocycline	–	–	Decrease	Inconsistent. Studies underway
N-Acetyl-cysteine (NAC) (Tardiolo et al. 2018)	–	–	Decrease	Inconsistent, but positive evidence in bipolar depression. Effects take a long time to become apparent
Co-enzyme Q10	Increase	?	Decrease	Evidence in bipolar depression. Effects take a long time to become apparent
Melatonin	Increase	Increase	Decrease	Evidence in unipolar and bipolar depression
Ebselen	–	–	Decrease	Studies currently underway
Mangosteen	Increase	–	Decrease	Improvement in psychotic and affective symptoms of schizophrenia
Ketogenic diet	Increase	Increase	Decrease	Animal models and human case reports
Resveratrol and pterostilbene	Increase	Increase	Decrease	Animal models show effectiveness. Human studies underway for resveratrol
Taurine	Increase	?	Decrease	Animal models only

References

A Pilot Study Investigating the Efficacy of Minocycline and N-Acetyl Cysteine for Bipolar Depression (2016). https://clinicaltrials.gov/ct2/show/NCT02719392

A Study by ChromaDex to Assess the Effects of TRU NIAGEN on Cognitive Function, Mood and Sleep in Older Adults (2018). https://clinicaltrials.gov/ct2/show/NCT03562468

Abeysundera H, Gill R (2018) Possible SAMe-induced mania. BMJ Case Rep 2018

Acuna-Castroviejo D, Escames G, Rodriguez MI et al (2007) Melatonin role in the mitochondrial function. Front Biosci 12:947–963

Akarsu S, Bolu A, Aydemir E et al (2018) The relationship between the number of manic episodes and oxidative stress indicators in bipolar disorder. Psychiatry Investig 15:514–519

Alcazar-Fabra M, Trevisson E, Brea-Calvo G (2018) Clinical syndromes associated with coenzyme Q10 deficiency. Essays Biochem 62:377–398

Allen PJ (2012) Creatine metabolism and psychiatric disorders: does creatine supplementation have therapeutic value? Neurosci Biobehav Rev 36:1442–1462

Almeida OP, Ford AH, Flicker L (2015) Systematic review and meta-analysis of randomized placebo-controlled trials of folate and vitamin B12 for depression. Int Psychogeriatr 27:727–737

Altunsoy N, Yuksel RN, Cingi Yirun M et al (2018) Exploring the relationship between vitamin D and mania: correlations between serum vitamin D levels and disease activity. Nord J Psychiatry 72:221–225

Andreazza AC, Shao L, Wang JF et al (2010) Mitochondrial complex I activity and oxidative damage to mitochondrial proteins in the prefrontal cortex of patients with bipolar disorder. Arch Gen Psychiatry 67:360–368

Ashok AH, Marques TR, Jauhar S et al (2017) The dopamine hypothesis of bipolar affective disorder: the state of the art and implications for treatment. Mol Psychiatry 22:666–679

Ashton MM, Dean OM, Walker AJ et al (2019) The therapeutic potential of mangosteen pericarp as an adjunctive therapy for bipolar disorder and schizophrenia. Front Psych 10:115

Baek JH, Bernstein EE, Nierenberg AA (2013) One-carbon metabolism and bipolar disorder. Aust N Z J Psychiatry 47:1013–1018

Baltan S (2015) Can lactate serve as an energy substrate for axons in good times and in bad, in sickness and in health? Metab Brain Dis 30:25–30

Baranov SV, Baranova OV, Yablonska S et al (2019) Mitochondria modulate programmed neuritic retraction. Proc Natl Acad Sci U S A 116:650–659

Bauer IE, Green C, Colpo GD et al (2018) A double-blind, randomized, placebo-controlled study of aspirin and N-acetylcysteine as adjunctive treatments for bipolar depression. J Clin Psychiatry 80

Berk M, Copolov DL, Dean O et al (2008) N-acetyl cysteine for depressive symptoms in bipolar disorder--a double-blind randomized placebo-controlled trial. Biol Psychiatry 64:468–475

Berk M, Turner A, Malhi GS et al (2019) A randomised controlled trial of a mitochondrial therapeutic target for bipolar depression: mitochondrial agents, N-acetylcysteine, and placebo. BMC Med 17:18

Berridge MJ (2014) Calcium signalling and psychiatric disease: bipolar disorder and schizophrenia. Cell Tissue Res 357:477–492

Bezafibrate Treatment for Bipolar Depression: A Proof of Concept Study (2015). https://clinicaltrials.gov/ct2/show/NCT02481245

Blacker CJ, Lewis CP, Frye MA et al (2017) Metabotropic glutamate receptors as emerging research targets in bipolar disorder. Psychiatry Res 257:327–337

Blazey T, Snyder AZ, Goyal MS et al (2018) A systematic meta-analysis of oxygen-to-glucose and oxygen-to-carbohydrate ratios in the resting human brain. PLoS One 13:e0204242

Boden WE, Sidhu MS, Toth PP (2014) The therapeutic role of niacin in dyslipidemia management. J Cardiovasc Pharmacol Ther 19:141–158

Bonora M, Patergnani S, Rimessi A et al (2012) ATP synthesis and storage. Purinergic Signal 8:343–357

Brennan BP, Jensen JE, Hudson JI et al (2013) A placebo-controlled trial of acetyl-L-carnitine and alpha-lipoic acid in the treatment of bipolar depression. J Clin Psychopharmacol 33:627–635
Brietzke E, Mansur RB, Subramaniapillai M et al (2018) Ketogenic diet as a metabolic therapy for mood disorders: evidence and developments. Neurosci Biobehav Rev 94:11–16
Brown NC, Andreazza AC, Young LT (2014) An updated meta-analysis of oxidative stress markers in bipolar disorder. Psychiatry Res 218:61–68
Bustillo JR, Jones T, Qualls C et al (2019) Proton magnetic resonance spectroscopic imaging of gray and white matter in bipolar-I and schizophrenia. J Affect Disord 246:745–753
Callaly E, Walder K, Morris G et al (2015) Mitochondrial dysfunction in the pathophysiology of bipolar disorder: effects of pharmacotherapy. Mini Rev Med Chem 15:355–365
Campbell IH, Campbell H (2019) Ketosis and bipolar disorder: controlled analytic study of online reports. BJPsych Open 5:e58
Carney MW, Chary TK, Bottiglieri T et al (1989) The switch mechanism and the bipolar/unipolar dichotomy. Br J Psychiatry 154:48–51
Cataldo AM, McPhie DL, Lange NT et al (2010) Abnormalities in mitochondrial structure in cells from patients with bipolar disorder. Am J Pathol 177:575–585
Chen S, Owens GC, Edelman DB (2008) Dopamine inhibits mitochondrial motility in hippocampal neurons. PLoS One 3:e2804
Chen WT, Huang TL, Tsai MC (2015) Bcl-2 associated with severity of manic symptoms in bipolar patients in a manic phase. Psychiatry Res 225:305–308
Chen SD, Wu CL, Hwang WC et al (2017) More insight into BDNF against neurodegeneration: anti-apoptosis, anti-oxidation, and suppression of autophagy. Int J Mol Sci 18:545
Chinnery PF, Hudson G (2013) Mitochondrial genetics. Br Med Bull 106:135–159
Chiu CT, Wang Z, Hunsberger JG et al (2013) Therapeutic potential of mood stabilizers lithium and valproic acid: beyond bipolar disorder. Pharmacol Rev 65:105–142
Clarke DDS (1999) Regulation of cerebral metabolic rate. In: Basic neurochemistry: molecular cellular and medical aspects, 6th edn. Lippincott-Raven, Philadelphia
Colle R, de Larminat D, Rotenberg S et al (2017) PPAR-gamma agonists for the treatment of major depression: a review. Pharmacopsychiatry 50:49–55
Conze D, Brenner C, Kruger CL (2019) Safety and metabolism of long-term administration of NIAGEN (nicotinamide riboside chloride) in a randomized, double-blind, placebo-controlled clinical trial of healthy overweight adults. Sci Rep 9:9772
Cuomo A, Maina G, Bolognesi S et al (2019) Prevalence and correlates of vitamin D deficiency in a sample of 290 inpatients with mental illness. Front Psych 10:167
Cuperfain AB, Zhang ZL, Kennedy JL et al (2018) The complex interaction of mitochondrial genetics and mitochondrial pathways in psychiatric disease. Mol Neuropsychiatry 4:52–69
Czerniczyniec A, Bustamante J, Lores-Arnaiz S (2007) Dopamine enhances mtNOS activity: implications in mitochondrial function. Biochim Biophys Acta 1767:1118–1125
Data-Franco J, Singh A, Popovic D et al (2017) Beyond the therapeutic shackles of the monoamines: new mechanisms in bipolar disorder biology. Prog Neuro-Psychopharmacol Biol Psychiatry 72:73–86
de Oliveira MR (2015) Vitamin A and retinoids as mitochondrial toxicants. Oxidative Med Cell Longev 2015:140267
de Oliveira GS, Cereser KM, Fernandes BS et al (2009) Decreased brain-derived neurotrophic factor in medicated and drug-free bipolar patients. J Psychiatr Res 43:1171–1174
de Sousa RT, Machado-Vieira R, Zarate CA Jr et al (2014) Targeting mitochondrially mediated plasticity to develop improved therapeutics for bipolar disorder. Expert Opin Ther Targets 18:1131–1147
de Sousa RT, Streck EL, Zanetti MV et al (2015) Lithium increases leukocyte mitochondrial complex I activity in bipolar disorder during depressive episodes. Psychopharmacology 232:245–250

de Sousa CNS, da Silva Leite CMG, da Silva Medeiros I et al (2019) Alpha-lipoic acid in the treatment of psychiatric and neurological disorders: a systematic review. Metab Brain Dis 34:39–52
Depeint F, Bruce WR, Shangari N et al (2006) Mitochondrial function and toxicity: role of the B vitamin family on mitochondrial energy metabolism. Chem Biol Interact 163:94–112
Dhitavat S, Ortiz D, Rogers E et al (2005) Folate, vitamin E, and acetyl-L-carnitine provide synergistic protection against oxidative stress resulting from exposure of human neuroblastoma cells to amyloid-beta. Brain Res 1061:114–117
Dienel GA (2012) Brain lactate metabolism: the discoveries and the controversies. J Cereb Blood Flow Metab 32:1107–1138
Dogan AE, Yuksel C, Du F et al (2018) Brain lactate and pH in schizophrenia and bipolar disorder: a systematic review of findings from magnetic resonance studies. Neuropsychopharmacology 43:1681–1690
Droge W (2002) Free radicals in the physiological control of cell function. Physiol Rev 82:47–95
Du F, Yuksel C, Chouinard VA et al (2018) Abnormalities in high-energy phosphate metabolism in first-episode bipolar disorder measured using (31)P-magnetic resonance spectroscopy. Biol Psychiatry 84:797–802
Dudley JA, Lee JH, Durling M et al (2015) Age-dependent decreases of high energy phosphates in cerebral gray matter of patients with bipolar I disorder: a preliminary phosphorus-31 magnetic resonance spectroscopic imaging study. J Affect Disord 175:251–255
Dudley J, DelBello MP, Weber WA et al (2016) Tissue-dependent cerebral energy metabolism in adolescents with bipolar disorder. J Affect Disord 191:248–255
Ear PH, Chadda A, Gumusoglu SB et al (2019) Maternal nicotinamide riboside enhances postpartum weight loss, juvenile offspring development, and neurogenesis of adult offspring. Cell Rep 26:969–983.e964
Ebselen as an add-on Treatment in Hypo/Mania (2017). https://clinicaltrials.gov/ct2/show/NCT03013400
Effectiveness of Ascorbic Acid and Tocopherol for Depression in Elderly (2016). https://clinicaltrials.gov/ct2/show/NCT02793648
Efficacy of Resveratrol in Depression (2017). https://clinicaltrials.gov/ct2/show/NCT03384329
Ende G, Cackowski S, van Eijk J et al (2016) Impulsivity and aggression in female BPD and ADHD patients: association with ACC glutamate and GABA concentrations. Neuropsychopharmacology 41:410–418
Erecinska M, Silver IA (1989) ATP and brain function. J Cereb Blood Flow Metab 9:2–19
Fan L, Feng Y, Chen GC et al (2017) Effects of coenzyme Q10 supplementation on inflammatory markers: a systematic review and meta-analysis of randomized controlled trials. Pharmacol Res 119:128–136
Flippo KH, Strack S (2017) Mitochondrial dynamics in neuronal injury, development and plasticity. J Cell Sci 130:671–681
Fonseca AC, Moreira PI, Oliveira CR et al (2015) Amyloid-beta disrupts calcium and redox homeostasis in brain endothelial cells. Mol Neurobiol 51:610–622
Forester BP, Harper DG, Georgakas J et al (2015) Antidepressant effects of open label treatment with coenzyme Q10 in geriatric bipolar depression. J Clin Psychopharmacol 35:338–340
Forlenza OV, De-Paula VJ, Diniz BS (2014) Neuroprotective effects of lithium: implications for the treatment of Alzheimer's disease and related neurodegenerative disorders. ACS Chem Neurosci 5:443–450
Garrido-Maraver J, Cordero MD, Oropesa-Avila M et al (2014) Clinical applications of coenzyme Q10. Front Biosci (Landmark Ed) 19:619–633
Geddes JR, Miklowitz DJ (2013) Treatment of bipolar disorder. Lancet 381:1672–1682
Geoffroy PA, Etain B, Lajnef M et al (2016) Circadian genes and lithium response in bipolar disorders: associations with PPARGC1A (PGC-1alpha) and RORA. Genes Brain Behav 15:660–668

Ghaleiha A, Davari H, Jahangard L et al (2016) Adjuvant thiamine improved standard treatment in patients with major depressive disorder: results from a randomized, double-blind, and placebo-controlled clinical trial. Eur Arch Psychiatry Clin Neurosci 266:695–702

Gladden LB (2004) Lactate metabolism: a new paradigm for the third millennium. J Physiol 558:5–30

Gomes MB, Negrato CA (2014) Alpha-lipoic acid as a pleiotropic compound with potential therapeutic use in diabetes and other chronic diseases. Diabetol Metab Syndr 6:80

Gowda U, Mutowo MP, Smith BJ et al (2015) Vitamin D supplementation to reduce depression in adults: meta-analysis of randomized controlled trials. Nutrition 31:421–429

Grant JE, Odlaug BL, Chamberlain SR et al (2014) A randomized, placebo-controlled trial of N-acetylcysteine plus imaginal desensitization for nicotine-dependent pathological gamblers. J Clin Psychiatry 75:39–45

Greer PL, Greenberg ME (2008) From synapse to nucleus: calcium-dependent gene transcription in the control of synapse development and function. Neuron 59:846–860

Grings M, Moura AP, Parmeggiani B et al (2017) Bezafibrate prevents mitochondrial dysfunction, antioxidant system disturbance, glial reactivity and neuronal damage induced by sulfite administration in striatum of rats: implications for a possible therapeutic strategy for sulfite oxidase deficiency. Biochim Biophys Acta Mol basis Dis 1863:2135–2148

Grob CS, Danforth AL, Chopra GS et al (2011) Pilot study of psilocybin treatment for anxiety in patients with advanced-stage cancer. Arch Gen Psychiatry 68:71–78

Guimaraes-Ferreira L (2014) Role of the phosphocreatine system on energetic homeostasis in skeletal and cardiac muscles. Einstein (Sao Paulo) 12:126–131

Guo C, Sun L, Chen X et al (2013) Oxidative stress, mitochondrial damage and neurodegenerative diseases. Neural Regen Res 8:2003–2014

Hagen TM, Liu J, Lykkesfeldt J et al (2002) Feeding acetyl-L-carnitine and lipoic acid to old rats significantly improves metabolic function while decreasing oxidative stress. Proc Natl Acad Sci U S A 99:1870–1875

Haghighatdoost F, Hariri M (2019) The effect of alpha-lipoic acid on inflammatory mediators: a systematic review and meta-analysis on randomized clinical trials. Eur J Pharmacol 849:115–123

Hammerling U (2016) Retinol as electron carrier in redox signaling, a new frontier in vitamin a research. Hepatobiliary Surg Nutr 5:15–28

Hardeland R (2017) Melatonin and the electron transport chain. Cell Mol Life Sci 74:3883–3896

Harwood AJ (2005) Lithium and bipolar mood disorder: the inositol-depletion hypothesis revisited. Mol Psychiatry 10:117–126

Haybaeck J, Postruznik M, Miller CL et al (2015) Increased expression of retinoic acid-induced gene 1 in the dorsolateral prefrontal cortex in schizophrenia, bipolar disorder, and major depression. Neuropsychiatr Dis Treat 11:279–289

Herlihy WC (2011) Repligen reports phase 2b results for RG2417 for bipolar depression. https://www.businesswire.com/news/home/20110307005868/en/Repligen-Reports-Phase-2b-Results-RG2417-Bipolar

Holper L, Ben-Shachar D, Mann JJ (2019) Multivariate meta-analyses of mitochondrial complex I and IV in major depressive disorder, bipolar disorder, schizophrenia, Alzheimer disease, and Parkinson disease. Neuropsychopharmacology 44:837–849

Howell N, Smejkal CB, Mackey DA et al (2003) The pedigree rate of sequence divergence in the human mitochondrial genome: there is a difference between phylogenetic and pedigree rates. Am J Hum Genet 72:659–670

Huang Y, Powers C, Moore V et al (2017) The PPAR pan-agonist bezafibrate ameliorates cardiomyopathy in a mouse model of Barth syndrome. Orphanet J Rare Dis 12:49

Ioannou N, Hargreaves IP, Allen G et al (2010) Bezafibrate induced increase in mitochondrial electron transport chain complex IV activity in human astrocytoma cells: implications for mitochondrial cytopathies and neurodegenerative diseases. Biofactors 36:468–473

Issac TG, Soundarya S, Christopher R et al (2015) Vitamin B12 deficiency: an important reversible co-morbidity in neuropsychiatric manifestations. Indian J Psychol Med 37:26–29
Jakaria M, Azam S, Haque ME et al (2019) Taurine and its analogs in neurological disorders: focus on therapeutic potential and molecular mechanisms. Redox Biol 24:101223
Jamilian H, Amirani E, Milajerdi A et al (2019) The effects of vitamin D supplementation on mental health, and biomarkers of inflammation and oxidative stress in patients with psychiatric disorders: a systematic review and meta-analysis of randomized controlled trials. Prog Neuro-Psychopharmacol Biol Psychiatry 94:109651
Jardim FR, de Rossi FT, Nascimento MX et al (2018) Resveratrol and brain mitochondria: a review. Mol Neurobiol 55:2085–2101
Javadov S, Chapa-Dubocq X, Makarov V (2018) Different approaches to modeling analysis of mitochondrial swelling. Mitochondrion 38:58–70
Jensen JE, Daniels M, Haws C et al (2008) Triacetyluridine (TAU) decreases depressive symptoms and increases brain pH in bipolar patients. Exp Clin Psychopharmacol 16:199–206
Jia ZQ, Li SQ, Qiao WQ et al (2018) Ebselen protects mitochondrial function and oxidative stress while inhibiting the mitochondrial apoptosis pathway after acute spinal cord injury. Neurosci Lett 678:110–117
Jou MJ (2011) Melatonin preserves the transient mitochondrial permeability transition for protection during mitochondrial Ca(2+) stress in astrocyte. J Pineal Res 50:427–435
Juarez Olguin H, Calderon Guzman D, Hernandez Garcia E et al (2016) The role of dopamine and its dysfunction as a consequence of oxidative stress. Oxidative Med Cell Longev 2016:9730467
Kannan K, Jain SK (2004) Effect of vitamin B6 on oxygen radicals, mitochondrial membrane potential, and lipid peroxidation in H2O2-treated U937 monocytes. Free Radic Biol Med 36:423–428
Kato T (2019) Current understanding of bipolar disorder: toward integration of biological basis and treatment strategies. Psychiatry Clin Neurosci 73:526–540
Kato T, Kato N (2000) Mitochondrial dysfunction in bipolar disorder. Bipolar Disord 2:180–190
Kato TM, Kubota-Sakashita M, Fujimori-Tonou N et al (2018) Ant1 mutant mice bridge the mitochondrial and serotonergic dysfunctions in bipolar disorder. Mol Psychiatry 23:2039–2049
Kim DJ, Lyoo IK, Yoon SJ et al (2007) Clinical response of quetiapine in rapid cycling manic bipolar patients and lactate level changes in proton magnetic resonance spectroscopy. Prog Neuro-Psychopharmacol Biol Psychiatry 31:1182–1188
Kim HW, Rapoport SI, Rao JS (2010) Altered expression of apoptotic factors and synaptic markers in postmortem brain from bipolar disorder patients. Neurobiol Dis 37:596–603
Kim Y, Santos R, Gage FH et al (2017) Molecular mechanisms of bipolar disorder: progress made and future challenges. Front Cell Neurosci 11:30
Kim Y, Vadodaria KC, Lenkei Z et al (2019) Mitochondria, metabolism, and redox mechanisms in psychiatric disorders. Antioxid Redox Signal 31:275–317
Kishi T, Nomura I, Sakuma K et al (2019) Melatonin receptor agonists-ramelteon and melatonin-for bipolar disorder: a systematic review and meta-analysis of double-blind, randomized, placebo-controlled trials. Neuropsychiatr Dis Treat 15:1479–1486
Kocot J, Luchowska-Kocot D, Kielczykowska M et al (2017) Does vitamin C influence neurodegenerative diseases and psychiatric disorders? Nutrients 9:659
Kondo DG, Sung YH, Hellem TL et al (2011) Open-label uridine for treatment of depressed adolescents with bipolar disorder. J Child Adolesc Psychopharmacol 21:171–175
Kontush K, Schekatolina S (2004) Vitamin E in neurodegenerative disorders: Alzheimer's disease. Ann N Y Acad Sci 1031:249–262
Kovacs Z, D'Agostino DP, Diamond D et al (2019) Therapeutic potential of exogenous ketone supplement induced ketosis in the treatment of psychiatric disorders: review of current literature. Front Psych 10:363
Kuang H, Duong A, Jeong H et al (2018) Lactate in bipolar disorder: a systematic review and meta-analysis. Psychiatry Clin Neurosci 72:546–555

Kubli DA, Gustafsson AB (2012) Mitochondria and mitophagy: the yin and yang of cell death control. Circ Res 111:1208–1221

Kumar P, Efstathopoulos P, Millischer V et al (2018) Mitochondrial DNA copy number is associated with psychosis severity and anti-psychotic treatment. Sci Rep 8:12743

Kumari S, Mehta SL, Milledge GZ et al (2016) Ubisol-Q10 prevents glutamate-induced cell death by blocking mitochondrial fragmentation and permeability transition pore opening. Int J Biol Sci 12:688–700

Lagopoulos J, Hermens DF, Tobias-Webb J et al (2013) In vivo glutathione levels in young persons with bipolar disorder: a magnetic resonance spectroscopy study. J Psychiatr Res 47:412–417

Lange KW, Li S (2018) Resveratrol, pterostilbene, and dementia. Biofactors 44:83–90

Lee A, Hirabayashi Y, Kwon SK et al (2018) Emerging roles of mitochondria in synaptic transmission and neurodegeneration. Curr Opin Physiol 3:82–93

Li XB, Gu JD, Zhou QH (2015) Review of aerobic glycolysis and its key enzymes – new targets for lung cancer therapy. Thorac Cancer 6:17–24

Liberti MV, Locasale JW (2016) The Warburg effect: how does it benefit cancer cells? Trends Biochem Sci 41:211–218

Liguori I, Russo G, Curcio F et al (2018) Oxidative stress, aging, and diseases. Clin Interv Aging 13:757–772

Lin DD, Crawford TO, Barker PB (2003) Proton MR spectroscopy in the diagnostic evaluation of suspected mitochondrial disease. AJNR Am J Neuroradiol 24:33–41

Loebl T, Raskin S (2013) A novel case report: acute manic psychotic episode after treatment with niacin. J Neuropsychiatry Clin Neurosci 25:E14

Lutz PE, Kieffer BL (2013) Opioid receptors: distinct roles in mood disorders. Trends Neurosci 36:195–206

Lyoo IK, Kong SW, Sung SM et al (2003) Multinuclear magnetic resonance spectroscopy of high-energy phosphate metabolites in human brain following oral supplementation of creatine-monohydrate. Psychiatry Res 123:87–100

MacDonald ML, Naydenov A, Chu M et al (2006) Decrease in creatine kinase messenger RNA expression in the hippocampus and dorsolateral prefrontal cortex in bipolar disorder. Bipolar Disord 8:255–264

Machado-Vieira R, Pivovarova NB, Stanika RI et al (2011) The Bcl-2 gene polymorphism rs956572AA increases inositol 1,4,5-trisphosphate receptor-mediated endoplasmic reticulum calcium release in subjects with bipolar disorder. Biol Psychiatry 69:344–352

Maes M, Mihaylova I, Kubera M et al (2009) Lower plasma coenzyme Q10 in depression: a marker for treatment resistance and chronic fatigue in depression and a risk factor to cardiovascular disorder in that illness. Neuro Endocrinol Lett 30:462–469

Manfredi G, Kwong JQ, Oca-Cossio JA et al (2003) BCL-2 improves oxidative phosphorylation and modulates adenine nucleotide translocation in mitochondria of cells harboring mutant mtDNA. J Biol Chem 278:5639–5645

Mantle D, Hargreaves I (2019) Coenzyme Q10 and degenerative disorders affecting longevity: an overview. Antioxidants (Basel) 8:44

Markham A, Bains R, Franklin P et al (2014) Changes in mitochondrial function are pivotal in neurodegenerative and psychiatric disorders: how important is BDNF? Br J Pharmacol 171:2206–2229

Marsh WK, Penny JL, Rothschild AJ (2017) Vitamin D supplementation in bipolar depression: a double blind placebo controlled trial. J Psychiatr Res 95:48–53

Masaki C, Sharpley AL, Cooper CM et al (2016) Effects of the potential lithium-mimetic, ebselen, on impulsivity and emotional processing. Psychopharmacology 233:2655–2661

Massaad CA, Klann E (2011) Reactive oxygen species in the regulation of synaptic plasticity and memory. Antioxid Redox Signal 14:2013–2054

Matschke V, Theiss C, Matschke J (2019) Oxidative stress: the lowest common denominator of multiple diseases. Neural Regen Res 14:238–241

McCann RF, Ross DA (2018) So happy together: the storied marriage between mitochondria and the mind. Biol Psychiatry 83:e47–e49

McCarthy MJ, Le Roux MJ, Wei H et al (2016) Calcium channel genes associated with bipolar disorder modulate lithium's amplification of circadian rhythms. Neuropharmacology 101:439–448

Mehrpooya M, Yasrebifar F, Haghighi M et al (2018) Evaluating the effect of coenzyme Q10 augmentation on treatment of bipolar depression: a double-blind controlled clinical trial. J Clin Psychopharmacol 38:460–466

Mertens J, Wang QW, Kim Y et al (2015) Differential responses to lithium in hyperexcitable neurons from patients with bipolar disorder. Nature 527:95–99

Minocycline for Bipolar Depression (2012). https://clinicaltrials.gov/ct2/show/NCT01514422

Modica-Napolitano JS, Renshaw PF (2004) Ethanolamine and phosphoethanolamine inhibit mitochondrial function in vitro: implications for mitochondrial dysfunction hypothesis in depression and bipolar disorder. Biol Psychiatry 55:273–277

Moffett JR, Arun P, Ariyannur PS et al (2013) N-Acetylaspartate reductions in brain injury: impact on post-injury neuroenergetics, lipid synthesis, and protein acetylation. Front Neuroenerg 5:11

Monzani E, Nicolis S, Dell'Acqua S et al (2019) Dopamine, oxidative stress and protein-quinone modifications in Parkinson's and other neurodegenerative diseases. Angew Chem Int Ed Engl 58:6512–6527

Moore A, Beidler J, Hong MY (2018) Resveratrol and depression in animal models: a systematic review of the biological mechanisms. Molecules 23:2197

Morris G, Anderson G, Berk M et al (2013) Coenzyme Q10 depletion in medical and neuropsychiatric disorders: potential repercussions and therapeutic implications. Mol Neurobiol 48:883–903

Morris G, Walder K, McGee SL et al (2017) A model of the mitochondrial basis of bipolar disorder. Neurosci Biobehav Rev 74:1–20

Murphy AN, Bredesen DE, Cortopassi G et al (1996) Bcl-2 potentiates the maximal calcium uptake capacity of neural cell mitochondria. Proc Natl Acad Sci U S A 93:9893–9898

Murphy BL, Stoll AL, Harris PQ et al (2012) Omega-3 fatty acid treatment, with or without cytidine, fails to show therapeutic properties in bipolar disorder: a double-blind, randomized add-on clinical trial. J Clin Psychopharmacol 32:699–703

Nascimento C, Kim HK, Young LT et al (2015) Glutathione-mediated effects of lithium in decreasing protein oxidation induced by mitochondrial complex I dysfunction. J Neural Transm (Vienna) 122:741–746

Naydenov AV, MacDonald ML, Ongur D et al (2007) Differences in lymphocyte electron transport gene expression levels between subjects with bipolar disorder and normal controls in response to glucose deprivation stress. Arch Gen Psychiatry 64:555–564

Naziroglu M, Cig B, Ozgul C (2013) Neuroprotection induced by N-acetylcysteine against cytosolic glutathione depletion-induced Ca2+ influx in dorsal root ganglion neurons of mice: role of TRPV1 channels. Neuroscience 242:151–160

Neergheen V, Chalasani A, Wainwright L, Yubero D, Montero R, Artuch R, Hargreaves I (2017) Coenzyme Q10 in the treatment of mitochondrial disease. J Inborn Errors Metabol Screen 5

Ngamchuea K, Batchelor-McAuley C, Williams C et al (2018) Salivary glutathione in bipolar disorder: a pilot study. J Affect Disord 238:277–280

Nierenberg AA, Kansky C, Brennan BP et al (2013) Mitochondrial modulators for bipolar disorder: a pathophysiologically informed paradigm for new drug development. Aust N Z J Psychiatry 47:26–42

Novakova M, Prasko J, Latalova K et al (2015) The circadian system of patients with bipolar disorder differs in episodes of mania and depression. Bipolar Disord 17:303–314

Odeya D, Galila A, Lilah T (2018) The observed alteration in BCL2 expression following lithium treatment is influenced by the choice of normalization method. Sci Rep 8:6399

Okereke OI, Cook NR, Albert CM et al (2015) Effect of long-term supplementation with folic acid and B vitamins on risk of depression in older women. Br J Psychiatry 206:324–331

Onukwufor JO, Berry BJ, Wojtovich AP (2019) Physiologic implications of reactive oxygen species production by mitochondrial complex I reverse electron transport. Antioxidants (Basel) 8:285
Orrenius S (2004) Mitochondrial regulation of apoptotic cell death. Toxicol Lett 149:19–23
Ozkul A, Sair A, Akyol A et al (2014) Effects of lithium and lamotrigine on oxidative-nitrosative stress and spatial learning deficit after global cerebral ischemia. Neurochem Res 39:853–861
Papakostas GI, Mischoulon D, Shyu I et al (2010) S-adenosyl methionine (SAMe) augmentation of serotonin reuptake inhibitors for antidepressant nonresponders with major depressive disorder: a double-blind, randomized clinical trial. Am J Psychiatry 167:942–948
Pattij T, Vanderschuren LJ (2008) The neuropharmacology of impulsive behaviour. Trends Pharmacol Sci 29:192–199
Pehlivan FE (2017) Vitamin C: an antioxidant agent. Vitamin C. https://doi.org/10.5772/intechopen.69660
Pei L, Wallace DC (2018) Mitochondrial etiology of neuropsychiatric disorders. Biol Psychiatry 83:722–730
Pereira C, Chavarria V, Vian J et al (2018) Mitochondrial agents for bipolar disorder. Int J Neuropsychopharmacol 21:550–569
Peters A, Schweiger U, Pellerin L et al (2004) The selfish brain: competition for energy resources. Neurosci Biobehav Rev 28:143–180
Petrov B, Aldoori A, James C et al (2018) Bipolar disorder in youth is associated with increased levels of vitamin D-binding protein. Transl Psychiatry 8:61
Pham-Huy LA, He H, Pham-Huy C (2008) Free radicals, antioxidants in disease and health. Int J Biomed Sci 4:89–96
Pizzino G, Irrera N, Cucinotta M et al (2017) Oxidative stress: harms and benefits for human health. Oxidative Med Cell Longev 2017:8416763
Pullar JM, Carr AC, Bozonet SM et al (2018) High vitamin C status is associated with elevated mood in male tertiary students. Antioxidants (Basel) 7:91
Raha S, Robinson BH (2000) Mitochondria, oxygen free radicals, disease and ageing. Trends Biochem Sci 25:502–508
Reiter RJ, Rosales-Corral S, Tan DX et al (2017) Melatonin as a mitochondria-targeted antioxidant: one of evolution's best ideas. Cell Mol Life Sci 74:3863–3881
Rex A, Schickert R, Fink H (2004) Antidepressant-like effect of nicotinamide adenine dinucleotide in the forced swim test in rats. Pharmacol Biochem Behav 77:303–307
Ricca C, Aillon A, Bergandi L et al (2018) Vitamin D receptor is necessary for mitochondrial function and cell health. Int J Mol Sci 19:1672
Rizvi S, Raza ST, Ahmed F et al (2014) The role of vitamin e in human health and some diseases. Sultan Qaboos Univ Med J 14:e157–e165
Robertson OD, Coronado NG, Sethi R et al (2019) Putative neuroprotective pharmacotherapies to target the staged progression of mental illness. Early Interv Psychiatry 13:1032–1049
Roitman S, Green T, Osher Y et al (2007) Creatine monohydrate in resistant depression: a preliminary study. Bipolar Disord 9:754–758
Rosa AR, Singh N, Whitaker E et al (2014) Altered plasma glutathione levels in bipolar disorder indicates higher oxidative stress; a possible risk factor for illness onset despite normal brain-derived neurotrophic factor (BDNF) levels. Psychol Med 44:2409–2418
Ross S, Bossis A, Guss J et al (2016) Rapid and sustained symptom reduction following psilocybin treatment for anxiety and depression in patients with life-threatening cancer: a randomized controlled trial. J Psychopharmacol 30:1165–1180
Sabir MS, Haussler MR, Mallick S et al (2018) Optimal vitamin D spurs serotonin: 1,25-dihydroxyvitamin D represses serotonin reuptake transport (SERT) and degradation (MAO-A) gene expression in cultured rat serotonergic neuronal cell lines. Genes Nutr 13:19
Sahlin K, Harris RC (2011) The creatine kinase reaction: a simple reaction with functional complexity. Amino Acids 40:1363–1367

Sahraian A, Ghanizadeh A, Kazemeini F (2015) Vitamin C as an adjuvant for treating major depressive disorder and suicidal behavior, a randomized placebo-controlled clinical trial. Trials 16:94

Samuni Y, Goldstein S, Dean OM et al (2013) The chemistry and biological activities of N-acetylcysteine. Biochim Biophys Acta 1830:4117–4129

Sanoobar M, Eghtesadi S, Azimi A et al (2015) Coenzyme Q10 supplementation ameliorates inflammatory markers in patients with multiple sclerosis: a double blind, placebo, controlled randomized clinical trial. Nutr Neurosci 18:169–176

Sarris J, Papakostas GI, Vitolo O et al (2014) S-adenosyl methionine (SAMe) versus escitalopram and placebo in major depression RCT: efficacy and effects of histamine and carnitine as moderators of response. J Affect Disord 164:76–81

Sarris J, Price LH, Carpenter LL et al (2015) Is S-adenosyl methionine (SAMe) for depression only effective in males? A re-analysis of data from a randomized clinical trial. Pharmacopsychiatry 48:141–144

Sarris J, Byrne GJ, Bousman C et al (2018) Adjunctive S-adenosylmethionine (SAMe) in treating non-remittent major depressive disorder: an 8-week double-blind, randomized, controlled trial. Eur Neuropsychopharmacol 28:1126–1136

Sarris J, Byrne GJ, Stough C et al (2019) Nutraceuticals for major depressive disorder- more is not merrier: an 8-week double-blind, randomised, controlled trial. J Affect Disord 245:1007–1015

Scaini G, Barichello T, Fries GR et al (2019) TSPO upregulation in bipolar disorder and concomitant downregulation of mitophagic proteins and NLRP3 inflammasome activation. Neuropsychopharmacology 44:1291–1299

Sena LA, Chandel NS (2012) Physiological roles of mitochondrial reactive oxygen species. Mol Cell 48:158–167

Sharpley AL, Hockney R, McPeake L et al (2014) Folic acid supplementation for prevention of mood disorders in young people at familial risk: a randomised, double blind, placebo controlled trial. J Affect Disord 167:306–311

Shults CW, Oakes D, Kieburtz K et al (2002) Effects of coenzyme Q10 in early Parkinson disease: evidence of slowing of the functional decline. Arch Neurol 59:1541–1550

Shultz RB, Zhong Y (2017) Minocycline targets multiple secondary injury mechanisms in traumatic spinal cord injury. Neural Regen Res 12:702–713

Sigitova E, Fisar Z, Hroudova J et al (2017) Biological hypotheses and biomarkers of bipolar disorder. Psychiatry Clin Neurosci 71:77–103

Sikoglu EM, Navarro AA, Starr D et al (2015) Vitamin D3 supplemental treatment for mania in youth with bipolar spectrum disorders. J Child Adolesc Psychopharmacol 25:415–424

Simon J, Geddes JR, Gardiner A et al (2018) Comparative economic evaluation of quetiapine plus lamotrigine combination vs quetiapine monotherapy (and folic acid vs placebo) in patients with bipolar depression (CEQUEL). Bipolar Disord 20:733–745

Siwek M, Sowa-Kucma M, Styczen K et al (2016) Thiobarbituric acid-reactive substances: markers of an acute episode and a late stage of bipolar disorder. Neuropsychobiology 73:116–122

Slusarczyk W, Olakowska E, Larysz-Brysz M et al (2019) Use of ebselen as a neuroprotective agent in rat spinal cord subjected to traumatic injury. Neural Regen Res 14:1255–1261

Soeiro-de-Souza MG, Pastorello BF, Leite Cda C et al (2016) Dorsal anterior cingulate lactate and glutathione levels in euthymic bipolar I disorder: 1H-MRS study. Int J Neuropsychopharmacol 19:pyw032

Soeiro-de-Souza MG, Otaduy MCG, Machado-Vieira R et al (2018a) Anterior cingulate cortex glutamatergic metabolites and mood stabilizers in euthymic bipolar I disorder patients: a proton magnetic resonance spectroscopy study. Biol Psychiatry Cogn Neurosci Neuroimaging 3:985–991

Soeiro-de-Souza MG, Otaduy MCG, Machado-Vieira R et al (2018b) Lithium-associated anterior cingulate neurometabolic profile in euthymic bipolar I disorder: a (1)H-MRS study. J Affect Disord 241:192–199

Srivastava R, Faust T, Ramos A et al (2018) Dynamic changes of the mitochondria in psychiatric illnesses: new mechanistic insights from human neuronal models. Biol Psychiatry 83:751–760

Stacey D, Schubert KO, Clark SR et al (2018) A gene co-expression module implicating the mitochondrial electron transport chain is associated with long-term response to lithium treatment in bipolar affective disorder. Transl Psychiatry 8:183

Steckert AV, Valvassori SS, Moretti M et al (2010) Role of oxidative stress in the pathophysiology of bipolar disorder. Neurochem Res 35:1295–1301

Stern S, Santos R, Marchetto MC et al (2018) Neurons derived from patients with bipolar disorder divide into intrinsically different sub-populations of neurons, predicting the patients' responsiveness to lithium. Mol Psychiatry 23:1453–1465

Szczepankiewicz A (2013) Evidence for single nucleotide polymorphisms and their association with bipolar disorder. Neuropsychiatr Dis Treat 9:1573–1582

Tardiolo G, Bramanti P, Mazzon E (2018) Overview on the effects of N-acetylcysteine in neurodegenerative diseases. Molecules 23:3305

Taurine as an Anti-Manic Agent: A Double-Blind, Placebo-Controlled Study (2005). https://clinicaltrials.gov/ct2/show/NCT00217165

Toniolo RA, Silva M, Fernandes FBF et al (2018) A randomized, double-blind, placebo-controlled, proof-of-concept trial of creatine monohydrate as adjunctive treatment for bipolar depression. J Neural Transm (Vienna) 125:247–257

Traina G (2016) The neurobiology of acetyl-L-carnitine. Front Biosci (Landmark Ed) 21:1314–1329

Tsai MC, Huang TL (2015) Thiobarbituric acid reactive substances (TBARS) is a state biomarker of oxidative stress in bipolar patients in a manic phase. J Affect Disord 173:22–26

Vaarmann A, Kovac S, Holmstrom KM et al (2013) Dopamine protects neurons against glutamate-induced excitotoxicity. Cell Death Dis 4:e455

Valvassori SS, Rezin GT, Ferreira CL et al (2010) Effects of mood stabilizers on mitochondrial respiratory chain activity in brain of rats treated with d-amphetamine. J Psychiatr Res 44:903–909

Valvassori SS, Bavaresco DV, Feier G et al (2018) Increased oxidative stress in the mitochondria isolated from lymphocytes of bipolar disorder patients during depressive episodes. Psychiatry Res 264:192–201

van Enkhuizen J, Janowsky DS, Olivier B et al (2015) The catecholaminergic-cholinergic balance hypothesis of bipolar disorder revisited. Eur J Pharmacol 753:114–126

Vellekkatt F, Menon V (2019) Efficacy of vitamin D supplementation in major depression: a meta-analysis of randomized controlled trials. J Postgrad Med 65:74–80

Wang D, Li Z, Liu W et al (2018) Differential mitochondrial DNA copy number in three mood states of bipolar disorder. BMC Psychiatry 18:149

Weber WA, Dudley J, Lee JH et al (2013) A pilot study of alterations in high energy phosphoryl compounds and intracellular pH in unmedicated adolescents with bipolar disorder. J Affect Disord 150:1109–1113

Wilson C, Lee MD, Heathcote HR et al (2019) Mitochondrial ATP production provides long-range control of endothelial inositol trisphosphate-evoked calcium signaling. J Biol Chem 294:737–758

Wu GF, Ren S, Tang RY et al (2017) Antidepressant effect of taurine in chronic unpredictable mild stress-induced depressive rats. Sci Rep 7:4989

Wurtman RJ, Regan M, Ulus I et al (2000) Effect of oral CDP-choline on plasma choline and uridine levels in humans. Biochem Pharmacol 60:989–992

Yamaki N, Otsuka I, Numata S et al (2018) Mitochondrial DNA copy number of peripheral blood in bipolar disorder: the present study and a meta-analysis. Psychiatry Res 269:115–117

Yang L, Ran Y, Quan Z et al (2019) Pterostilbene, an active component of the dragon's blood extract, acts as an antidepressant in adult rats. Psychopharmacology 236:1323–1333

Yoon SJ, Lyoo IK, Haws C et al (2009) Decreased glutamate/glutamine levels may mediate cytidine's efficacy in treating bipolar depression: a longitudinal proton magnetic resonance spectroscopy study. Neuropsychopharmacology 34:1810–1818

Yu W, Greenberg ML (2016) Inositol depletion, GSK3 inhibition and bipolar disorder. Future Neurol 11:135–148

Yuksel C, Du F, Ravichandran C et al (2015) Abnormal high-energy phosphate molecule metabolism during regional brain activation in patients with bipolar disorder. Mol Psychiatry 20:1079–1084

Zhang J, Wang X, Vikash V et al (2016) ROS and ROS-mediated cellular signaling. Oxidative Med Cell Longev 2016:4350965

Zheng W, Zhu XM, Zhang QE et al (2019) Adjunctive minocycline for major mental disorders: a systematic review. J Psychopharmacol 33:1215–1226

Zverova M, Hroudova J, Fisar Z et al (2019) Disturbances of mitochondrial parameters to distinguish patients with depressive episode of bipolar disorder and major depressive disorder. Neuropsychiatr Dis Treat 15:233–240

Intracellular Signaling Cascades in Bipolar Disorder

Gregory H. Jones, Carola Rong, Aisha S. Shariq, Abhinav Mishra, and Rodrigo Machado-Vieira

Contents

Abstract Bipolar spectrum disorders carry a significant public health burden. Disproportionately high rates of suicide, incarceration, and comorbid medical conditions necessitate an extraordinary focus on understanding the intricacies of this disease. Elucidating granular, intracellular details seems to be a necessary preamble to advancing promising therapeutic opportunities. In this chapter, we review a wide range of intracellular mechanisms including mitochondrial energetics, calcium

G. H. Jones, C. Rong, and R. Machado-Vieira (✉)
Department of Psychiatry, University of Texas Health Science Center at Houston, Houston, TX, USA
e-mail: Rodrigo.MachadoVieira@uth.tmc.edu

A. S. Shariq
Department of Psychiatry, Texas Tech University Health Science Center, El Paso, TX, USA

Texas Tech University Health Science Center, Paul L. Foster School of Medicine, El Paso, TX, USA

A. Mishra
Texas Tech University Health Science Center, Paul L. Foster School of Medicine, El Paso, TX, USA

Curr Topics Behav Neurosci (2021) 48: 101–132
https://doi.org/10.1007/7854_2020_157
Published Online: 29 August 2020

signaling, neuroinflammation, the microbiome, neurotransmitter metabolism, glycogen synthase kinase 3-beta (GSK3β), protein kinase C (PKC) and diacylglycerol (DAG), and neurotrophins (especially BDNF), as well as the glutamatergic, dopaminergic, purinergic, and neurohormonal systems. Owing to the relative lack of understanding and effective therapeutic options compared to the rest of the spectrum, special attention is paid in the chapter to the latest developments in bipolar depression. Likewise, from a therapeutic standpoint, special attention should be paid to the pervasive mechanistic actions of lithium as a means of amalgamating numerous, disparate cascades into a digestible cognitive topology.

Keywords Brain-derived neurotrophic factor (BDNF) · Glycogen synthase kinase 3-beta (GSK3β) · Lithium · Protein kinase C (PKC) and diacylglycerol (DAG)

1 Introduction

BD has a lifetime prevalence of approximately 4% in the general population and imposes substantial medical and psychosocial morbidity on patients and families (Kessler et al. 2005). Bipolar disorder (BD) has been associated with a wide range of neurobiological models whose pathophysiological mechanisms have not yet been fully characterized. As our attempts to understand and treat this disease evolve, the clearer it becomes that BD is a multifactorial disease comprised of many dynamic, interdependent systems. Findings in recent decades have encouraged a transition from research predominantly centered around monoamine neurotransmission to intracellular signaling, synaptic/neural plasticity, and other cellular mechanisms. Cumulative data also demonstrates that conventional mood stabilizers and other efficacious compounds work by modulating many of these pathways. In this present chapter, we provide updated descriptions of the prominent pathophysiologic models as well as the latest therapeutic developments in the treatment of BD.

2 Mitochondrial Dysfunction

Dysfunction in brain-energy metabolism has been considered a key mechanism in the pathophysiology of BD (Zuccoli et al. 2017). Mitochondria primarily regulate intracellular energy metabolism, and abnormalities in both mitochondrial structure and function have been associated with the development of BD. In that regard, cells from individuals with BD demonstrate distinct abnormalities in mitochondrial structure (Cataldo et al. 2010). Moreover, some reports have indicated that patients with various mitochondrial diseases demonstrate up to a 20-fold higher incidence of BD than the general population (Goodwin and Jamison 2007). Accumulated deletions in mitochondrial DNA (ΔmtDNA) of transgenic mice have been shown to result in recurrent, spontaneous depression-like episodes which are both prevented SSRI therapy and worsened by lithium withdrawal (Kato 2017). Postmortem analysis of brain tissue from BD patients also demonstrates increased rates of ΔmtDNA (Kato et al. 1997; Sequeira et al. 2012).

Mitochondria regulate energy production in the cell, and their dysfunction can result in neuronal damage via multiple pathways: oxidative damage, decreased ATP production, abnormal calcium sequestration, and apoptosis via activation of caspase proteases. During times of cellular stress and increased energy demand, ATP production through the anaerobic pathway is upregulated. This results in an increased buffering burden for neuronal lactate, forming more lactic acid and decreasing intracellular pH. This process also enhances the production of various reactive oxygen species, which can lead to further free-radical damage. Due to its relatively higher energy demand and limited buffering capacity, the brain is far more susceptible to this paradigm than other body systems. In line with this theory, there have been several magnetic resonance spectroscopy (MRS) studies in BD patients showing increased neuronal lactate and decreased pH levels.

Furthermore, patients with BD in a current major depressive episode have shown a significant decrease in cingulate cortex lactate after being treated with a 6-week course of lithium monotherapy (Machado-Vieira et al. 2017b; Clay et al. 2011; Dager et al. 2004). Another indication of mitochondrial dysfunction is the decreased levels of N-acetyl-aspartate (NAA) seen in BD patients compared to normal controls (Kubo et al. 2017). NAA is produced in the mitochondria and is one of the most abundant brain metabolites, also serving as an important marker of neuronal viability (Rosso et al. 2017). In a clinical trial assessing NAA levels using MRS, BD patients were shown to have significantly lower NAA levels than controls. Furthermore, they had significant NAA increases after treatment with a 12-week course of lamotrigine (Croarkin et al. 2015). In addition to mitochondrial dysfunction, elevations of peripheral markers of oxidative stress have been described in BD patients, which correlate with a longer duration and earlier onset of illness (Machado-Vieira et al. 2007; Brown et al. 2014). De Sousa et al. have further demonstrated that a reactive increase in antioxidant enzymes occurs early in the developmental course of BD, especially during depressive phases. Additionally, BD patients have been shown to exhibit a decrease in lipid peroxidation following lithium treatment (de Sousa et al. 2014). Thus, oxidative damage appears to play a significant inciting role in the development of BD, subsequently interfering with endogenous repair mechanisms if left untreated.

Within the mitochondria itself, the electron transport chain (ETC) serves as a major mediator of oxidation/reduction and overall mitochondrial function. Postmortem analysis of cerebral tissue in patients with bipolar disorder demonstrates a decrease in ETC complex I concentration (Andreazza et al. 2010). Furthermore, lithium has been shown to increase mitochondrial complex I activity in BD patients significantly, and post-treatment complex I activity positively correlates with plasma lithium levels (de Sousa et al. 2015a, b). This strongly suggests that beyond its acute therapeutic mechanisms (discussed elsewhere), lithium provides a long-term, neuroprotective benefit to patients by reducing overall oxidative damage to neurons.

Converging lines of this hypothesis have proffered several substances as mitochondrial activity modulators: omega-3 fatty acids, coenzyme Q10, acetyl-L-carnitine, N-acetylcysteine (NAC), S-adenosylmethionine (SAM), alpha-lipoic acid, creatine monohydrate, melatonin, L-tryptophan, magnesium, folic acid, and branched-chain amino acids (BCAAs) (Nierenberg et al. 2013; Sarris et al. 2011). Overall, NAC has demonstrated the largest effect size in treating bipolar depression. Mixed, but mostly positive, evidence also supports the use of omega-3 fatty acids. Significant reductions in mania have also been demonstrated with the use of L-tryptophan, magnesium, folic acid, and BCAAs (Sarris et al. 2011). While some of these compounds may demonstrate efficacy as monotherapies, ultimately, this class may prove most effective as augmentation therapy for conventional BD treatments (Sarris et al. 2009; Mischoulon 2009). Larger, more-robust clinical trials are needed before their use becomes commonplace.

3 Calcium Channel Modulators

Ca2+ signaling is a tightly regulated process that plays an important role in a variety of cellular processes, including neuronal excitability, neurotransmitter synthesis/release, synaptogenesis, and plasticity. Even small perturbations in the balance between intracellular and extracellular calcium can trigger cell-death programs which are unresponsive to molecular rescue. Two cellular organelles, mitochondria and endoplasmic reticula (ER), are critical in maintaining a delicate balance of intracellular Ca2+ concentrations through sequestration, buffering, and mobilization. As previously discussed, increasing evidence suggests that mitochondrial function plays a key role in governing the synaptic properties of neuronal circuits that govern complex human behaviors. Mitochondrial dysfunction has also been implicated in a wide array of neurological and psychiatric disorders (Kato 2017).

In 1922, Weston and Howard observed that individuals with mania had lower spinal fluid concentrations of Ca2+ compared to depressed individuals (Weston 1922). This realization was subsequently obscured by decades of controversial findings regarding Ca2+ homeostasis in BD. However, more recent studies have revived this hypothesis by demonstrating that, compared to healthy controls, BD patients have elevated Ca2+ levels in platelets and lymphocytes, in addition to increased Ca2+ ATPase activity in red blood cells (Dubovsky et al. 1992; Warsh et al. 2004). As mitochondrial structure plays a significant role in Ca2+ homeostasis, these findings are in accordance with an ion-imbalance hypothesis in BD (Clay et al. 2011).

Various lines of pharmacologic evidence also support the relationship between calcium signaling and mood disorders. In animal and human models, lithium has been shown to significantly reduce NMDA receptor-stimulated Ca2+ responses (Nonaka et al. 1998). Moreover, both lithium and valproate are potent enhancers of Bcl-2 expression (Chen and Chuang 1999). The Bcl-2 family of proteins promote neuronal survival in part by inhibiting Ca2+ release from the ER (Rong and Distelhorst 2008). SNPs of the Bcl-2 gene, which are associated with developing BD, are also associated with elevated basal calcium levels and increased cytosolic Ca2+ release (Machado-Vieira et al. 2011). This alteration in membrane excitability may disrupt several neural components responsible for controlling mood and behavior.

Additionally, through separate pathways, lithium and valproate both cause depletion of inositol, a key calcium signaling intermediary. Lithium inhibits inositol monophosphatase (IMPase) involved in phosphatidylinositol 4,5-biphosphate (PIP2) turnover, which in turn reduces calcium entry into the cell and thus overall neuronal excitability. This may partially explain the efficacy of lithium in correcting neurotransmission, which has been shown to be altered in BD (Berridge 2014). Furthermore, inositol depletion has been shown to increase synapse formation between hippocampal neurons in vitro (Kim and Thayer 2009). Thus, modulation of calcium signaling appears to have multiple benefits, acutely stabilizing neurotransmission as well as the long-term potentiation of synaptogenesis.

As follows, researchers have investigated various calcium channel antagonists (CCAs) as possible therapeutic targets for BD. Studies indicate that only slow-gated L-type channels (LTCCs) are sensitive to CCAs in neurons. In that regard, three antihypertensive medications with LTCC activity, verapamil, nimodipine, and diltiazem were initially tested, with mixed results. Verapamil was first to be compared to lithium and placebo to treat mania. Both compounds had similar anti-manic effects, which were superior to placebo (Giannini et al. 1984). Despite this initial promise, recent reviews have shown inconsistent findings. Janicak et al. (1998) found no benefit over placebo in treating mania (Janicak et al. 1998). Keck et al. (2000) found both positive and negative results, suggesting the need for further investigation (Keck et al. 2000). Giannini et al. (2000) found that verapamil combined with magnesium oxide was superior to verapamil monotherapy (Giannini et al. 2000). More recently, verapamil was compared to lithium in a double-blind continuation study for acute mania. Individuals unresponsive to lithium (phase 1) were randomized to either continue lithium or switch to verapamil (phase 2). Non-responders in phase 2 were further allocated to verapamil-lithium combination treatment (phase 3). Authors found some efficacy for verapamil as monotherapy, but no significant difference compared to lithium and no superiority in combination (Mallinger et al. 2008).

In 1998, Pazzaglia et al. conducted a double-blind placebo-controlled study investigating the efficacy of nimodipine in 30 patients with refractory affective disorders (23 with bipolar and 7 with unipolar depression). One-third of the patients showed a significant response to nimodipine monotherapy. Non-responders ($n = 14$) added carbamazepine, with roughly another third achieving remission. Authors further noted that individuals with rapid-cycling BD were more apt to respond to the carbamazepine-nimodipine combination therapy (Pazzaglia et al. 1998). Subsequent studies were conducted around this time, most with modest results, and some evidence suggesting that low baseline CSF somatostatin may predict response to nimodipine (Frye et al. 2003).

Lastly, investigators have attempted to demonstrate the efficacy of diltiazem as add-on therapy in BD. In 2000, researchers conducted an open-label study with eight women in either a manic or depressive episode, generating promising findings. In comparing the 6 months before and after starting the add-on therapy, they observed a significant improvement in manic/depressive symptoms (Silverstone and Birkett 2000). However, several studies have reported that verapamil and diltiazem may enhance lithium excretion, increase carbamazepine levels, and create a synergistic, neurotoxic effect with either mood stabilizer (Brodie and MacPhee 1986; Bahls et al. 1991; Price and James Giannini 1986).

Despite a lack of definitive clinical outcomes and potential adverse side effects, recent preclinical research has reinforced the importance of aberrant calcium signaling in BD (Harrison 2016; Heyes et al. 2015). Specifically, genomic data shows that several LTCC-subunit-encoding genes, especially the CACNA1C locus, are associated with BD, schizophrenia, and MDD (Craddock and Sklar 2013). Additionally, neuron-like cells derived from BD patients show increased CACNA1C gene expression and enhanced calcium signaling (Yoshimizu et al. 2015). Mertens et al. further found that induced stem-cell-derived (iPSC) hippocampal neurons from patients with BD display a characteristic, hyperexcitable phenotype in relation to calcium signaling. Importantly, this phenotype was selectively reversed by lithium in vitro, only in neurons derived from patients that responded to lithium clinically (Mertens et al. 2015).

In summary, despite contradictory findings in the early twentieth century, more recent preclinical evidence has strongly implicated alterations in mitochondrial function and calcium signaling in the pathogenesis of BD. Elevated endogenous calcium concentrations in bipolar patients and reductions calcium hyperexcitability caused by lithium clearly demonstrate promise for this line of investigation. Building on this hypothesis, conventional CCAs (verapamil, diltiazem, nimodipine) demonstrated some initial promise. However, subsequent, more-rigorously designed studies failed to replicate these findings, in addition to revealing several adverse side effects. Despite this obstacle, the large quantity of in vitro evidence underscores a need for further proof-of-concept trials, potentially involving other CCAs. Given the wide distribution of LTCCs throughout the body, future clinical interventions should focus on developing more-targeted and genome-directed compounds with greater blood-brain barrier penetration and higher specificity for CNS-LTCC ligands. Most importantly, genomic evidence strongly suggests that future patients should be selectively enrolled in trails based on genetic and phenotypic amenability to such therapies.

4 Intracellular Signaling and Inflammation

Numerous studies support the role of the immune dysfunction in the development of BD (Rosenblat and McIntyre 2015; Munkholm et al. 2013; Goldstein et al. 2009; Leboyer et al. 2012). Bipolar disorder has been implicated in increased rates (1.5–3 times) of premature death due to medical comorbidities, higher prevalence of metabolic syndrome, and greater cardiovascular mortality than that in the adult general population (Leboyer et al. 2012). Furthermore, there is evidence that patients with BD have a reduced life expectancy (9–14 years) and exhibit shortened telomeres compared to the general population (Rizzo et al. 2013; Chang et al. 2015). BD has also been associated with numerous other immunological conditions: systemic lupus erythematosus, autoimmune thyroiditis, psoriasis, rheumatoid arthritis, multiple sclerosis, inflammatory bowel disease, and others (Rosenblat and McIntyre 2015). Though causality in this regard has yet to be established, recent research indicates that the relationship is likely bidirectional (Perugi et al. 2014).

These associations are reinforced by several cross-sectional studies which have demonstrated that, compared to healthy controls, patients with BD have significant elevations in proinflammatory cytokines such as TNF-α, IL-1β, IL-4, IL-6, and others. These studies have even suggested a state-dependent association with various inflammatory markers. For example, during manic episodes, BD patients demonstrate increased expression of TNF-α, IL-4, IL-6, CXCL10, and other inflammatory molecules (Barbosa et al. 2014; Modabbernia et al. 2013). Many of these markers have not yet been robustly investigated in bipolar depression; however evidence from MDD patients suggests that similar cytokine alterations also occur during depressive episodes (Rosenblat et al. 2014; Felger and Lotrich 2013). A recent meta-analysis of 32 inflammation studies found TNF-α to be the only biomarker elevated in both mania and bipolar depression (Rowland et al. 2018). Though further longitudinal research is necessary to fully validate a causal link in this regard, mounting evidence suggests that BD patients experience a chronic, low-grade inflammatory state which may be enhanced during acute mood episodes (Rosenblat et al. 2014; Fillman et al. 2014).

It is important to note that the majority of the aforementioned studies measure peripheral cytokine levels. While the direct observation of CNS inflammation may be more desirable, doing so is typically a highly invasive and cost-prohibitive endeavor. As with most neuroinflammatory conditions, altered blood-brain barrier (BBB) permeability allows for the bidirectional passage of peripheral and central immune mediators. Additionally, recent paradigm-shifting animal research has demonstrated a contiguous link between systemic lymphatic drainage and dural sinus vasculature, further allowing for the amalgamation of CNS and systemic immune constituents (Louveau et al. 2015). As follows, numerous BD studies have demonstrated a correlation between peripheral and central cytokine expression (Munkholm et al. 2013; Barbosa et al. 2014).

Several mechanisms have been proposed as to the specific, deleterious effects of inflammation in mood disorders. Under normal conditions, CNS macrophages (microglia) play several vital roles: immune surveillance, scavenging unwanted cellular products, and pruning of underutilized synaptic pathways. However, in states of elevated inflammation, microglia disproportionately adopt a proinflammatory phenotype leading to unwanted pruning, increased BBB permeability, and resultant influx of peripheral immune cells (Stertz et al. 2013). Evidence for this mechanism is supported by both PET imaging and postmortem tissue analysis, demonstrating increased microglial activity in the hippocampus and prefrontal cortex of BD patients (Haarman et al. 2014a, b; Rao et al. 2010). Proinflammatory cytokines have also been shown to alter the concentration of monoamines, for example, by decreasing the conversion of tryptophan to serotonin, accelerating serotonin breakdown, and increasing levels of depressogenic tryptophan metabolites (Capuron et al. 2003; Zhang et al. 2001). This process has been shown to be mediated in part by HPA axis over-activation and resultant hypercortisolemia (Maes et al. 2011). A detailed explanation of the HPA axis and cortisol overproduction in BD is discussed elsewhere in this chapter. In short, cortisol upregulates hepatic tryptophan 2,3-dioxygenase activity, causing unwanted monoamine catabolism. Additionally, proinflammatory cytokines interfere with hypothalamic glucocorticoid receptor function, disrupting normal feedback inhibition on cortisol production (Pace and Miller 2009).

With regard to systemic inflammation, significant attention has recently been directed towards elucidating a connection between the enteric nervous system (ENS) and the CNS, otherwise known as the "gut-brain axis." Numerous in vitro and animal studies have suggested that the mucosal microbiome may exert substantial influence on behavior and cognition via alterations in immune function and neurotransmitter metabolism. Though microbiome research in psychiatry is in its relative infancy, there has been a long-standing association between various gastrointestinal pathologies and psychiatric disorders (Severance et al. 2015; Lee et al. 2015). Recent research suggests that an imbalance in gut microflora (dysbiosis) may cause luminal inflammation which facilitates microbial translocation into systemic circulation (Dickerson et al. 2017).

Recently, researchers have developed antibodies to diagnose Crohn's disease, which detect the presence of normal gut microbes in systemic circulation (Desplat-Jego et al. 2007). Elevated antibody production against one such yeast (*Saccharomyces cerevisiae)* has been detected in patients with both schizophrenia and BD. Importantly, titers were significantly higher in individuals experiencing recent-onset of their disorder and those who were antipsychotic-naïve, suggesting an inciting mechanism that cannot be solely attributed to medication usage (Severance et al. 2012, 2014). Dickerson et al. have conducted several longitudinal observation studies which demonstrated that, compared to healthy controls, BD patients in acute manic episodes have increased production of antibodies directed against gliadin (a marker of gluten sensitivity), NR2 (an NMDA receptor peptide), Mason-Pfizer monkey virus 24, and *Toxoplasma gondii*. These findings appeared to be state-dependent, as titers did not differ from controls at 6-month follow-up.

Furthermore, elevated inflammation scores during initial manic episodes were predictive of psychiatric re-hospitalization during the follow-up period (Dickerson et al. 2012a, b, 2013). Though initial findings are intriguing, more research is needed to delineate specific microbial-inflammatory mechanisms. We must also assess the influence of confounding variables such as environment, medication usage, and smoking. Furthermore, there is a relative paucity of evidence regarding the interplay between gut dysbiosis and neurotransmitter-mediated, retrograde communication between the ENS and CNS. In that regard, more information is needed to bolster therapeutic efforts, as initial studies with general probiotics have yet to yield impactful results (Dickerson et al. 2017).

Current knowledge gaps notwithstanding, specifically targeting immune dysfunction in BD, is highly desirable, especially given that doing so can potentially ameliorate numerous associated comorbidities. Emerging evidence suggests that some antipsychotics and lithium downregulate the expression of inflammatory genes in the cells of BD patients (Haarman et al. 2014a, b). However, the exact mechanism by which these agents exert their anti-inflammatory effects is unknown. As previously mentioned, the anti-inflammatory/antioxidant compound N-acetylcysteine (NAC) has shown significant promise as an adjunct treatment in BD. Multiple DBRPCTs have demonstrated significant reductions in depressive symptoms compared to placebo in bipolar patients (Berk et al. 2008, 2012). A small number of trials have also investigated the efficacy of the anti-inflammatory agents like celecoxib (Arabzadeh et al. 2015) and aspirin (Berk et al. 2013) as adjunctive treatment in BD, with mixed results. Specifically, Arabzadeh et al. found celecoxib 400 mg to be superior to placebo as adjunctive therapy to sodium valproate in treating bipolar mania without psychotic features.

Pioglitazone is a peroxisome proliferator-activated receptor-gamma agonist with well-established, potent anti-inflammatory and anti-hyperglycemic properties. In a 2015 double-blind RCT of BD patients in a current major depressive episode (MDE), Zeinoddini et al. found that pioglitazone caused significant reductions in depressive symptoms after 6 weeks (Zeinoddini et al. 2015). Further research has shown it to be particularly effective in reducing depressive symptoms when used as an adjunct therapy in BD patients with comorbid metabolic dysfunction (Kemp et al. 2014). Importantly, in this study, elevated baseline levels of the proinflammatory cytokine IL-6 correlated with greater reductions in depression severity. Likewise, a recent 12-week, randomized, double-blind, placebo-controlled, parallel-group trial of 60 participants investigated the use of adjunctive infliximab in the treatment of adults with bipolar I/II depression. Despite being a negative trial overall infliximab significantly reduced depressive symptoms compared to placebo in patients with baseline elevations in serum CRP and TNF-α.

Preclinical evidence supports an intricate association between immune function, the microbiome, BD, and many of its comorbid conditions. Immune cells and inflammatory signaling molecules have been shown to produce local damage, prevent repair, and affect downstream neurotransmitter metabolism through a myriad of interconnected pathways. Beyond demonstrating therapeutic potential, several trials of anti-inflammatory compounds have importantly shown enhanced efficacy in patients with elevated levels of pre-treatment inflammation. This underscores the significant need for further investigation in this area, both in developing targeted therapies and screening/selecting patients based on their inflammatory profile. Numerous immune modulators have been approved for various systemic conditions over the last two decades, only a handful of which have been evaluated in the treatment of psychiatric disorders. Thus, given the ability to bypass significant startup costs and regulatory obstacles inherent to developing novel agents, it would seem prudent going forward to dedicate more resources towards clinical trials which investigate the efficacy of approved but untested medications in this class.

5 Glycogen Synthase Kinase 3-Beta (GSK3β)

Several lines of investigation have implicated the enzyme GSK3β in the parthenogenesis of BD. In 1996, GSK3β was discovered to be a target of lithium, specifically mediated by a magnesium-competitive inhibition mechanism (Klein and Melton 1996). Since that time, researchers have uncovered nearly 50 substrates of GSK3β that are key modulators in several processes related to neuronal function. The most prominent group of these compounds are known as Wnt molecules. Throughout early development and adult life, Wnt ligands mediate cell patterning, differentiation, proliferation, neuronal morphology, and cellular-integration into established neuronal circuits. The Wnt pathway facilitates these processes by first inhibiting the constituently active GSK3β enzyme, leading to the nuclear translocation of β-catenin and subsequent activation of several enhancers/transcription factors involved in neurotrophic generation (BDNF), circadian regulation, and inflammatory modulation (Valvezan and Klein 2012). Indeed, GSK3β overactivity has been linked to dysfunction in all of these domains, causing decreased BDNF production (Machado-Vieira et al. 2009), prevention of medication-mediated mood-stabilization and neurogenesis (Hussaini et al. 2014), increased expression of pro-inflammatory cytokines such as IL-6 and TNF-α (Ajmone-Cat et al. 2016), and lengthening of the circadian rhythm (McCarthy et al. 2013).

Various lines of animal and human evidence further support the role of aberrant Wnt/GSK3β activity in BD. Transgenic mice that over-express GSK3β develop hyperactive symptoms akin to human mania, whereas haplo-insufficient mice ($GSK3^{+/-}$) with decreased GSK3β activity exhibit features that replicate chronic lithium treatment. Moreover, induced GSK3β overexpression in mice nullifies the effect of chronic lithium treatment. Importantly, lithium is a well-validated GSK3β inhibitor (Cole 2013).

Patients with BD exhibit decreased levels of β-catenin mRNA and protein in the dorsolateral-prefrontal and temporal regions compared to those with schizophrenia and healthy controls (Pandey et al. 2015). Moreover, several SNPs within the promoter region for GSK3β are associated with an earlier age of onset and sensitivity to lithium therapy (Cole 2013).

Researchers have also demonstrated decreased GSK3β inhibition in the peripheral blood mononuclear cells (PBMCs) of bipolar individuals. Notably, in this small cohort study, a lower magnitude of GSK3β inhibition significantly correlated with the severity of manic/depressive episodes (Polter et al. 2010). More recent evidence has bolstered these findings by demonstrating that treatment with lithium causes an increase in PBMC-GSK3β inhibition, which also correlates with symptomatic improvement (de Sousa et al. 2015a, b).

Apart from lithium, other psychoactive compounds with efficacy in BD also demonstrate GSK3β pathway activity. Conflicting evidence exists regarding the GSK3 activity of anticonvulsant mood stabilizers such as valproic acid and carbamazepine (Muneer 2017). However, second-generation antipsychotics (SGAs), which are becoming increasingly important in the treatment of all phases of BD, have well-documented GSK3β inhibitory properties (Kalinichev and Dawson 2011; Pandey et al. 2015). Biased agonists such as aripiprazole have also been shown to increase the transcription of Wnt-related genes in the β-catenin pathway (de Bartolomeis et al. 2015). Lastly, conventional antidepressants have been shown to interact with several Wnt/GSK3 cellular cascades. GSK3β-mediated alterations in neuronal plasticity occur at the level of genetic transcription and are thus expected to take weeks before subjective improvement in symptoms occurs. This timeline thus concurs with the natural treatment course of SSRI therapies (Pilar-Cúellar et al. 2014).

Given the pervasive involvement of this pathway in a wide range of psychiatric disorders and medication classes, GSK3β inhibition stands as an attractive, albeit illusive therapeutic target. Over the past two decades, AstraZeneca has been at the forefront of GSK3β inhibitor development, with six novel candidates, only one of which progressed to phase II trials. Due to concerns over toxicity, safety margins, as well as various other issues, progress in this area has been slow (Bhat et al. 2018). Building on these efforts, Bhat et al. note that future GSK3 drug development should focus on increasing kinase selectivity as well as using novel conjugated-peptide technology to enhance tissue specificity.

6 Protein Kinase C (PKC) and Diacylglycerol (DAG)

Protein kinase C (PKC) is a family of calcium and phospholipid-dependent enzymes implicated in mood regulation. More specifically, PKC in the brain is present in high levels at presynaptic terminals and thought to play a role in the regulation of neuronal excitability, neurotransmitter release, and neuroplasticity (Zarate and Manji 2009). Conventional PKC (cPKC) isoforms (α, βI, βII, γ) require calcium and diacylglycerol (DAG) for activation, while novel PKC isoforms (δ, ε, η, θ, μ) only require DAG for activation (Zarate and Manji 2009). In CNS, cPKC isoforms are the most common and are highly expressed in several structures classically associated with mood regulation (prefrontal cortex, amygdala, and hippocampus) (Naik et al. 2000). This family of enzymes modulate neuronal transmission at several levels: short term (neurotransmitter/ion flux), intermediate (receptor regulation), and long term (synaptic remodeling, cell proliferation, genetic expression) (Amadio et al. 2006). PKC signaling underlies many of the pathologic mechanisms discussed throughout this chapter: neuronal excitability (Pahl et al. 2014), neurotransmitter release (Zarate et al. 2006), glutamate signaling (Zarate et al. 2003), neuroinflammation (Jun et al. 2014), neuroplasticity (Chu et al. 2014), and mitochondrial dysfunction, especially as it pertains oxidative stress and apoptosis (Nam et al. 2015).

Multiple lines of animal and human evidence support the pervasive involvement of this family of enzymes in BD. PKC inhibition reduces manic-like behaviors and hippocampal cell degeneration in rat sleep deprivation mania models (Abrial et al. 2013). Compared to healthy controls, BD patients demonstrate increased levels of both central (cortical) and peripheral (platelet) PKC activity (Wang and Friedman 1996; Wang et al. 1999). Lastly, a meta-analysis of 8,700 patients with both unipolar and bipolar depression found that suicidality was significantly associated with the genetic locus for PKCε (Saxena et al. 2017).

In that regard, chronic lithium treatment has been shown to decrease PKC levels in platelets of BD patients (Soares et al. 2000). Both lithium and valproic acid inhibit PKC activity in vitro and in vivo (Zarate and Manji 2009; Chen et al. 1994, 2000). Chronic administration of quercetin, a non-specific PKC inhibitor, prevents methylphenidate-induced hyperlocomotion and lipid peroxidation in mice (Kanazawa et al. 2017). Tamoxifen, another potent PKC inhibitor, has demonstrated efficacy in acute manic or mixed episodes of BD as both adjunct and monotherapy (Talaei et al. 2016; Yildiz et al. 2016, 2008; Amrollahi et al. 2011).

7 Neurotrophins

Neurotrophins are cellular growth factors involved in synaptic plasticity, neurogenesis, cell survival, and long-term memory formation (Grande et al. 2010). Within the neurotrophin family, brain-derived neurotrophic factor (BDNF) is the most extensively studied in BD (Fernandes et al. 2011; Rowland et al. 2018). Data from six meta-analyses have consistently demonstrated that BD patients have significantly lower levels of plasma BDNF compared to healthy controls, as do individuals with schizophrenia and unipolar depression (Rowland et al. 2018; Fernandes et al. 2015; Molendijk et al. 2014). Furthermore, BD patients in current manic or depressive episodes both demonstrate reductions in peripheral BDNF levels, unlike their euthymic counterparts (Fernandes et al. 2015; Rowland et al. 2018). The effect size is particularly large in bipolar depression (SMD – 0.86) and moderate in mania (SMD – 0.54) (Rowland et al. 2018).

BDNF plays a significant role in both GSK3β/Wnt and PKC activity (discussed in more detail in their respective sections). Briefly, BDNF transcription can be increased by various GSK3β inhibitors (including lithium) and through GSK3 gene silencing. Moreover, BDNF itself can inhibit GSK3β activity (Machado-Vieira et al. 2009). BDNF also mediates some of the genetic expression associated with synaptic plasticity via PKC signaling (Arevalo and Wu 2006). Beyond acting as a second messenger, PKC itself can also influence the expression of BDNF and other neurotrophins (Xu et al. 2013, 2015). The reciprocity and interdependence involved in both of these pathways underscores the pervasive and complex role that neurotrophins play in modulating a wide array of neuronal signaling mechanisms.

Some studies have suggested an association between polymorphisms in BDNF promoter genes and vulnerability for developing BD (D'Addario et al. 2012). Using a systems model of the human prefrontal transcriptional network, researchers have highlighted the importance of the early growth response gene 3 (EGR3) gene, which was previously shown to be modulated by BDNF (Pfaffenseller et al. 2016). A current proposal suggests a positive feedback loop wherein BDNF signaling dysfunction leads to reduced EGR3 expression, thus impairing neuroplasticity and resilience, increasing the vulnerability to stress, and further lowering BDNF expression (Pfaffenseller et al. 2018). Moreover, using a multisystem analysis of BDNF genes and other loci, researchers were able to identify bipolar patients with a sensitivity of 73% and a specificity of 71% (Munkholm et al. 2019). Building on this success, recent findings indicate that measuring BDNF in conjunction with TNF-α may provide even greater fidelity in distinguishing manic/depressed individuals from euthymic patients (Rowland et al. 2018). All of these studies suggest that BD pathogenesis is highly connected to many BDNF-related pathways. As with the "triple test" in Down syndrome and diagnostic tests for thyroid-related illnesses, BDNF measurements may be of greatest use when combined with other known biomarkers in BD. Given its pervasive influence on neuronal activity, BDNF may also have great potential as future therapeutic target.

Currently, no pharmacological agent used in BD directly targets the neurotrophins; however, antidepressants, lithium, electroconvulsive therapy (ECT), and glutamatergic agents like ketamine and memantine all increase peripheral levels of BDNF in patients with major depressive disorder (MDD) (Molendijk et al. 2014; Duncan and Zarate 2013; Haile et al. 2014; Lu et al. 2012). The pervasive, reciprocal interactions of neurotrophins with many other pathways present significant confounding obstacles for investigators. For now, BDNF may be of greatest utility as biomarker for disease severity and treatment response.

8 Glutamatergic System

Alterations in the glutamatergic system have been extensively implicated in the pathophysiology of BD. Likewise, the glutamate-modulating capability of conventional BD treatments such as lithium, valproate, lamotrigine, and antidepressants is well-documented (Machado-Vieira et al. 2012). Glutamate is the most abundant excitatory neurotransmitter in the brain. It acts in three different cellular compartments – pre/postsynaptic neurons and glia – characterized as the "tripartite glutamatergic synapse."

Particularly through its action at ionotropic AMPA and NMDA receptors, glutamatergic signaling plays a crucial role in excitatory neurotransmission, synaptic function, neuroplasticity, and neurogenesis. Indeed, AMPA (GluA1 subunit) knockout mice provide a successful model for depression, and NMDA receptor antagonists have repeatedly demonstrated antidepressant efficacy (Papp and Moryl 1994; Chourbaji et al. 2008). Interestingly, Du et al. (2010) showed that the GSK3 inhibitor

AR-A014418 regulates AMPA-induced GluR1 and GluR2 internalization via phosphorylation of kinesin light chain 2 (KLC2), the key molecule of the kinesin cargo delivery system (Du et al. 2010). As previously discussed, neurotrophins modulate the activity of GSK3 (Bartzokis 2012). Thus, glutamate seemingly acts as a molecular intermediary between many of the signaling pathways discussed throughout this chapter.

Candidate gene and genome-wide association studies (GWAS) have also implicated glutamate signaling in BD pathophysiology, albeit with some conflicting results. In a recent systematic review, de Sousa et al. found a positive association between BD and glutamate-related genes in 12 of 34 (35%) studies (Nurnberger et al. 2014; de Sousa et al. 2017). Overall, these genetic findings warrant deeper investigation, especially for the glutamatergic genes which have repeatedly demonstrated associations with BD (*GRIA3*, *GRIK2*, *GRIK4*, and *GRM7*). Moreover, researchers have demonstrated alterations in mRNA expression as well as induction of depressive-like behaviors in genetically altered animal models, further implicating these candidate genes in the disease process (Beneyto et al. 2007; Catches et al. 2012; Cryan et al. 2003; Duric et al. 2013).

Separate genetic analysis has also uncovered a significant association between BD and the glia-astrocyte pathway (Duncan et al. 2014). In addition to being the primary mediators of neuroinflammation, glial cells are integral to the glutamate recycling process. In that regard, investigation into glutamate/glial modulators has substantially increased in recent years. Ketamine, a N-methyl-D-aspartate (NMDA) receptor antagonist, has been placed in the spotlight due to its rapid antidepressant action in both unipolar and bipolar depression (Sanacora et al. 2017). While previous data did not demonstrate effective remission rates in bipolar depression (McCloud et al. 2016), a more recent double-blind, randomized placebo-control study phase 1 trial found that repeated ketamine infusions were effective in treating bipolar depression (Chen et al. 2019). Research suggests that ketamine's NMDA receptor antagonism increases BDNF-mediated synaptic protein synthesis via downstream glutamate-stimulation of AMPA receptors (Duman et al. 2012). Ketamine has also demonstrated a significant degree of interaction with glia-astrocyte-mediated inflammatory signaling (Miller 2013). Thus, beyond its therapeutic efficacy, ketamine potentially represents a paradigm-shifting convergence of the inflammatory and glutamate hypotheses of depression.

Approved by the FDA for the treatment of amyotrophic lateral sclerosis, riluzole (2-amino-6-trifluoromethoxy benzothiazole), is a glutamatergic modulator with both neuroprotective and anticonvulsant properties. Riluzole inhibits voltage-dependent sodium channels in neurons with subsequent inhibition of glutamate release, enhancing AMPA trafficking and membrane insertion of GluR1 and GluR2 and also increasing glutamate reuptake (Bellingham 2011). While early studies showed promise of riluzole in bipolar depression (Machado-Vieira et al. 2012), a more recent double-blind, placebo-controlled trial failed to show efficacy (Park et al. 2017). Another NMDA agent, memantine, is approved for dementia; it acts as a low-affinity NMDA receptor antagonist (Rammes et al. 2008). Like riluzole, initial clinical trials and case reports generated relatively positive results as an add-on to mood stabilizers in both mania (Koukopoulos et al. 2010; Serra et al. 2013) and bipolar depression (Stevens et al. 2013). However, a recent meta-analysis of randomized, double-blind controlled trials failed to demonstrate efficacy (Zheng et al. 2019). Especially given recent findings in relation to NMDA signaling, targeted modulation of glutamate neurotransmission will likely be an area of great clinical potential in the near future.

9 Dopaminergic System

Dopaminergic neurotransmission and its dysregulation in the development of BD are shown at different levels of evidence: behavioral, biomarker, and pharmacological proof-of-concept studies. Increasing dopamine activity is known to produce mania-like clinical presentations (Cousins et al. 2009). Specifically, amphetamine inhibits dopamine transporter-mediated reuptake and increases dopamine synaptic stimulation in the striatum, nucleus accumbens, and the frontal cortex. Predictably, it induces mood elevation, goal-directed activities, and motor hyperactivity and decreases the need for sleep. Likewise, L-dopa, a dopamine precursor, can induce manic or hypomanic states when used as a treatment for Parkinson's disease (PD). Homovanillic acid (HVA), a metabolite of dopamine, is decreased in the CSF of untreated individuals with bipolar depression. Conversely, treated and manic subjects have normal or increased HVA (Zarate et al. 2004). A recent meta-analysis of 26 studies supports this claim that decreased HVA is found in those with depressive disorders (Ogawa et al. 2018).

Measurements of in vivo availability of dopamine receptor binding using single-photon emission computed tomography (SPECT) radiotracer [^{123}I] iodobenzamide (IBZM) have generated mixed results. One study demonstrated reduced D_1 receptor binding potentials in the frontal cortex when compared to healthy controls, while D_2 receptor density was normal in all phases of non-psychotic BD individuals (Gonul et al. 2009). Using SPECT radiotracer [^{99m}Tc], a different study found increased dopamine transporter (DAT) availability in euthymic bipolar patients when compared to healthy subjects (Chang et al. 2010). Furthermore, another study used positron emission tomography (PET) and found that unmedicated BD subjects had significantly lower DAT availability and possibly higher dopamine concentrations in the dorsal caudate (Anand et al. 2011). In contrast, a different study found greater DAT density in unipolar and bipolar depressed vs non-depressed subjects (Amsterdam et al. 2012). More studies are needed to elucidate the role of dopamine receptor binding in mania and depression.

As increasing evidence reinforces the role of dopamine in BD pathophysiology, prescriptions of atypical antipsychotics are increasing. One cohort of 343 patients with bipolar depression who received olanzapine monotherapy significantly improved from their symptoms when compared to 171 bipolar depression patients receiving placebo in a randomized, double-blind placebo-controlled trial (Tohen et al. 2012). A separate randomized, double-blind placebo-controlled trial of

802 subjects with acute bipolar depression, which compared two dosages of quetiapine with lithium and placebo, found quetiapine more effective than both (Young et al. 2010). Recently, another randomized, double-blind placebo-controlled trial found cariprazine, a different atypical antipsychotic, to be effective, well-tolerated, and safe in depressive symptoms in adults with bipolar I (Earley et al. 2019). One proposed theory is that the atypical antipsychotics have 5-HT_{2A} antagonism and some D_2 receptor antagonism/partial D_2 agonism, which balances the dopamine signaling and leads to anti-manic effects (Kato 2019). This could partially explain the aforementioned mixed findings regarding dopamine receptor activity in BD.

Pramipexole, an aminothiazole derivative D_2/D_3 agonist, approved for use in PD, has been employed in proof-of-concept clinical trials in bipolar depression. Its theoretical utility as an antidepressant is derived in part from its D_3 receptor activity, which modulates several neuronal circuits implicated in depressive states (Zarate et al. 2004). It also increases anti-apoptotic Bcl-2 expression (Inden et al. 2009), suggesting an added neuroprotective benefit (Shaltiel et al. 2007). Previously, a cohort of 23 patients were followed, and 60.9% of patients showed sustained remission of a depressive episode (defined as a $>$ 50% on the Montgomery-Asberg Depression Rating Scale score) (Cassano et al. 2004). Using similar criteria, a subsequent meta-analysis showed that pramipexole had a response rate of 52.5%, a long-term response rate of 62.1%, and a long-term remission rate of 39.6% in 504 patients (Tundo et al. 2019). Randomized clinical trials have also demonstrated superiority compared to placebo in major depressive episodes (Tundo et al. 2019). While pramipexole showed beneficial as a therapeutic option for depression in BD patients, the authors suggest having an RCT of a higher power to confirm its benefit (Tundo et al. 2019).

10 Neurohormonal System

Circadian abnormalities have been recently associated with BD (Melo et al. 2017). Rhythm disruption can both precipitate and exacerbate mood episodes (Hirata et al. 2007). It is suggested that causality of this phenomenon is bidirectional, with circadian disruptions also being a consequence of affective disorders themselves (Bechtel 2015). Even in the euthymic state, BD patients have more alterations in their sleeping patterns than healthy controls (Ng et al. 2015). Circadian rhythms are regulated primarily by the secretion of melatonin from the suprachiasmatic nucleus (SCN) and subsequent binding to MT1 and MT2 receptors in the posterior hypothalamus. Normally, melatonin secretion reaches its peak during the night and trough during the day. BD patients characteristically show a delayed onset and lower evening peak of melatonin secretion compared to healthy controls (Robillard et al. 2013; Nurnberger et al. 2000). Mutant mice, deficient in the MT1 receptor encoding gene, display behavioral alterations that resemble some symptoms of depression (Comai et al. 2015).

Corticotropin-releasing hormone (CRH), secreted by the hypothalamus, stimulates the release of adrenocorticotropic hormone (ACTH) from the pituitary gland. ACTH then enables the secretion of cortisol by the adrenal glands. Circulating cortisol exerts a negative feedback effect at multiple levels within the hypothalamic-pituitary-adrenal (HPA) axis. In healthy individuals, cortisol displays the opposite circadian pattern to melatonin: peaking early in the morning and lowest at night. In addition to secreting melatonin, the SCN also modulates cortisol secretion and cognitive performance, both of which are altered in mania and depression (Linkowski et al. 1994). Both hypercortisolemia and a flattening of circadian cortisol variation are seen in depressed individuals (Carroll et al. 1976). Elevated cortisol response to dexamethasone suppression and CRH (DEX/CRH test) at the time of discharge from an inpatient treatment is also predictive of relapse in depressed individuals (Zobel et al. 2001). As with depression, newly diagnosed bipolar patients display elevated baseline cortisol (Coello et al. 2019).

Additionally, changes in CRH secretion may have potential as a future biomarker, as they can be detected before the clinically observable onset of manic or hypomanic symptoms (Daban et al. 2005). Previous research indicates that cortisol secretion is elevated in the manic and the depressed phases of bipolar disorder, suggesting that aberrant stress responsiveness plays a role in both phases of the disease (Cervantes et al. 2001). Higher cortisol concentrations have also been demonstrated in both remitted and non-remitted bipolar patients, implying a persistence of HPA axis abnormalities beyond symptomatic control (Watson et al. 2004). Ellenbogen et al. found that offspring of parents with bipolar disorder have elevated cortisol compared to those with no parental history of mental illness (Ellenbogen et al. 2004). The heritability of basal cortisol secretion level is estimated to be 60% (Bartels et al. 2003). Taken together, this implicates hormonal axis dysfunction as a core mechanism underlying the pathogenesis and transmission of BD. Aberrancies precede, are passed on, and may persist beyond a patient's detectable symptomatic course.

In addition to hormone secretion, the glucocorticoid receptor (GR) function plays an intricate role in the pathophysiology of mood disorders. Cortisol diffuses through cellular membranes and binds to both GR and mineralocorticoid (MR) intracellular receptors. The human stress response increases glucocorticoid secretion, preferentially causing MR saturation first. GR binding thus becomes the primary mediator of feedback inhibition on the HPA axis (Pariante et al. 2001). Both glucocorticoids and stress have been shown to impair neurogenesis and promote atrophy of dendritic processes in the hippocampus (Kempermann 2002; McEwen 2005). GR antagonists block the detrimental effects of hypercortisolism and increase the expression of GRs, exerting a more powerful negative feedback effect on the HPA axis.

The extensive involvement of the HPA axis in affective disorders makes this system an attractive target for therapeutic interventions. Tricyclic antidepressants increase glucocorticoid receptor protein binding and capacity in vivo. This effect was also seen in rats when submitted to lithium therapy or electroconvulsive therapy. In humans, elevated diurnal cortisol is dampened with chronic lithium prophylaxis (Colla et al. 2009). Thus, it is suggested that part of the overall therapeutic mechanism in the treatment of BD may involve a gradual reinstatement of appropriate glucocorticoid receptor function and some degree of HPA axis normalization.

As follows, several other therapeutic interventions have been attempted, targeting various aspects of HPA/circadian homeostasis. Exogenous melatonin has well-documented hypnotic properties and has been shown to improve sleep disturbances in the context of depression (Satyanarayanan et al. 2018). However, administering melatonin on its own to patients fails to display any antidepressant efficacy. Conversely, agomelatine, a MT1 and MT2 agonist but also a 5-HT2c antagonist, displayed clear antidepressant activity in animal models of depression (Thomas et al. 2016). A single dose of agomelatine can restore sleep architecture in severely depressed individuals. The compound also induces a faster symptomatic improvement in depressed patients than the SSRI sertraline (Kasper and Hamon 2009). This finding was bolstered by a subsequent systematic review which concluded that agomelatine is a viable treatment for depression (Fornaro et al. 2010). With regard to bipolar disorder, an open-label study for agomelatine enrolled 21 patients with bipolar I depression. Patients received 25 mg/day for 6–46 weeks, with 81% of subjects showing substantial improvement (>50% improvement in HAM-D scores) (Calabrese et al. 2007).

Researchers have also investigated the efficacy of the GR antagonist mifepristone (RU-486) in proof-of-concept trials. In one double-blind, placebo-controlled crossover study, patients with psychotic depression experienced rapid improvement of their symptoms following 4 days of mifepristone therapy (Belanoff et al. 2001). Authors replicated this result in a larger study which also found that elevated dosages (>600 mg/day) may be the optimal therapeutic range (Belanoff et al. 2002). Young et al. conducted a mifepristone trial (600 mg/day vs placebo) in 20 patients with bipolar depression. Spatial recognition/memory, verbal fluency, and HAM-D rating scores all significantly improved in the intervention group. Notably, improvements in cognition were inversely correlated with baseline cortisol levels, further implicating an anti-glucocorticoid effect as the therapeutic mechanism (Young et al. 2004). Gallagher et al. conducted a placebo-controlled, 7-day mifepristone trial of 19 individuals with bipolar disorder and 20 with schizophrenia. There was a significant elevation in cortisol secretion directly following mifepristone administration. However, at 21 days post-treatment, cortisol levels decreased significantly, falling below baseline level in both treatment subgroups. Conversely, cortisol levels for both bipolar and schizophrenic patients in the placebo group remained unchanged. This reinforces the notion that GR antagonists cause normalization of the HPA axis through long-term, receptor-mediated negative feedback. More recently, Watson et al. conducted a RDBPCT evaluating a 7-day course of mifepristone (600 mg/day) vs placebo as an adjunct therapy in 60 patients with bipolar depression. Though patients experienced no antidepressant efficacy, spatial working memory showed substantial improvement, which was sustained 7 weeks after cessation of treatment. Consistent with previous findings, the magnitude of cortisol response to mifepristone directly correlated with the magnitude of memory improvement (Watson et al. 2012).

Glucocorticoid synthesis inhibitors such as ketoconazole and metyrapone have also shown some antidepressant efficacy in clinical and preclinical models. In a double-blind, randomized, controlled trial for MDD, metyrapone was superior to placebo as add-on therapy to conventional antidepressants and accelerated the onset of symptomatic relief (Jahn et al. 2004). Gallagher et al. evaluated five trials conducted with ketoconazole in either MDD or BD, finding a significant difference in HAM-D scores favoring treatment (Gallagher et al. 2008). However, the risks associated with long-term use of these agents may limit their applicability in mood disorders. Nevertheless, these proof-of-concept studies show promise for more-targeted approaches in the future.

The neurohormonal modulator, liraglutide, a glucagon-like peptide 1(GLP-1) analog, has already shown promise in preclinical models of Alzheimer's disease, Parkinson's disease, and stroke. Weina et al. found that liraglutide successfully decreased behavioral symptoms of depression and anxiety in a corticosterone-induced mouse model of depression. Importantly, these findings were associated with a reduction in ACTH-mediated stress responsiveness. The drug was shown to preserve neuronal plasticity and increase the density of immature neurons in the hippocampus (Weina et al. 2018). Further investigation regarding its clinical efficacy in humans is highly warranted, especially given the strong association between metabolic disease, insulin resistance, and BD.

Non-pharmacologic interventions targeting the neurohormonal/circadian system, including sleep manipulation/deprivation and light therapy, have also been investigated. Total sleep deprivation and partial sleep deprivation during the second half of the night have shown the most antidepressant efficacy (Wirz-Justice et al. 2005). Sleep deprivation, in combination with SSRI, has been proven to accelerate the improvement of depressive symptoms compared to SSRI alone (Benedetti et al. 1997). Light therapy has also shown promise in certain mood disorders, especially seasonal depression. Like sleep deprivation, best results have been demonstrated when used in combination with antidepressants (Lanfumey et al. 2013). Tseng et al. conducted a meta-analysis regarding the use of light therapy and sleep deprivation in patients with bipolar depression. The results of 9 studies, including 489 patients, indicated that adjunct light therapy significantly reduced disease severity in medicated patients (Tseng et al. 2016).

As the primary arbiters of circadian rhythm and stress responses, neurohormones such as cortisol and melatonin play a significant role in the development and transmission of BD. Whether through genetic, epigenetic, or direct signaling mechanisms, circadian and HPA axis dysfunction seems to underlie most mood disorders, making the pathway an attractive target for therapeutic interventions going forward. Successful treatment of BD seems to entail a certain degree of normalization both in GR function and the HPA axis overall. Supporting this hypothesis, therapeutic efficacy, especially in the cognitive domain, is consistently associated with beneficial alterations in cortisol secretion. Currently, however, therapeutic interventions have been mostly limited to proof-of-concept studies and non-pharmacologic modalities, which may not be broadly applicable in real-world clinical settings. Given the significant association between metabolic syndrome and BD, further investigation of hormonal modulators, particularly those which affect insulin signaling, is warranted.

11 Purinergic System and Mania

In 1921, Kraepelin indirectly implicated the purinergic system in the parthenogenesis of BD by suggesting an association between manic symptoms, hyperuricemia, and gout (Ketter 2009). Clinicians have since observed that patients being treated with lithium see improvements not only in their mood disturbances but also in their gout/hyperuricemia (Shorter 2009). This association has subsequently been confirmed by several studies showing elevated uric acid (UA) levels in drug-naïve patients experiencing their first manic episode, as well as nationwide, population-based data indicating an increased risk of gout in patients with BD (Salvadore et al. 2010; Chung et al. 2010). Uric acid modulates a variety of endogenous functions, including sleep, appetite, cognition, memory, motor activity, and social interaction (Machado-Vieira et al. 2002; Lorenzi et al. 2010). Elevated UA levels have been associated with specific manic traits such as impulsivity, irritability, increased drive, disinhibition, and hyperthymia (Machado-Vieira et al. 2002; Lorenzi et al. 2010; Sutin et al. 2014). Manic patients demonstrate increased serum UA compared to those with bipolar depression or MDD (Bartoli et al. 2017b). Higher serum UA is also predictive of bipolar conversion in currently depressed patients (Oliveira et al. 2019). Indeed, hyperuricemia appears to be highly selective for mania in BD (Berardis and De Berardis 2008), whereas low UA levels have been linked to depressive mood scores in BD, independent of current disease phase (Albert et al. 2015; Muti et al. 2015). Reductions in mania are associated with a concurrent decrease in serum uric acid levels, and enhanced urinary excretion of UA has been observed during lithium-induced remission from hypomania (Machado-Vieira et al. 2017a).

Furthermore, plasma uric acid level abnormalities are typically absent during euthymic periods (Berardis and De Berardis 2008). This evidence strongly suggests that purinergic system dysfunction plays an important role, especially early in the development of BD (Machado-Vieira et al. 2008). Given the positive association between peripheral and central uric acid levels (Bowman et al. 2010), UA holds significant potential as a state-dependent biomarker and outcome indicator in BD.

The purine adenosine is decreased in the serum of euthymic bipolar patients when compared to healthy controls, and lower levels confer a higher degree of functional impairment (Gubert et al. 2016). Adenosine receptor agonists characteristically display anti-aggressive, anticonvulsant, and antipsychotic properties, whereas antagonists (caffeine, theophylline) can enhance irritability, anxiety, and insomnia (Lara et al. 2006). This suggests that, beyond acting as surrogate markers for other cellular processes (i.e., nucleic acid turnover), purinergic constituents independently contribute to the overall disease process in BD. Machado-Vieira et al. previously demonstrated the efficacy of allopurinol in treatment-resistant mania associated with hyperuricemia (Machado-Vieira et al. 2001). By inhibiting xanthine oxidase, it is suggested that allopurinol may increase CNS levels of adenosine and other purine metabolites in addition to decreasing UA formation (Marro et al. 2006). Several trials have subsequently demonstrated the efficacy of allopurinol as an adjunct to standard therapy in improving manic symptoms (Machado-Vieira et al. 2002, 2008; Akhondzadeh et al. 2006). Moreover, a 2017 systematic review and meta-analysis of 5 RCTs using adjuvant purine-modulators confirmed their anti-manic efficacy compared to placebo (Bartoli et al. 2017a). Of note, allopurinol can also effectively treat aggressive behavior in patients with dementia, further solidifying the association between UA and the aggression/impulsivity (Lara et al. 2003).

Genetic studies have recently revealed a specific SNP in the P2X7 gene, which encodes a CNS-expressed purine receptor associated with both BD and MDD in animal models (Sperlagh et al. 2012). P2X7 expression is also increased by sleep deprivation and rapid cycling in humans (Backlund et al. 2012). P2X7 receptor-associated neuroinflammation has also been implicated in the parthenogenesis of BD (Masuch et al. 2016). Recently, significant attention has been directed towards developing targeted therapies for P2X7 in BD and MDD, though robust clinical trials have not yet been executed (Bhattacharya et al. 2013; Dodd et al. 2015).

In conclusion, beyond acting as surrogate markers, evidence strongly implicates purinergic system constituents as a priori disease modulators in BD. The diversity of metabolites and downstream interactions associated with this pathway espouse significant potential for future therapeutic and biomarker development.

12 Conclusion

Our evolving understanding of BD weaves an ever-expanding web of conceptual integration. Complexities notwithstanding, several key entities (cortisol, BDNF, PKC, GSK3, calcium, glutamate, and inflammatory cytokines) seem to underlie and connect many of the seemingly disparate concepts discussed in this chapter. Through these elements, we can see how oxidation and mitochondrial function, for example, ultimately affect neurotransmission, plasticity, apoptosis, and even the inflammatory milieu. Despite our limited understanding of the microbiome, we can already appreciate this same dynamic. Small alterations in the gut seem to affect everything from neurotransmitter metabolism, to neuroinflammation, and even the HPA axis. From a therapeutic standpoint, lithium perfectly exemplifies this concept. Its efficacy stems from its ability to modulate so many aspects of this disease from oxidative stress, mitochondrial function, apoptosis, and calcium signaling to inflammation, hormonal activity, glutamate signaling, and even the purinergic system. Likewise, we see how novel therapies like ketamine simultaneously integrate both the glutamate and inflammatory hypotheses of depression through final common pathways. It is no coincidence that emerging biomarkers such as BDNF, TNF-α, UA, and others seem to act as molecular conduits between so many different pathways and systems. Moving forward, clinicians and scientists must build on this ethic: making connections that emphasize a holistic context for every specific discovery, increasing patient screening with objective markers, and implementing multi-modal, tailored approaches to treatment.

References

Abrial E, Etievant A, Bétry C, Scarna H, Lucas G, Haddjeri N, Lambás-Señas L (2013) Protein kinase C regulates mood-related behaviors and adult hippocampal cell proliferation in rats. Prog Neuropsychopharmacol Biol Psychiatry 43:40–48

Ajmone-Cat MA, D'Urso MC, di Blasio G, Brignone MS, De Simone R, Minghetti L (2016) Glycogen synthase kinase 3 is part of the molecular machinery regulating the adaptive response to LPS stimulation in microglial cells. Brain Behav Immun 55:225–235

Akhondzadeh S, Milajerdi MR, Amini H et al (2006) Allopurinol as an adjunct to lithium and haloperidol for treatment of patients with acute mania: a double-blind, randomized, placebo-controlled trial. Bipolar Disord 8:485–489

Albert U et al (2015) Increased uric acid levels in bipolar disorder subjects during different phases of illness. J Affect Disord 173:170–175

Amadio M, Battaini F, Pascale A (2006) The different facets of protein kinases C: old and new players in neuronal signal transduction pathways. Pharmacol Res 54:317–325

Amrollahi Z, Rezaei F, Salehi B, Modabbernia AH, Maroufi A, Esfandiari GR, Naderi M, Ghebleh F, Ahmadi-Abhari SA, Sadeghi M, Tabrizi M, Akhondzadeh S (2011) Double-blind, randomized, placebo-controlled 6-week study on the efficacy and safety of the tamoxifen adjunctive to lithium in acute bipolar mania. J Affect Disord 129(1–3):327–331

Amsterdam JD et al (2012) Greater striatal dopamine transporter density may be associated with major depressive episode. J Affect Disord 141(2–3):425–431

Anand A et al (2011) Striatal dopamine transporter availability in unmedicated bipolar disorder. Bipolar Disord 13(4):406–413

Andreazza AC, Shao L, Wang JF, Young LT (2010) Mitochondrial complex I activity and oxidative damage to mitochondrial proteins in the prefrontal cortex of patients with bipolar disorder. Arch Gen Psychiatry 67:360–368

Arabzadeh S et al (2015) Celecoxib adjunctive therapy for acute bipolar mania: a randomized, double-blind, placebo-controlled trial. Bipolar Disord 17(6):606–614

Arevalo JC, Wu SH (2006) Neurotrophin signaling: many exciting surprises! Cell Mol Life Sci 63:1523–1537

Backlund L et al (2012) P2RX7: expression responds to sleep deprivation and associates with rapid cycling in bipolar disorder type 1. PLoS One 7(8):e43057

Bahls FH, Ozuna J, Ritchie DE (1991) Interactions between calcium channel blockers and the anticonvulsants carbamazepine and phenytoin. Neurology 41(5):740–742

Barbosa IG, Bauer ME, Machado-Vieira R et al (2014) Cytokines in bipolar disorder: paving the way for neuroprogression. Neural Plast 2014:360481

Bartels M et al (2003) Heritability of daytime cortisol levels in children. Behav Genet 33 (4):421–433

Bartoli F, Crocamo C et al (2017a) Purinergic system dysfunctions in subjects with bipolar disorder: a comparative cross-sectional study. Compr Psychiatry 73:1–6

Bartoli F, Carrà G, Clerici M (2017b) Purinergic dysfunction in bipolar disorder: Any role for the antioxidant uric acid as a trait and state biomarker? Psychiatry Clin Neurosci 71(6):417

Bartzokis G (2012) Neuroglialpharmacology: myelination as a shared mechanism of action of psychotropic treatments. Neuropharmacology 62(7):2137–2153

Bechtel W (2015) Circadian rhythms and mood disorders: are the phenomena and mechanisms causally related? Front Psych 6:118

Belanoff JK et al (2001) Rapid reversal of psychotic depression using mifepristone. J Clin Psychopharmacol 21(5):516–521
Belanoff JK et al (2002) An open label trial of C-1073 (mifepristone) for psychotic major depression. Biol Psychiatry 52(5):386–392
Bellingham MC (2011) A review of the neural mechanisms of action and clinical efficiency of riluzole in treating amyotrophic lateral sclerosis: what have we learned in the last decade? CNS Neurosci Ther 17(1):4–31
Benedetti F et al (1997) Sleep deprivation hastens the antidepressant action of fluoxetine. Eur Arch Psychiatry Clin Neurosci 247(2):100–103
Beneyto M, Kristiansen LV, Oni-Orisan A, McCullumsmith RE, Meador-Woodruff JH (2007) Abnormal glutamate receptor expression in the medial temporal lobe in schizophrenia and mood disorders. Neuropsychopharmacology 32:1888–1902
Berardis DD, De Berardis D (2008) The role of duloxetine in the treatment of anxiety disorders. Neuropsychiatr Dis Treat 1:929
Berk M, Copolov DL, Dean O et al (2008) N-acetyl cysteine for depressive symptoms in bipolar disorder–a double-blind randomized placebo-controlled trial. Biol Psychiatry 64:468–475
Berk M, Dean OM, Cotton SM et al (2012) Maintenance N-acetyl cysteine treatment for bipolar disorder: a double-blind randomized placebo controlled trial. BMC Med 10:91
Berk M et al (2013) Aspirin: a review of its neurobiological properties and therapeutic potential for mental illness. BMC Med 11(74):2013
Berridge MJ (2014) Calcium signalling and psychiatric disease: bipolar disorder and schizophrenia. Cell Tissue Res 357(2):477–492
Bhat RV, Andersson U, Andersson S, Kneer L, Bauer U, Sundgren-Andersson AK (2018) The conundrum of GSK3 inhibitors: is it the dawn of a new beginning? J Alzheimers Dis 64(1): S547–S554
Bhattacharya A et al (2013) Pharmacological characterization of a novel centrally permeable P2X7 receptor antagonist: JNJ-47965567. Br J Pharmacol 170(3):624–640
Bowman GL et al (2010) Uric acid as a CNS antioxidant. J Alzheimer's Dis 19(4):1331–1336
Brodie MJ, MacPhee GJ (1986) Carbamazepine neurotoxicity precipitated by diltiazem. Br Med J 292(6529):1170–1171
Brown NC, Andreazza AC, Trevor Young L (2014) An updated meta-analysis of oxidative stress markers in bipolar disorder. Psychiatry Res 218(1–2):61–68
Calabrese JR et al (2007) Agomelatine adjunctive therapy for acute bipolar depression: preliminary open data. Bipolar Disord 9(6):628–635
Capuron L, Neurauter G, Musselman DL et al (2003) Interferon-alpha-induced changes in tryptophan metabolism. Relationship to depression and paroxetine treatment. Biol Psychiatry 54:906–914
Carroll BJ, Curtis GC, Mendels J (1976) Cerebrospinal fluid and plasma free cortisol concentrations in depression. Psychol Med 6(2):235–244
Cassano P et al (2004) Pramipexole in treatment-resistant depression: an extended follow-up. Depress Anxiety 20(3):131–138
Cataldo AM, McPhie DL, Lange NT, Punzell S, Elmiligy S, Ye NZ, Froimowitz MP, Hassinger LC, Menesale EB, Sargent LW, Logan DJ, Carpenter AE, Cohen BM (2010) Abnormalities in mitochondrial structure in cells from patients with bipolar disorder. Am J Pathol 177 (2):575–585
Catches JS, Xu J, Contractor A (2012) Genetic ablation of the GluK4 kainate receptor subunit causes anxiolytic and antidepressant-like behavior in mice. Behav Brain Res 228:406–414
Cervantes P et al (2001) Circadian secretion of cortisol in bipolar disorder. J Psychiatry Neurosci 26 (5):411–416
Chang TT et al (2010) Higher striatal dopamine transporters in euthymic patients with bipolar disorder: a SPECT study with [Tc] TRODAT-1. Bipolar Disord 12(1):102–106

Chang K-C et al (2015) Estimation of life expectancy and the expected years of life lost among heroin users in the era of opioid substitution treatment (OST) in Taiwan. Drug Alcohol Depend 153:152–158

Chen RW, Chuang DM (1999) Long term lithium treatment suppresses p53 and Bax expression but increases Bcl-2 expression. A prominent role in neuroprotection against excitotoxicity. J Biol Chem 274(10):6039–6042

Chen G, Manji HK, Hawver DB, Wright CB, Potter WZ (1994) Chronic sodium valproate selectively decreases protein kinase C alpha and epsilon in vitro. J Neurochem 63:2361–2364

Chen G, Masana MI, Manji HK (2000) Lithium regulates PKC-mediated intracellular cross-talk and gene expression in the CNS in vivo. Bipolar Disord 2:217–236

Chen M-H et al (2019) Maintenance of antidepressant and antisuicidal effects by D-cycloserine among patients with treatment-resistant depression who responded to low-dose ketamine infusion: a double-blind randomized placebo-control study. Neuropsychopharmacology 44 (12):2112–2118

Chourbaji S, Vogt MA, Fumagalli F, Sohr R, Frasca A, Brandwein C et al (2008) AMPA receptor subunit 1 (GluR-A) knockout mice model the glutamate hypothesis of depression. FASEB J 22:3129–3134

Chu Y, Fioravante D, Leitges M, Regehr WG (2014) Calcium-dependent PKC isoforms have specialized roles in short-term synaptic plasticity. Neuron 82:859–871

Chung K-H, Huang C-C, Lin H-C (2010) Increased risk of gout among patients with bipolar disorder: a nationwide population-based study. Psychiatry Res 180(2–3):147–150

Clay HB, Sillivan S, Konradi C (2011) Mitochondrial dysfunction and pathology in bipolar disorder and schizophrenia. Int J Dev Neurosci 29(3):311–324

Coello K et al (2019) Hair cortisol in newly diagnosed bipolar disorder and unaffected first-degree relatives. Psychoneuroendocrinology 99:183–190

Cole AR (2013) Glycogen synthase kinase 3 substrates in mood disorders and schizophrenia. FEBS J 280(21):5213–5227

Colla M et al (2009) Glutamate as a spectroscopic marker of hippocampal structural plasticity is elevated in long-term euthymic bipolar patients on chronic lithium therapy and correlates inversely with diurnal cortisol. Mol Psychiatry 14(7):696–704, 647

Comai S et al (2015) Melancholic-Like behaviors and circadian neurobiological abnormalities in melatonin MT1 receptor knockout mice. Int J Neuropsychopharmacol 18(3):pyu075

Cousins DA, Butts K, Young AH (2009) The role of dopamine in bipolar disorder. Bipolar Disord 11(8):787–806

Craddock N, Sklar P (2013) Genetics of bipolar disorder. Lancet 381(9878):1654–1662

Croarkin PE et al (2015) N-acetylaspartate normalization in bipolar depression after lamotrigine treatment. Bipolar Disord 17(4):450–457

Cryan JF, Kelly PH, Neijt HC, Sansig G, Flor PJ, van Der Putten H (2003) Antidepressant and anxiolytic-like effects in mice lacking the group III metabotropic glutamate receptor mGluR7. Eur J Neurosci 17:2409–2417

D'Addario C et al (2012) Selective DNA Methylation of BDNF Promoter in Bipolar Disorder: Differences Among Patients with BDI and BDII. Neuropsychopharmacology 37(7):1647–1655

Daban C et al (2005) Hypothalamic-pituitary-adrenal axis and bipolar disorder. Psychiatr Clin North Am 28(2):469–480

Dager S, Friedman S, Parow A, Demopulos A, Stoll A, Lyoo I, Dunner D, Renshaw P (2004) Brain metabolic alterations in medication-free patients with bipolar disorder. Arch Gen Psychiatry 61 (5):450–458

de Bartolomeis A, Tomasetti C, Iasevoli F (2015) Update on the mechanism of action of aripiprazole: translational insights into antipsychotic strategies beyond dopamine receptor antagonism. CNS Drugs 29(9):773–799

de Sousa RT et al (2014) Oxidative stress in early stage bipolar disorder and the association with response to lithium. J Psychiatr Res 50:36–41 de Sousa RT, Streck EL, Zanetti MV et al (2015a) Lithium increases leukocyte mitochondrial complex I activity in bipolar disorder during depressive episodes. Psychopharmacology (Berl) 232(1):245

de Sousa RT, Zanetti MV, Talib LL, Serpa MH, Chaim TM, Carvalho AF, Brunoni AR, Busatto GF, Gattaz WF, Machado-Vieira R (2015b) Lithium increases platelet serine-9 phosphorylated GSK-3β levels in drug-free bipolar disorder during depressive episodes. J Psychiatr Res 62:78–83

de Sousa RT, Loch AA, Carvalho AF, Brunoni AR, Haddad MR, Henter ID, Zarate CA, Machado-Vieira R (2017) Genetic studies on the tripartite glutamate synapse in the pathophysiology and therapeutics of mood disorders. Neuropsychopharmacology 42(4):787–800

Desplat-Jego S, Johanet C, Escande A, Goetz J, Fabien N, Olsson N et al (2007) Update on anti-*Saccharomyces cerevisiae* antibodies, anti-nuclear associated anti-neutrophil antibodies and antibodies to exocrine pancreas detected by indirect immunofluorescence as biomarkers in chronic inflammatory bowel diseases: results of a multicenter study. World J Gastroenterol 13 (16):2312–2318

Dickerson F, Stallings C, Origoni A, Vaughan C, Khushalani S, Yolken R (2012a) Markers of gluten sensitivity in acute mania: a longitudinal study. Psychiatry Res 196(1):68–71

Dickerson F, Stallings C, Origoni A, Vaughan C, Khushalani S, Yolken R (2012b) Antibodies to the glutamate receptor in mania. Bipolar Disord 14(5):547–553

Dickerson F, Stallings C, Origoni A, Vaughan C, Khushalani S, Yolken R (2013) A combined marker of inflammation in individuals with mania. PLoS One 8(9):e73520

Dickerson F, Severance E, Yolken R (2017) The microbiome, immunity, and schizophrenia and bipolar disorder. Brain Behav Immun 62:46–52

Dodd S, Fernandes BS, Dean OM (2015) Future directions for pharmacotherapies for treatment-resistant bipolar disorder. Curr Neuropharmacol 13(5):656–662

Du J et al (2010) A kinesin signaling complex mediates the ability of GSK-3beta to affect mood-associated behaviors. Proc Natl Acad Sci U S A 107(25):11573–11578

Dubovsky SL et al (1992) Abnormal intracellular calcium ion concentration in platelets and lymphocytes of bipolar patients. Am J Psychiatry 149(1):118–120

Duman RS, Li N, Liu RJ, Duric V, Aghajanian G (2012) Signaling pathways underlying the rapid antidepressant actions of ketamine. Neuropharmacology 62:35–41

Duncan WC Jr, Zarate CA Jr (2013) Ketamine, sleep, and depression: current status and new questions. Curr Psychiatry Rep 15(9):394

Duncan LE, Holmans PA, Lee PH, O'Dushlaine CT, Kirby AW, Smoller JW et al (2014) Pathway analyses implicate glial cells in schizophrenia. PLoS One 9:e89441

Duric V, Banasr M, Stockmeier CA, Simen AA, Newton SS, Overholser JC et al (2013) Altered expression of synapse and glutamate related genes in post-mortem hippocampus of depressed subjects. Int J Neuropsychopharmacol 16:69–82

Earley W et al (2019) Cariprazine treatment of bipolar depression: a randomized double-blind placebo-controlled phase 3 study. Am J Psychiatry 176(6):439–448

Ellenbogen MA, Hodgins S, Walker C-D (2004) High levels of cortisol among adolescent offspring of parents with bipolar disorder: a pilot study. Psychoneuroendocrinology 29(1):99–106

Felger JC, Lotrich FE (2013) Inflammatory cytokines in depression: neurobiological mechanisms and therapeutic implications. Neuroscience 246:199–229

Fernandes BS et al (2011) Brain-derived neurotrophic factor as a state-marker of mood episodes in bipolar disorders: a systematic review and meta-regression analysis. J Psychiatr Res 45 (8):995–1004

Fernandes BS et al (2015) Peripheral brain-derived neurotrophic factor (BDNF) as a biomarker in bipolar disorder: a meta-analysis of 52 studies. BMC Med 13:289

Fillman SG, Sinclair D, Fung SJ, Webster MJ, Shannon Weickert C (2014) Markers of inflammation and stress distinguish subsets of individuals with schizophrenia and bipolar disorder. Transl Psychiatry 4:e365

Fornaro M et al (2010) A systematic, updated review on the antidepressant agomelatine focusing on its melatonergic modulation. Curr Neuropharmacol 8(3):287–304
Frye MA et al (2003) Low CSF somatostatin associated with response to nimodipine in patents with affective illness. Biol Psychiatry 53(2):180–183
Gallagher P et al (2008) Antiglucocorticoid treatments for mood disorders. Cochrane Database Syst Rev (1):CD005168
Giannini AJ et al (1984) Antimanic effects of verapamil. Am J Psychiatry 141(12):1602–1603
Giannini AJ et al (2000) Magnesium oxide augmentation of verapamil maintenance therapy in mania. Psychiatry Res 93(1):83–87
Goldstein BI, Kemp DE, Soczynska JK et al (2009) Inflammation and the phenomenology, pathophysiology, comorbidity, and treatment of bipolar disorder: a systematic review of the literature. J Clin Psychiatry 70:1078–1090
Gonul AS, Coburn K, Kula M (2009) Cerebral blood flow, metabolic, receptor, and transporter changes in bipolar disorder: the role of PET and SPECT studies. Int Rev Psychiatry 21 (4):323–335
Goodwin FK, Jamison K (2007) Manic-depressive illness, 2nd edn. Oxford University Press, New York
Grande I, Fries GR, Kunz M, Kapczinski F (2010) The role of BDNF as a mediator of neuroplasticity in bipolar disorder. Psychiatry Investig 7(4):243–250
Gubert C et al (2016) Peripheral adenosine levels in euthymic patients with bipolar disorder. Psychiatry Res 246:421–426
Haarman BCM et al (2014a) Relationship between clinical features and inflammation-related monocyte gene expression in bipolar disorder – towards a better understanding of psychoimmunological interactions. Bipolar Disord 16(2):137–150
Haarman BC, Riemersma-Van der Lek RF, de Groot JC et al (2014b) Neuroinflammation in bipolar disorder – A [(11)C]-(R)-PK11195 positron emission tomography study. Brain Behav Immun 40:219–225
Haile CN et al (2014) Plasma brain derived neurotrophic factor (BDNF) and response to ketamine in treatment-resistant depression. Int J Neuropsychopharmacol 17(2):331–336
Harrison PJ (2016) Molecular neurobiological clues to the pathogenesis of bipolar disorder. Curr Opin Neurobiol 36:1–6
Heyes S et al (2015) Genetic disruption of voltage-gated calcium channels in psychiatric and neurological disorders. Prog Neurobiol 134:36–54
Hirata FC et al (2007) Depression in medical school: the influence of morningness-eveningness. Chronobiol Int 24(5):939–946
Hussaini SM, Choi CI, Cho CH, Kim HJ, Jun H, Jang MH (2014) Wnt signaling in neuropsychiatric disorders: ties with adult hippocampal neurogenesis and behavior. Neurosci Biobehav Rev 47:369–383
Inden M et al (2009) Neuroprotective effect of the antiparkinsonian drug pramipexole against nigrostriatal dopaminergic degeneration in rotenone-treated mice. Neurochem Int 55 (8):760–767
Jahn H et al (2004) Metyrapone as additive treatment in major depression. Arch Gen Psychiatry 61 (12):1235
Janicak PG et al (1998) Verapamil for the treatment of acute mania: a double-blind, placebo-controlled trial. Am J Psychiatry 155(7):972–973
Jun C, Choi Y, Lim SM, Bae S, Hong YS, Kim JE, Lyoo IK (2014) Disturbance of the glutamatergic system in mood disorders. Exp Neurobiol 23:28–35
Kalinichev M, Dawson LA (2011) Evidence for antimanic efficacy of glycogen synthase kinase-3 (GSK3) inhibitors in a strain-specific model of acute mania. Int J Neuropsychopharmacol 14 (8):1051–1067
Kanazawa LK, Vecchia DD, Wendler EM, Hocayen PA, Berao PS Jr, de Melo ML, Dos Reis Livero FA, Corso CR, Stipp MC, Acco A, Andreatini R (2017) Effects of acute and chronic

quercetin administration on methylphenidate-induced hyperlocomotion and oxidative stress. Life Sci 171:1–8
Kasper S, Hamon M (2009) Beyond the monoaminergic hypothesis: agomelatine, a new antidepressant with an innovative mechanism of action. World J Biol Psychiatry 10(2):117–126
Kato T (2017) Neurobiological basis of bipolar disorder: mitochondrial dysfunction hypothesis and beyond. Schizophr Res 187:62–66
Kato T (2019) Current understanding of bipolar disorder: toward integration of biological basis and treatment strategies. Psychiatry Clin Neurosci 73(9):526–540
Kato T, Stine O, McMahon F, Crowe R (1997) Increased levels of a mitochondrial DNA deletion in the brain of patients with bipolar disorder. Biol Psychiatry 42(10):871–875
Keck PE Jr et al (2000) A review of randomized, controlled clinical trials in acute mania. J Affect Disord 59(1):S31–S37
Kemp DE, Schinagle M, Gao K et al (2014) PPAR-gamma agonism as a modulator of mood: proof-of-concept for pioglitazone in bipolar depression. CNS Drugs 28:571–581
Kempermann G (2002) Regulation of adult hippocampal neurogenesis – implications for novel theories of major depression. Bipolar Disord 4(1):17–33
Kessler RC, Berglund P, Demler O, Jin R, Merikangas KR, Walters EE (2005) Lifetime prevalence and age-of-onset distributions of DSM-IV disorders in the National Comorbidity Survey Replication. Arch Gen Psychiatry 62:593–602
Ketter TA (2009) Handbook of diagnosis and treatment of bipolar disorders. American Psychiatric Pub, Washington
Kim HJ, Thayer SA (2009) Lithium increases synapse formation between hippocampal neurons by depleting phosphoinositides. Mol Pharmacol 75(5):1021–1030
Klein PS, Melton DA (1996) A molecular mechanism for the effect of lithium on development. Proc Natl Acad Sci U S A 93(16):8455–8459
Koukopoulos A et al (2010) Antimanic and mood-stabilizing effect of memantine as an augmenting agent in treatment-resistant bipolar disorder. Bipolar Disord 12(3):348–349
Kubo H et al (2017) 1H-magnetic resonance spectroscopy study of glutamate-related abnormality in bipolar disorder. J Affect Disord 208:139–144
Lanfumey L, Mongeau R, Hamon M (2013) Biological rhythms and melatonin in mood disorders and their treatments. Pharmacol Ther 138(2):176–184
Lara DR et al (2003) Allopurinol for the treatment of aggressive behaviour in patients with dementia. Int Clin Psychopharmacol 18(1):53–55
Lara DR et al (2006) Involvement of adenosine in the neurobiology of schizophrenia and its therapeutic implications. Prog Neuropsychopharmacol Biol Psychiatry 30(4):617–629
Leboyer M et al (2012) Can bipolar disorder be viewed as a multi-system inflammatory disease? J Affect Disord 141(1):1–10
Lee YT, Hu LY, Shen CC, Huang MW, Tsai SJ, Yang AC et al (2015) Risk of psychiatric disorders following irritable bowel syndrome: a nationwide population-based cohort study. PLoS One 10 (7):e0133283
Linkowski P et al (1994) The 24-hour profiles of cortisol, prolactin, and growth hormone secretion in mania. Arch Gen Psychiatry 51(8):616–624
Lorenzi TM et al (2010) Association of serum uric acid levels with emotional and affective temperaments. J Affect Disord 121(1–2):161–164
Louveau A, Smirnov I, Keyes TJ et al (2015) Structural and functional features of central nervous system lymphatic vessels. Nature 523:337–341
Lu R-B et al (2012) Neuroprotective and neurogenesis agent for treating bipolar II disorder: add-on memantine to mood stabilizer works. Med Hypotheses 79(2):280–283
Machado-Vieira R et al (2001) Therapeutic efficacy of allopurinol in mania associated with hyperuricemia. J Clin Psychopharmacol 21(6):621–622
Machado-Vieira R et al (2002) Purinergic dysfunction in mania: an integrative model. Med Hypotheses 58(4):297–304

Machado-Vieira R et al (2007) Oxidative stress parameters in unmedicated and treated bipolar subjects during initial manic episode: a possible role for lithium antioxidant effects. Neurosci Lett 421(1):33–36

Machado-Vieira R et al (2008) A double-blind, randomized, placebo-controlled 4-week study on the efficacy and safety of the purinergic agents allopurinol and dipyridamole adjunctive to lithium in acute bipolar mania. J Clin Psychiatry 69(8):1237–1245

Machado-Vieira R, Manji HK, Zarate CA Jr (2009) The role of lithium in the treatment of bipolar disorder: convergent evidence for neurotrophic effects as a unifying hypothesis. Bipolar Disord:92–109. https://doi.org/10.1111/j.1399-5618.2009.00714.x

Machado-Vieira R et al (2011) The Bcl-2 gene polymorphism rs956572AA increases inositol 1,4,5-trisphosphate receptor-mediated endoplasmic reticulum calcium release in subjects with bipolar disorder. Biol Psychiatry 69(4):344–352

Machado-Vieira R et al (2012) Novel glutamatergic agents for major depressive disorder and bipolar disorder. Pharmacol Biochem Behav 100(4):678–687

Machado-Vieira R, Salem H et al (2017a) Convergent lines of evidence support the role of uric acid levels as a potential biomarker in bipolar disorder. Expert Rev Mol Diagn 17(2):107–108

Machado-Vieira R, Zanetti MV et al (2017b) Increased brain lactate during depressive episodes and reversal effects by lithium monotherapy in drug-naive bipolar disorder. J Clin Psychopharmacol 37(1):40–45

Maes M, Leonard BE, Myint AM et al (2011) The new '5-HT' hypothesis of depression: cell-mediated immune activation induces indoleamine 2,3-dioxygenase, which leads to lower plasma tryptophan and an increased synthesis of detrimental tryptophan catabolites (TRYCATs), both of which contribute to the onset of depression. Prog Neuro-Psychopharmacology Biol Psychiatry 35:702–721

Mallinger AG et al (2008) Verapamil augmentation of lithium treatment improves outcome in mania unresponsive to lithium alone: preliminary findings and a discussion of therapeutic mechanisms. Bipolar Disord 10(8):856–866

Marro PJ, Mishra OP, Delivoria-Papadopoulos M (2006) Effect of allopurinol on brain adenosine levels during hypoxia in newborn piglets. Brain Res 1073-1074:444–450

Masuch A et al (2016) Mechanism of microglia neuroprotection: involvement of P2X7, TNFα, and valproic acid. Glia 64(1):76–89

McCarthy MJ, Wei H, Marnoy Z, Darvish RM, McPhie DL, Cohen BM, Welsh DK (2013) Genetic and clinical factors predict lithium's effects on PER2 gene expression rhythms in cells from bipolar disorder patients. Transl Psychiatry 3:e318

McCloud TL et al (2016) Ketamine and other glutamate receptor modulators for depression in bipolar disorder in adults. BJ Psych Advances 22(4):215–215

McEwen BS (2005) Glucocorticoids, depression, and mood disorders: structural remodeling in the brain. Metab Clin Exp 54(5). Suppl 1:20–23

Melo MCA et al (2017) Chronotype and circadian rhythm in bipolar disorder: a systematic review. Sleep Med Rev 34:46–58

Mertens J et al (2015) Differential responses to lithium in hyperexcitable neurons from patients with bipolar disorder. Nature 527(7576):95–99

Miller A (2013) Conceptual confluence: the kynurenine pathway as a common target for ketamine and the convergence of the inflammation and glutamate hypotheses of depression. Neuropsychopharmacology 38:1607–1608

Mischoulon D (2009) Update and critique of natural remedies as antidepressant treatments. Obstet Gynecol Clin North Am 36:789–807

Modabbernia A, Taslimi S, Brietzke E et al (2013) Cytokine alterations in bipolar disorder: a meta-analysis of 30 studies. Biol Psychiatry 74:15–25

Molendijk ML et al (2014) Serum BDNF concentrations as peripheral manifestations of depression: evidence from a systematic review and meta-analyses on 179 associations (N=9484). Mol Psychiatry 19(7):791–800

Muneer A (2017) Wnt and GSK3 signaling pathways in bipolar disorder: clinical and therapeutic implications. Clin Psychopharmacol Neurosci 15(2):100–114
Munkholm K, Brauner JV, Kessing LV et al (2013) Cytokines in bipolar disorder vs. healthy control subjects: a systematic review and meta-analysis. J Psychiatr Res 47:1119–1133
Munkholm K et al (2019) A multisystem composite biomarker as a preliminary diagnostic test in bipolar disorder. Acta Psychiatr Scand 139(3):227–236
Muti M et al (2015) Serum uric acid levels and different phases of illness in bipolar I patients treated with lithium. Psychiatry Res 225(3):604–608
Naik MU, Benedikz E, Hernandez I, Libien J, Hrabe J, Valsamis M, Dow-Edwards D, Osman M, Sacktor TC (2000) Distribution of protein kinase Mzeta and the complete protein kinase C isoform family in rat brain. J Comp Neurol 426:243–258
Nam Y, Wie MB, Shin EJ, Nguyen TT, Nah SY, Ko SK, Jeong JH, Jang CG, Kim HC (2015) Ginsenoside Re protects methamphetamine-induced mitochondrial burdens and proapoptosis via genetic inhibition of protein kinase C delta in human neuroblastoma dopaminergic SH-SY5Y cell lines. J Appl Toxicol 35:927–944
Ng TH et al (2015) Sleep-wake disturbance in interepisode bipolar disorder and high-risk individuals: a systematic review and meta-analysis. Sleep Med Rev 20:46–58
Nierenberg AA et al (2013) Mitochondrial modulators for bipolar disorder: a pathophysiologically informed paradigm for new drug development. Aust N Z J Psychiatry 47(1):26–42
Nonaka S, Hough CJ, Chuang DM (1998) Chronic lithium treatment robustly protects neurons in the central nervous system against excitotoxicity by inhibiting N-methyl-D-aspartate receptor-mediated calcium influx. Proc Natl Acad Sci U S A 95(5):2642–2647
Nurnberger JI Jr et al (2000) Melatonin suppression by light in euthymic bipolar and unipolar patients. Arch Gen Psychiatry 57(6):572–579
Nurnberger JI Jr et al (2014) Identification of pathways for bipolar disorder: a meta-analysis. JAMA Psychiat 71(6):657–664
Ogawa S, Tsuchimine S, Kunugi H (2018) Cerebrospinal fluid monoamine metabolite concentrations in depressive disorder: a meta-analysis of historic evidence. J Psychiatr Res 105:137–146
Oliveira PMS et al (2019) Serum uric acid as a predictor of bipolarity in individuals with a major depressive episode. Bipolar Disord 21(3):235–243
Pace TW, Miller AH (2009) Cytokines and glucocorticoid receptor signaling. Relevance to major depression. Ann N Y Acad Sci 1179:86–105
Pahl S, Tapken D, Haering SC, Hollmann M (2014) Trafficking of kainate receptors. Membranes 4:565–595
Pandey GN, Rizavi HS, Tripathi M, Ren X (2015) Region-specific dysregulation of glycogen synthase kinase-3β and β-catenin in the postmortem brains of subjects with bipolar disorder and schizophrenia. Bipolar Disord 17:160–171
Papp M, Moryl E (1994) Antidepressant activity of non-competitive and competitive NMDA receptor antagonists in a chronic mild stress model of depression. Eur J Pharmacol 263:1–7
Pariante CM et al (2001) Antidepressants enhance glucocorticoid receptor function in vitro by modulating the membrane steroid transporters. Br J Pharmacol 134(6):1335–1343
Park LT et al (2017) A double-blind, placebo-controlled, pilot study of riluzole monotherapy for acute bipolar depression. J Clin Psychopharmacol 37(3):355–358
Pazzaglia PJ et al (1998) Nimodipine monotherapy and carbamazepine augmentation in patients with refractory recurrent affective illness. J Clin Psychopharmacol 18(5):404–413
Perugi G, Quaranta G, Belletti S et al (2014) General medical conditions in 347 bipolar disorder patients: clinical correlates of metabolic and autoimmune-allergic diseases. J Affect Disord 107C:95–103
Pfaffenseller B et al (2016) Differential expression of transcriptional regulatory units in the prefrontal cortex of patients with bipolar disorder: potential role of early growth response gene 3. Transl Psychiatry 6(5):e805–e805
Pfaffenseller B et al (2018) Immediate early gene and the brain-derived neurotrophic factor in bipolar disorder. Front Behav Neurosci 12:15

Pilar-Cúellar F, Vidal R, Díaz A, Castro E, dos Anjos S, Vargas V, Romero B, Valdizán EM (2014) Signaling pathways involved in antidepressant-induced cell proliferation and synaptic plasticity. Curr Pharm Des 20(23):3776–3794

Polter A, Beurel E, Yang S, Garner R, Song L, Miller CA, Sweatt JD, McMahon L, Bartolucci AA, Li X, Jope RS (2010) Deficiency in the inhibitory serine-phosphorylation of glycogen synthase kinase-3 increases sensitivity to mood disturbances. Neuropsychopharmacology 35 (8):1761–1774

Price WA, James Giannini A (1986) Neurotoxicity caused by lithium-verapamil synergism. J Clin Pharmacol 26(8):717–719

Rammes G, Danysz W, Parsons CG (2008) Pharmacodynamics of memantine: an update. Curr Neuropharmacol 6(1):55–78

Rao JS, Harry GJ, Rapoport SI et al (2010) Increased excitotoxicity and neuroinflammatory markers in postmortem frontal cortex from bipolar disorder patients. Mol Psychiatry 15:384–392

Rizzo LB et al (2013) Immunosenescence is associated with human cytomegalovirus and shortened telomeres in type I bipolar disorder. Bipolar Disord 15(8):832–838

Robillard R et al (2013) Sleep-wake cycle and melatonin rhythms in adolescents and young adults with mood disorders: comparison of unipolar and bipolar phenotypes. Eur Psychiatry 28 (7):412–416

Rong Y, Distelhorst CW (2008) Bcl-2 protein family members: versatile regulators of calcium signaling in cell survival and apoptosis. Annu Rev Physiol 70:73–91

Rosenblat JD, McIntyre RS (2015) Are medical comorbid conditions of bipolar disorder due to immune dysfunction? Acta Psychiatr Scand 132(3):180–191

Rosenblat JD, Cha DS, Mansur RB et al (2014) Inflamed moods: a review of the interactions between inflammation and mood disorders. Prog Neuropsychopharmacol Biol Psychiatry 53:23–34

Rosso IM, Crowley DJ, Silveri MM, Rauch SL, Jensen JE (2017) Hippocampus glutamate and N-acetyl aspartate markers of excitotoxic neuronal compromise in posttraumatic stress disorder. Neuropsychopharmacology 42(8):1698–1705

Rowland T, Perry BI, Upthegrove R, Barnes N, Chatterjee J, Gallacher D, Marwaha S (2018) Neurotrophins, cytokines, oxidative stress mediators and mood state in bipolar disorder: systematic review and meta-analyses. Br J Psychiatry 213(3):514–525

Salvadore G et al (2010) Increased uric acid levels in drug-naïve subjects with bipolar disorder during a first manic episode. Prog Neuropsychopharmacol Biol Psychiatry 34(6):819–821

Sanacora G et al (2017) A consensus statement on the use of ketamine in the treatment of mood disorders. JAMA Psychiat 74(4):399–405

Sarris J, Schoendorfer N, Kavanagh DJ (2009) Major depressive disorder and nutritional medicine: a review of monotherapies and adjuvant treatments. Nutr Rev 67:125–131

Sarris J, Mischoulon D, Schweitzer I (2011) Adjunctive nutraceuticals with standard pharmacotherapies in bipolar disorder: a systematic review of clinical trials. Bipolar Disord 13:454–465

Satyanarayanan SK et al (2018) Circadian rhythm and melatonin in the treatment of depression. Curr Pharm Des 24(22):2549–2555

Saxena A, Scaini G, Bavaresco DV et al (2017) Role of protein kinase C in bipolar disorder: a review of the current literature [published correction appears in Mol Neuropsychiatry. 2018 Feb;3(3):180. Valvassoria, Samira S [corrected to Valvassori, Samira S]]. Mol Neuropsychiatry 3(2):108–124

Sequeira A, Martin M, Rollins B, Moon E, Bunney W, Macciardi F, Lupoli S, Smith E, Kelsoe J, Magnan C, van Oven M, Baldi P, Wallace D, Vawter M (2012) Mitochondrial mutations and polymorphisms in psychiatric disorders. Front Genet 3:103

Serra G et al (2013) Antimanic and long-lasting mood stabilizing effect of memantine in bipolar I mood disorder: two case reports. J Clin Psychopharmacol 33(5):715–717

Severance EG, Alaedini A, Yang S, Halling M, Gressitt KL, Stallings CR et al (2012) Gastrointestinal inflammation and associated immune activation in schizophrenia. Schizophr Res 138 (1):48–53

Severance EG, Gressitt KL, Yang S, Stallings CR, Origoni AE, Vaughan C et al (2014) Seroreactive marker for inflammatory bowel disease and associations with antibodies to dietary proteins in bipolar disorder. Bipolar Disord 16(3):230–240
Severance EG, Prandovszky E, Castiglione J, Yolken RH (2015) Gastroenterology issues in schizophrenia: why the gut matters. Curr Psychiatry Rep 17(5):27
Shaltiel G, Chen G, Manji HK (2007) Neurotrophic signaling cascades in the pathophysiology and treatment of bipolar disorder. Curr Opin Pharmacol 7(1):22–26
Shorter E (2009) The history of lithium therapy. Bipolar Disord 11(2):4–9
Silverstone PH, Birkett L (2000) Diltiazem as augmentation therapy in patients with treatment-resistant bipolar disorder: a retrospective study. J Psychiatry Neurosci 25(3):276–280
Soares JC et al (2000) Concurrent measures of protein kinase C and phosphoinositides in lithium-treated bipolar patients and healthy individuals: a preliminary study. Psychiatry Res 95(2):109–118
Sperlagh B et al (2012) The role of purinergic signaling in depressive disorders. Neuropsychopharmacol Hung 14(4):231–238
Stertz L, Magalhaes PV, Kapczinski F (2013) Is bipolar disorder an inflammatory condition? The relevance of microglial activation. Curr Opin Psychiatry 26:19–26
Stevens J et al (2013) Bayesian model of Hamilton Depression Rating Score (HDRS) with memantine augmentation in bipolar depression. Br J Clin Pharmacol 75(3):791–798
Sutin AR et al (2014) Impulsivity is associated with uric acid: evidence from humans and mice. Biol Psychiatry 75(1):31–37
Talaei A et al (2016) Tamoxifen: a protein kinase C inhibitor to treat mania: a systematic review and meta-analysis of randomized, placebo-controlled trials. J Clin Psychopharmacol 36(3):272–275
Thomas J, Khanam R, Vohora D (2016) Augmentation of antidepressant effects of venlafaxine by agomelatine in mice are independent of kynurenine pathway. Neurochem Int 99:103–109
Tohen M et al (2012) Randomised, double-blind, placebo-controlled study of olanzapine in patients with bipolar I depression. Br J Psychiatry J Ment Sci 201(5):376–382
Tseng P-T et al (2016) Light therapy in the treatment of patients with bipolar depression: a meta-analytic study. Eur Neuropsychopharmacol 26(6):1037–1047
Tundo A, de Filippis R, De Crescenzo F (2019) Pramipexole in the treatment of unipolar and bipolar depression. A systematic review and meta-analysis. Acta Psychiatr Scand 140 (2):116–125
Valvezan AJ, Klein PS (2012) GSK-3 and Wnt signaling in neurogenesis and bipolar disorder. Front Mol Neurosci 5(1):1
Wang HY, Friedman E (1996) Enhanced protein kinase C activity and translocation in bipolar affective disorder brains. Biol Psychiatry 40:568–575
Wang HY, Markowitz P, Levinson D, Undie AS, Friedman E (1999) Increased membrane-associated protein kinase C activity and translocation in blood platelets from bipolar affective disorder patients. J Psychiatr Res 33:171–179
Warsh JJ, Andreopoulos S, Li PP (2004) Role of intracellular calcium signaling in the pathophysiology and pharmacotherapy of bipolar disorder: current status. Clin Neurosci Res 4 (3–4):201–213
Watson S et al (2004) Hypothalamic-pituitary-adrenal axis function in patients with bipolar disorder. Br J Psychiatry 184:496–502
Watson S et al (2012) A randomized trial to examine the effect of mifepristone on neuropsychological performance and mood in patients with bipolar depression. Biol Psychiatry 72 (11):943–949
Weina H et al (2018) Liraglutide attenuates the depressive- and anxiety-like behaviour in the corticosterone induced depression model via improving hippocampal neural plasticity. Brain Res 1694:55–62
Weston PG (1922) The determination of sodium, potassium, calcium, and magnesium in the blood and spinal fluid of patients suffering from manic depressive insanity. Arch Neurol Psychiatry 8 (2):179

Wirz-Justice A et al (2005) Chronotherapeutics (light and wake therapy) in affective disorders. Psychol Med 35(7):939–944
Xu P, Rosen KM, Hedstrom K, Rey O, Guha S, Hart C, Corfas G (2013) Nerve injury induces glial cell line-derived neurotrophic factor (GDNF) expression in Schwann cells through purinergic signaling and the PKC-PKD pathway. Glia 61:1029–1040
Xu H, Czerwinski P, Xia N, Forstermann U, Li H (2015) Downregulation of BDNF expression by PKC and by TNF-alpha in human endothelial cells. Pharmacology 96:1–10
Yildiz A et al (2008) Protein kinase C inhibition in the treatment of mania: a double-blind, placebo-controlled trial of tamoxifen. Arch Gen Psychiatry 65(3):255–263
Yildiz A et al (2016) Antimanic treatment with tamoxifen affects brain chemistry: a double-blind, placebo-controlled proton magnetic resonance spectroscopy study. Biol Psychiatry Cogn Neurosci Neuroimaging 1(2):125–131
Yoshimizu T et al (2015) Functional implications of a psychiatric risk variant within CACNA1C in induced human neurons. Mol Psychiatry 20(2):162–169
Young AH et al (2004) Improvements in neurocognitive function and mood following adjunctive treatment with mifepristone (RU-486) in bipolar disorder. Neuropsychopharmacology 29 (8):1538–1545
Young AH et al (2010) A double-blind, placebo-controlled study of quetiapine and lithium monotherapy in adults in the acute phase of bipolar depression (EMBOLDEN I). J Clin Psychiatry 71(2):150–162
Zarate CA Jr, Du J, Quiroz J, Gray NA, Denicoff KD, Singh J, Charney DS, Manji HK (2003) Regulation of cellular plasticity cascades in the pathophysiology and treatment of mood disorders: role of the glutamatergic system. Ann N Y Acad Sci 1003:273–291
Zarate CA Jr et al (2004) Pramipexole for bipolar II depression: a placebo-controlled proof of concept study. Biol Psychiatry 56(1):54–60
Zarate CA Jr, Singh J, Manji HK (2006) Cellular plasticity cascades: targets for the development of novel therapeutics for bipolar disorder. Biol Psychiatry 59:1006–1020
Zarate CA, Manji HK (2009) Protein kinase C inhibitors: rationale for use and potential in the treatment of bipolar disorder. CNS Drugs 23(7):569–582
Zeinoddini A, Sorayani M, Hassanzadeh E et al (2015) Pioglitazone adjunctive therapy for depressive episode of bipolar disorder: a randomized, double-blind, placebo-controlled trial. Depress Anxiety 32:167–173
Zhang J, Terreni L, De Simoni MG et al (2001) Peripheral interleukin-6 administration increases extracellular concentrations of serotonin and the evoked release of serotonin in the rat striatum. Neurochem Int 38:303–308
Zheng W et al (2019) Adjunctive memantine for major mental disorders: a systematic review and meta-analysis of randomized double-blind controlled trials. Schizophr Res 209:12–21
Zobel AW et al (2001) Cortisol response in the combined dexamethasone/CRH test as predictor of relapse in patients with remitted depression. a prospective study. J Psychiatr Res 35(2):83–94
Zuccoli GS et al (2017) The energy metabolism dysfunction in psychiatric disorders postmortem brains: focus on proteomic evidence. Front Neurosci 11:493

Sleep and Circadian Rhythm Disorder in Bipolar Affective Disorder

Attia Ahmad, Kirstie N. Anderson, and Stuart Watson

Contents

Abstract Symptoms of affective disorders encompass a range of changes to biological processes such as sleep and appetite. These processes are regulated over a 24-h cycle known as the circadian rhythm. Sleep is a particularly useful marker of this rhythm as it is readily measurable and functionally significant. Sleep disturbance is common in bipolar affective disorder and may act as a marker, and precipitant, of relapse. Circadian rhythms are modulated by environmental and social cues and

A. Ahmad (✉)
Cumbria, Northumberland, Tyne and Wear NHS Foundation Trust, Wolfson Institute, Campus for Aging and Vitality, Newcastle upon Tyne, UK
e-mail: AttiaAhmad@cntw.nhs.uk

K. N. Anderson
Newcastle University, Translational and Cultural Research Institute, Newcastle upon Tyne, UK

Faculty of Medical Science, Newcastle upon Tyne NHS Foundation Trust, Newcastle upon Tyne, UK

S. Watson
Newcastle University, Translational and Cultural Research Institute, Newcastle upon Tyne, UK

Faculty of Medical Science, Cumbria, Northumberland, Tyne and Wear NHS Foundation Trust, Wolfson Institute, Campus for Aging and Vitality, Newcastle upon Tyne, UK

Curr Topics Behav Neurosci (2021) 48: 133–148
https://doi.org/10.1007/7854_2020_150
Published Online: 6 February 2021

have been shown to be influenced by treatment in BPAD. As such understanding of circadian rhythms may lead to a better understanding of the pathophysiology of BPAD and its treatment. This chapter will explore the neurobiology of the circadian clock and the putative role of circadian rhythm dysregulation in the pathophysiology and treatment of bipolar affective disorder (BPAD).

Keywords Bipolar · Chronobiology · Circadian rhythm · Melatonin · Sleep · Social rhythm

1 Bipolar Affective Disorder

A diagnosis of BPAD requires at least one episode of hypomania or mania (World Health Organisation 1993). The clinical features of manic episodes include changes in numerous biological rhythms (Schnell et al. 2014) including an increase in energy, a change in appetite and a decreased need for sleep (World Health Organisation 1993). A manic episode is further characterised by euphoria, irritability or expansiveness with rapid and pressured speech, disordered thought such as a flight of ideas, increased self-esteem or grandiosity, distractibility and impulsive or reckless behaviour.

Although BPAD is defined by mania, patients typically spend more time depressed (Judd et al. 2017). A depressive episode is characterised by almost daily depressed mood or diminished interest lasting at least 2 weeks accompanied by other symptoms, which include poor concentration, psychomotor agitation or retardation, feelings of worthlessness or excessive or inappropriate guilt, hopelessness and recurrent thoughts of death or suicide. Rhythm disruptions are revealed by decreased energy, fatigue, change in appetite (increased or decreased) and change in sleep (increased or decreased) (World Health Organisation 1993). These changes in biological rhythms, particularly sleep, can be utilised as early markers of relapse.

2 Circadian Rhythm

Circadian rhythms synchronise physiological activity to the solar day through biological processes that oscilate over a 24-h period. This allows us to maintain harmony between physiological activities and our external environment and to predict and manage change in the environment. A master clock, the suprachiasmatic

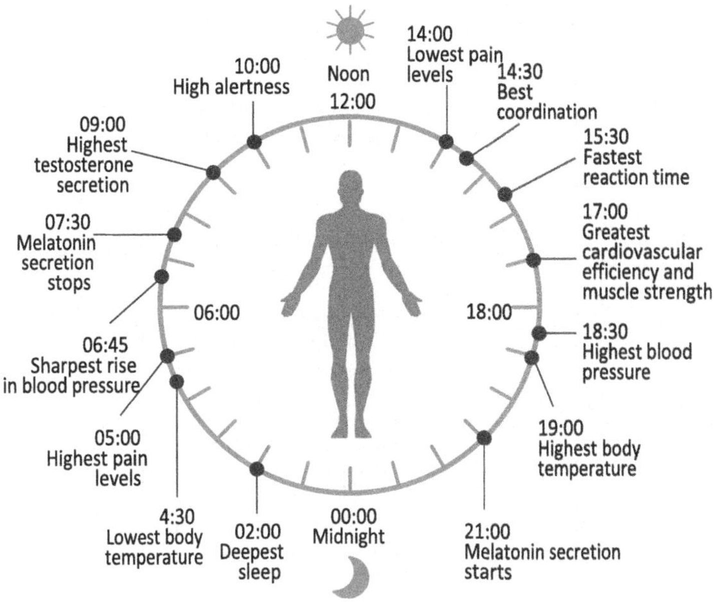

Fig. 1 Biological clock illustrates changing functions over the course of the day

nucleus (SCN), is located within the hypothalamus. Light intensity entrains it with a specific subset of melanopsin-expressing, intrinsically photosensitive retinal ganglion cells (ipRGCs) that signal directly to the hypothalamus via the retinohypothalamic tract (Dibner et al. 2010). Whilst the SCN is influenced by other external cues, such as blood pressure, locomotion, food and melatonin levels, which are known as zeitgebers or timekeepers, the variation in light intensity remains the most important driver of all of our circadian rhythms. The SCN maintains the alignment of all of the separate, independent molecular peripheral clocks within the tissues via the rhythmic downstream expression of various proteins (hormones, neurotransmitters, etc.) (Reppert and Weaver 2001) which ensure that physiological functions, for example, DNA repair, occur at the optimal time over the course of the solar day (Kang et al. 2009) (Fig. 1 – *the biological clock here*). Without a regular exposure to light and the other zeitgebers, there is "free-running" of the internal clock with a rhythm in humans that is slightly longer than 24 h (Czeisler et al. 1999). Disruptions to external social zeitgebers such as variable sleep and wake time, or meal times effects the internal biological zeitgebers. This can then cause disruption to the circadian rhythm as posited in the social zeitgeber theory (Grandin et al. 2006) (Fig. 2 *social zeitgebers* here).

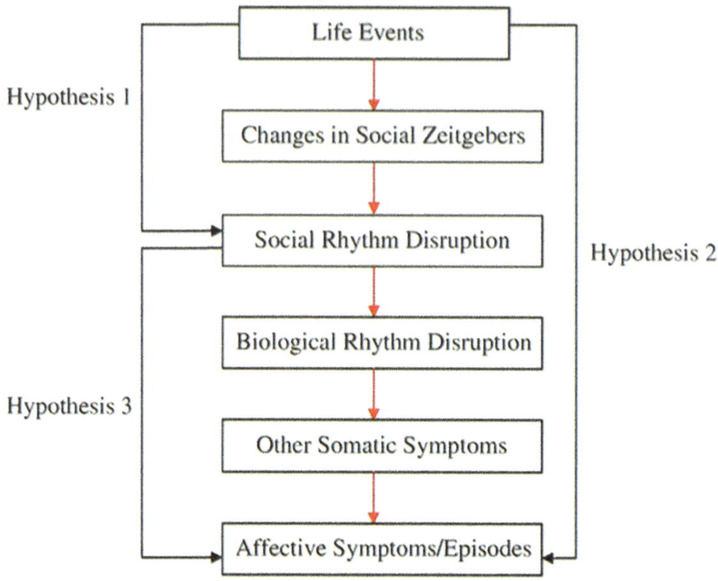

Fig. 2 Social zeitgeber theory posits life events influence external zeitgebers resulting in biological and social disruptions leading to mood episode (Reppert and Weaver 2001)

3 Circadian Rhythm and Bipolar Affective Disorder

A significant disruption in processes regulated by the circadian clock, including sleep-wake patterns and hormone production, has been demonstrated in bipolar patients. Indeed, sleep disturbance forms part of the diagnostic criteria for both bipolar manic and depressive episodes, and is commonly seen between episodes, with actigraphic data demonstrating differences in sleep latency, sleep duration, wake after sleep onset and sleep efficiency (reviewed by Bellivier et al. 2015). Overt sleep disturbance also heralds relapse, particularly of mania (Jackson et al. 2003). In the hypo-/manic phase, the most commonly reported sleep disturbance is "decreased need for sleep" cited for 69–99% of patients, whereas in depressed phases, insomnia is more commonly reported (40–100%) than hypersomnia (23–78%) (Steardo et al. 2019).

Primary sleep disorders such as obstructive sleep apnoea and restless legs syndrome are treatable and are common in BPAD patients. The putative role of poor sleep in precipitating relapse (Wehr et al. 1979; Francesca 1983) highlights the need to screen for, and treat, sleep disorders in BPAD.

Phase advance, measured through actigraphic data, salivary cortisol and gene expressions, has been shown during manic episodes (approximately 7 h) and phase delay during depressive and mixed states (approximately 7 h), both of which resolve with effective treatment (Moon et al. 2016). A more recent work in a BPAD population, many of whom were not in episode, has revealed that sleep and circadian rhythm disorders are common in BPAD (Bradley et al. 2017) suggesting these features as accessible direct points of intervention.

4 Circadian Rhythm Sleep-Wake Disorders

Circadian rhythm sleep-wake disorders (CRSWDs) are characterised by a loss of synchronicity between the sleep-wake cycle and the day-night cycle. As such, they represent defined circadian rhythm disruption to sleep. There are six distinct CRSWDs: delayed sleep-wake phase disorder; advanced sleep-wake phase disorder; irregular sleep-wake rhythm disorder; non-24-h sleep-wake rhythm disorder (free-running or non-entrained disorder); shift work disorder; and jet lag disorder (Darien 2014). CRSWDs, particularly delayed sleep-wake phase disorder, are more common in BPAD (Talih et al. 2018), with an estimated prevalence of 35% (Takaesu et al. 2016). The presence of these disorders has been associated with shorter time to relapse (Takaesu et al. 2018) and as a predictor of subsequent mania in people with a diagnosis of major depressive disorder (Takaesu et al. 2017). A suggested framework for questions to screen for circadian rhythm sleep-wake disorders is within Table 1.

4.1 Delayed Sleep Phase Disorder

Delayed sleep phase disorder (DSPD) has been estimated to account for 5–10% of presentations to a sleep medicine clinic in those aged under 25 with complaints of insomnia (Anderson 2018). DSPD is characterised by habitual sleep-wake times that

Table 1 Screening questions for a circadian rhythm disorder (Darien 2014)

Specific history to screen for a circadian rhythm disorder
History
"Take me through your typical 24-h day". lights out and lights on time
Include specific patterns and rotas for any shift work, time out of the house, light exposure during the day, differences between weekends and weekdays
"If you are allowed to sleep when you feel the need to, do you sleep well?" insomniacs will typically say "no", but those with circadian rhythm disorders more often describe longer periods of consolidated sleep
"Do you go out of the house at least once a day?" lack of natural light exposure will contribute to circadian rhythm disruption
Investigations
Sleep diaries. Ideally 2 weeks long (include work and non-work days)
Actigraphy. Physical activity data acts as a proxy for sleep-wake data. An established research tool and used within the diagnostic criteria for circadian rhythm disorders albeit without accepted validated normal ranges

are at least 2 h later than conventional or socially acceptable patterns and often delayed by 3–6 h. If subjects are allowed to sleep to their preferred schedule, sleep is typical of normal duration. Typical patients will have evening chronotypes and will show a delayed phase in all the circadian markers, including melatonin and temperature, parameters not routinely evaluated in clinical practice. The diagnosis can be made with well-completed sleep diaries, preferably for 2–3 weeks, ideally supported by actigraphy. Treatment can be rapidly effective, particularly for those who have acquired a DSPD later in life rather than those with abnormal sleep timing from early childhood. The American Academy of Sleep Medicine practice parameters have produced treatment guidelines recommending precisely timed melatonin administration before bed alongside morning light exposure but do acknowledge a lack of consensus in important details such as melatonin dose in the published studies (Auger et al. 2015). Giving melatonin at a dose of 1–2 mg, 6 h before a consistent sleep diary recorded sleep onset can be effective, especially alongside exposure to a light box at around 8 a.m. (at least 2,500 lx and at least 1 h in most studies). How long to continue melatonin is not clear. There have not been any trials using modified-release preparations of melatonin, although anecdotal reports are positive.

5 Genetics

The machinery of the circadian rhythm at a cellular level is encoded by a number of identified genes (Fig. 3 *here, cellular clock*). Some of the core genes involved include Circadian Locomotor Output Cycles Kaput (CLOCK), Aryl hydrocarbon Receptor Nuclear Translocator-Like protein 1 (ARNTL), Period (PER) and Timeless Circadian Regulator (TIMELESS).

CLOCK knockout mice have been shown to demonstrate manic-like behaviours (McClung 2007; Mukherjee et al. 2010) with an increase in depression-like behaviour in one of these studies (Mukherjee et al. 2010). A single-nucleotide polymorphism (SNP) in the CLOCK gene associates with a higher recurrence rate of bipolar episodes (Benedetti et al. 2003) and with greater insomnia and decreased need for sleep in bipolar patients (Serretti et al. 2003).

In bipolar disorder, polymorphisms of TIMELESS associate with a family history of suicide, violent suicide and multiple suicide attempts (Pawlak et al. 2015), the ratio of hypo-/manic to depressive episodes and the number and frequency of depressive episodes (Pawlak et al. 2017). PER3 SNPs (Brasil Rocha et al. 2017) and tandem repeat polymorphisms (Benedetti et al. 2008; Karthikeyan et al. 2014) have been inconsistently (see (Brasil Rocha et al. 2017) shown to associate with the onset and diagnosis of BPAD. ARNTL plays a role in transcription of monoamine-oxidase A (MAO-A) amino acid. An altered methylation of ARNTL in bipolar patients compared to controls has been found. Due to its connection in MAO-A, this is proposed by its authors to suggest a mechanistic clue to underlying circadian rhythms and mood swings in bipolar disorder (Bengesser et al. 2016).

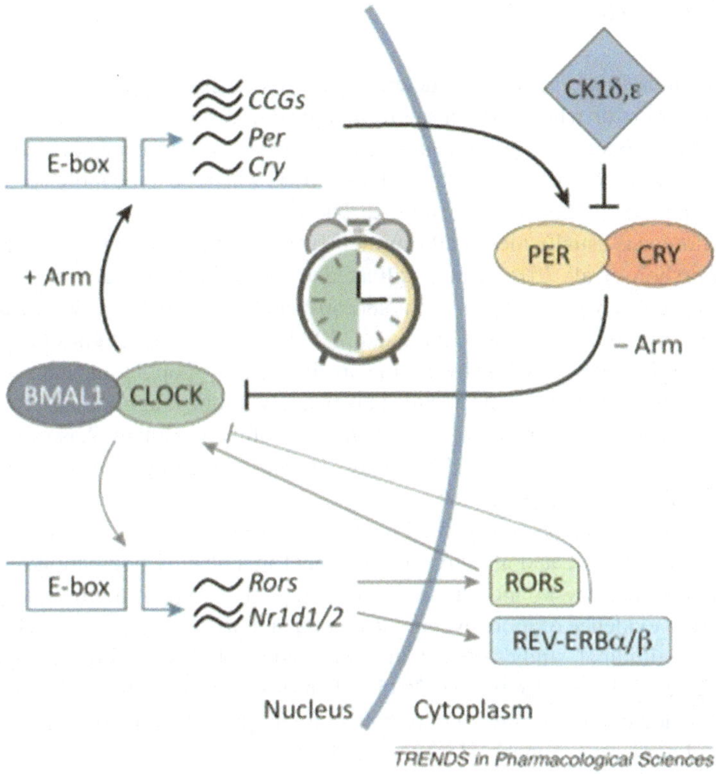

Fig. 3 The molecular clock. The molecular feedback loop drives approximately 24-h oscillations in core clock protein expression. The loop is composed of a positive arm (CLOCK and BMAL1) that binds to E-box consensus sequences driving the expression of PER and CRY, components of the negative arm of the loop. PER and CRY inhibit the ability of CLOCK and BMAL1 to bind onto DNA, thereby leading to a gradual decline of PER and CRY levels, allowing CLOCK and BMAL1 to once again restart the positive drive of the loop. Many of these circadian clock genes drive the rhythmic expression of other output genes or clock-controlled genes (CCGs) that are involved in a variety of cellular processes. Abbreviations: +/–, positive and negative arms; BMAL1, brain and muscle ARNT-like 1/Arntl; Clock, Circadian locomotor Output Cycles Kaput; CK1δ,ε, casein kinases 1δ,ε; CRY, cryptochrome; PER, period; Nr1d1/2, nuclear receptor subfamily 1, group D, members 1,2 (REV-ERBα/β); ROR, retinoid acid receptor (RAR)-related orphan receptor (Schroeder and Colwell 2013)

In general, whilst genetic manipulation of highly penetrant clock genes in animal models yields neuropsychiatric consequences, in humans, genome-wide association studies (GWAS) have not yet identified the associations of common genetic variants of the core clock genes with either BPAD or schizophrenia (McCarthy 2018). Candidate gene studies, similarly, have not shown a clear, convincing, replicated association with bipolar disorder but do reveal associations of common clock gene mutations with phenotypes related to BPAD, such as seasonal affective disorder (Zhang et al. 2016) and DSPD (Patke et al. 2017).

6 Chronotype

There are individual, genetically determined, preferences in phase position of the circadian oscillator which impact upon biological and social rhythms. This preference is our chronotype. People can be characterised as morning larks, early risers who are more active early in the day and night owls who tend to be more active in the evening and thus demonstrate evening chronotypes.

Chronotype is usually measured from self-report correlates, with objective assessment (Gershon et al. 2018) using validated interviewer tools such as Biological Rhythms Interview of Assessment in Neuropsychiatry (BRIAN) which measures a number of rhythmic variables (Giglio et al. 2009). Longitudinal studies have shown that this preference is stable over time in the healthy population and so single measures can be used to reflect circadian phase without longitudinal assessment. As such chronotype can be used in the estimation of the circadian phase simply and reliably, although it is influenced by factors such as age and gender (Randler et al. 2016; Randler and Engelke 2019).

Associations have been identified between polymorphisms of circadian rhythm genes and chronotype and affective temperament in a healthy population of university students (Jankowski and Dmitrzak-Weglarz 2017). Within the bipolar population, there is evidence of more eveningness compared to healthy controls (Baek et al. 2016). Longitudinal studies reveal the persistence of this preference in BPAD, including remission (Seleem et al. 2015) where it has been shown to be associated with worsened quality of life, an impairment in interpersonal relationships (Ng et al. 2016) and a poorer prognosis (Melo et al. 2019).

7 Melatonin and Cortisol

Melatonin is a hormone synthesised from serotonin and secreted by the pineal gland. It has been long identified as a marker of circadian phase (Lewy and Sack 1989). Soon after sundown secretion of this hormone begins, it peaks in the middle of the night (80–120 pg/mL) and then falls through the remainder of the night with low levels through the day (10–20 pg/mL). Bright light inhibits its production, with information about light being transmitted to the pineal gland via the modulated activity of the SCN. Melatonin receptors are found throughout the body, and its interaction with these receptors provides a temporal cue for peripheral tissues and so influences oscillations of peripheral activities (Tordjman et al. 2017).

Differences in the level of melatonin, as well as the pattern of its production, have been identified in those with BPAD across mood states with lower melatonin levels identified in recovered patients (Nurnberger et al. 2000). In bipolar depression, melatonin production was found to be reduced and delayed to a greater degree as compared to those with unipolar depression (Robillard et al. 2013). Morning melatonin levels in the cerebrospinal fluid of bipolar patients have been found to be significantly lower than healthy controls and patients with unipolar depression potentially again suggesting phase shift in this population (Bumb et al. 2016). Compared to the depressive phase, melatonin levels have been found to be higher and onset of production earlier in those with mania (Novakova et al. 2015). The suppression of melatonin in studies is suggestive of hypersensitivity of melatonin to light suppression in those with BPAD. However, this has not been found by all (Bradley et al. 2017; Nurnberger et al. 2000).

The hypothalamic-pituitary-adrenal axis activity also follows a clear circadian variation and appears to differ between those with and without the bipolar disorder (Watson et al. 2004).

8 Social Zeitgeber Theory

This theory proposed in 1988 by Ehlers, Frank and Kupfer attempted to explain the connection between life events, disruption of biological rhythms and depression (Ehlers et al. 1988). Irregular social rhythms are more common in bipolar and cyclothymic disorders (Shen et al. 2008) and appear to increase the vulnerability to depressive relapse (Alloy et al. 2015; Ehlers et al. 1988; Shen et al. 2008) perhaps by impacting biological circadian rhythms.

The integrated reward/circadian rhythm model is a development of the social zeitgeber theory and posits that the activation or deactivation of the behavioural approach system by goal achievement or by failure, respectively, sensitises or desensitises the reward system. Sensitisation of the reward system induces increased goal striving, appetitive motivation and initiation, leading to a neglect of social routines and rhythms and consequent disruption of circadian rhythms (Alloy et al. 2015) and enhanced vulnerability to relapse.

9 Circadian Rhythm and Bipolar Disorder

The interpersonal and social rhythm therapy (IPRST) utilises interpersonal psychotherapy and behaviour technique to help patients regulate their daily routines, minimise interpersonal problems and adhere to medications (Frank et al. 2000). It has demonstrated efficacy in individual (Frank et al. 2005; Miklowitz et al. 2007; Swartz et al. 2018) and group settings (Hoberg et al. 2013) including the young/adolescent populations (Inder et al. 2015; Goldstein et al. 2018).

For those with insomnia disorder, there is robust evidence for insomnia-specific CBT (CBTi) (reviewed by Krystal 2019). This is a multicomponent therapy with techniques including psychoeducation about sleep and circadian rhythm, sleep scheduling, sleep hygiene and cognitive control and typically requires an ongoing completion of daily sleep diaries. There are potential concerns about the elements of the programme that restrict sleep with a risk of precipitating mania. However, a pilot

RCT in bipolar disorder demonstrated improvements in insomnia without mood relapse (Harvey et al. 2015).

The use of light therapy has an evidence base for the acute treatment of seasonal affective disorder (Golden et al. 2005) and meta-analysis evidence of efficacy in bipolar depression (Tseng et al. 2016). A narrative review of the literature as pertains to all mood disorders (seasonal and non-seasonal) also supports the effective use of this therapy in bipolar depression both in the acute phase and in the management of sleep disorder in those with remitted bipolar disorder (Maruani and Geoffroy 2019). Concerns about the risk of manic switching have not been borne out by the evidence, although this is limited by the variability in methodological approach, including timing of light exposure, duration of light exposure and intensity of light used (Benedetti 2018). Dark therapy has also been used in the treatment of mania. Two single-patient case studies involving patients with rapid cycling presentations and a small pilot including inpatients with mania have provided evidence of a beneficial impact of total darkness – for an extended period of 14 h – as an adjunct in the treatment of mania (Wehr et al. 1998; Wirz-Justice et al. 1999; Barbini et al. 2005). A more recent RCT took a more nuanced approach of blocking blue light, the spectrum of light associated with daylight. This was achieved through use of blue light blocking yellow-tinted glasses creating "virtual darkness" without using the actual total darkness. Young Mania Rating Scale scores decreased more significantly over the 7 days of trials in the treatment groups, and the use of glasses was acceptable to patients (Henriksen et al. 2016). Chronotherapeutics remain safe and well-tolerated and are therefore worthy of further study (Gottlieb et al. 2019). There are ongoing specific trials into appropriate protocols of administering light therapy in bipolar depression (Geoffroy et al. 2018).

Lithium remains one of the most widely used and effective therapies for BPAD (Goodwin et al. 2016; Yatham et al. 2018). Part of its mechanism of action may be an effect on the circadian system. Chronotype and the intrinsic circadian rhythm of a patient have been shown to predict the clinical response to lithium therapy (McCarthy et al. 2019). In animal models of mania with CLOCK gene mutations displaying manic behaviour, the administration of lithium reduced the activity to wild-type levels (Roybal et al. 2007). Glycogen synthase kinase 3 beta (GSK3 beta) has been shown to play an important role in the regulation of the mammalian circadian clock; lithium is a known inhibitor of this enzyme. Exposing this enzyme in mice to lithium has been demonstrated to delay its phosphorylation activity and so delay the phase of clock gene expression (Iitaka et al. 2005). A more recent study demonstrated that lithium lengthens the circadian period in a number of brain regions, hypothesising that this aids in resynchronising the circadian clock (Yoshikawa and Honma 2016).

10 Conclusion

Bipolar disorder is characterised by significant disruptions to sleep and circadian rhythm with any disruption to circadian rhythm and poor sleep, both predicting a worse long-term outcome. The fundamental difference at a genetic level may underpin these differences.

References

Alloy LB, Nusslock R, Boland EM (2015) The development and course of bipolar spectrum disorders: an integrated reward and circadian rhythm dysregulation model. Annu Rev Clin Psychol 11:213–250

Anderson KN (2018) Circadian rhythm sleep disorders. In: Sleep disorders in neurology. Wiley, Hoboken, pp 169–181. https://onlinelibrary.wiley.com/doi/abs/10.1002/9781118777251.ch11

Auger RR, Burgess HJ, Emens JS, Deriy LV, Thomas SM, Sharkey KM (2015) Clinical practice guideline for the treatment of intrinsic circadian rhythm sleep-wake disorders: advanced sleep-wake phase disorder (ASWPD), delayed sleep-wake phase disorder (DSWPD), Non-24-hour sleep-wake rhythm disorder (N24SWD), and irregular sleep-W. J Clin sleep Med JCSM Off Publ Am Acad Sleep Med 11(10):1199–1236

Baek JH, Kim JS, Kim MJ, Ryu S, Lee K, Ha K et al (2016) Lifetime characteristics of evening-preference and irregular bed-rise time are associated with lifetime seasonal variation of mood and behavior: comparison between individuals with bipolar disorder and healthy controls. Behav Sleep Med 14(2):155–168

Barbini B, Benedetti F, Colombo C, Dotoli D, Bernasconi A, Cigala-Fulgosi M et al (2005) Dark therapy for mania: a pilot study. Bipolar Disord 7(1):98–101

Bellivier F, Geoffroy P-A, Etain B, Scott J (2015) Sleep- and circadian rhythm-associated pathways as therapeutic targets in bipolar disorder. Expert Opin Ther Targets 19(6):747–763

Benedetti F (2018) Rate of switch from bipolar depression into mania after morning light therapy: a historical review. Psychiatry Res 261:351–356

Benedetti F, Serretti A, Colombo C, Barbini B, Lorenzi C, Campori E et al (2003 Nov) Influence of CLOCK gene polymorphism on circadian mood fluctuation and illness recurrence in bipolar depression. Am J Med Genet B Neuropsychiatr Genet 123B(1):23–26

Benedetti F, Dallaspezia S, Colombo C, Pirovano A, Marino E, Smeraldi E (2008 Nov) A length polymorphism in the circadian clock gene Per3 influences age at onset of bipolar disorder. Neurosci Lett 445(2):184–187

Bengesser SA, Reininghaus EZ, Lackner N, Birner A, Fellendorf FT, Platzer M et al (2016) Is the molecular clock ticking differently in bipolar disorder? Methylation analysis of the clock gene ARNTL. World J Biol Psychiatry:1–9

Bradley AJ, Webb-Mitchell R, Hazu A, Slater N, Middleton B, Gallagher P et al (2017) Sleep and circadian rhythm disturbance in bipolar disorder. Psychol Med 47(9)

Brasil Rocha PM, Campos SB, Neves FS, da Silva Filho HC (2017) Genetic association of the PERIOD3 (Per3) clock gene with bipolar disorder. Psychiatry Investig 14(5):674–680

Bumb JM, Enning F, Mueller JK, van der List T, Rohleder C, Findeisen P et al (2016) Differential melatonin alterations in cerebrospinal fluid and serum of patients with major depressive disorder and bipolar disorder. Compr Psychiatry 68:34–39

Czeisler CA, Duffy JF, Shanahan TL, Brown EN, Mitchell JF, Rimmer DW et al (1999 Jun) Stability, precision, and near-24-hour period of the human circadian pacemaker. Science 284 (5423):2177–2181

Darien I (2014) International classification of sleep disorders, 3rd edn. American Academy of Sleep Medicine, Darien

Dibner C, Schibler U, Albrecht U (2010) The mammalian circadian timing system: organization and coordination of central and peripheral clocks. Annu Rev Physiol 72:517–549

Ehlers CL, Frank E, Kupfer DJ (1988) Social zeitgebers and biological rhythms. A unified approach to understanding the etiology of depression. Arch Gen Psychiatry 45(10):948–952

Francesca B (1983) Circadian rhythms of hormones in primary affective disorders. Cephalalgia 3 (Suppl 1):105–110

Frank E, Swartz HA, Kupfer DJ (2000 Sep) Interpersonal and social rhythm therapy: managing the chaos of bipolar disorder. Biol Psychiatry 48(6):593–604

Frank E, Kupfer DJ, Thase ME, Mallinger AG, Swartz HA, Fagiolini AM et al (2005 Sep) Two-year outcomes for interpersonal and social rhythm therapy in individuals with bipolar I disorder. Arch Gen Psychiatry 62(9):996–1004

Geoffroy PA, El Abbassi EMB, Maruani J, Etain B, Lejoyeux M, Amad A et al (2018) Bright light therapy in the morning or at mid-day in the treatment of non-seasonal bipolar depressive episodes (LuBi): study protocol for a dose research phase I / II trial. Psychiatry Investig 15 (12):1188–1202

Gershon A, Kaufmann CN, Depp CA, Miller S, Do D, Zeitzer JM et al (2018) Subjective versus objective evening chronotypes in bipolar disorder. J Affect Disord 225:342

Giglio LMF, da Magalhaes PVS, Andreazza AC, Walz JC, Jakobson L, Rucci P et al (2009) Development and use of a biological rhythm interview. J Affect Disord 118(1–3):161–165

Golden RN, Gaynes BN, Ekstrom RD, Hamer RM, Jacobsen FM, Suppes T et al (2005) The efficacy of light therapy in the treatment of mood disorders: a review and meta-analysis of the evidence. Am J Psychiatry 162(4):656–662

Goldstein TR, Merranko J, Krantz M, Garcia M, Franzen P, Levenson J et al (2018) Early intervention for adolescents at-risk for bipolar disorder: a pilot randomized trial of interpersonal and social rhythm therapy (IPSRT). J Affect Disord 235:348–356

Goodwin GM, Haddad PM, Ferrier IN, Aronson JK, Barnes T, Cipriani A et al (2016) Evidence-based guidelines for treating bipolar disorder: revised third edition recommendations from the British Association for Psychopharmacology. J Psychopharmacol 30(6):495–553

Gottlieb JF, Benedetti F, Geoffroy PA, Henriksen TEG, Lam RW, Murray G et al (2019) The chronotherapeutic treatment of bipolar disorders: a systematic review and practice recommendations from the ISBD task force on chronotherapy and chronobiology. Bipolar Disord

Grandin LD, Alloy LB, Abramson LY (2006 Oct) The social zeitgeber theory, circadian rhythms, and mood disorders: review and evaluation. Clin Psychol Rev 26(6):679–694

Harvey AG, Soehner AM, Kaplan KA, Hein K, Lee J, Kanady J et al (2015) Treating insomnia improves mood state, sleep, and functioning in bipolar disorder: a pilot randomized controlled trial. J Consult Clin Psychol 83(3):564–577

Henriksen TE, Skrede S, Fasmer OB, Schoeyen H, Leskauskaite I, Bjorke-Bertheussen J et al (2016) Blue-blocking glasses as additive treatment for mania: a randomized placebo-controlled trial. Bipolar Disord 18(3):221–232

Hoberg AA, Ponto J, Nelson PJ, Frye MA (2013) Group interpersonal and social rhythm therapy for bipolar depression. Perspect Psychiatr Care 49(4):226–234

Iitaka C, Miyazaki K, Akaike T, Ishida N (2005) A role for glycogen synthase kinase-3beta in the mammalian circadian clock. J Biol Chem 280(33):29397–29402

Inder ML, Crowe MT, Luty SE, Carter JD, Moor S, Frampton CM et al (2015) Randomized, controlled trial of interpersonal and social rhythm therapy for young people with bipolar disorder. Bipolar Disord 17(2):128–138

Jackson A, Cavanagh J, Scott J (2003 May) A systematic review of manic and depressive prodromes. J Affect Disord 74(3):209–217

Jankowski KS, Dmitrzak-Weglarz M (2017) ARNTL, CLOCK and PER3 polymorphisms - links with chronotype and affective dimensions. Chronobiol Int 34(8):1105–1113

Judd LL, Akiskal HS, Schettler PJ, Endicott J, Maser J, Solomon DA et al (2017) The long-term natural history of the weekly symptomatic status of bipolar I disorder. https://doi.org/10.1001/archpsyc.59.6.530

Kang T-H, Reardon JT, Kemp M, Sancar A (2009) Circadian oscillation of nucleotide excision repair in mammalian brain. Proc Natl Acad Sci U S A 106(8):2864–2867

Karthikeyan R, Marimuthu G, Ramasubramanian C, Arunachal G, BaHammam AS, Spence DW et al (2014) Association of Per3 length polymorphism with bipolar I disorder and schizophrenia. Neuropsychiatr Dis Treat 10:2325–2330

Krystal AD, Prather AA, Ashbrook LH (2019) The assessment and management of insomnia: an update. World Psychiatry 18(3):337–352

Lewy AJ, Sack RL (1989) The dim light melatonin onset as a marker for circadian phase position. Chronobiol Int 6(1):93–102

Maruani J, Geoffroy PA (2019) Bright light as a personalized precision treatment of mood disorders. Front Psych 10:85

McCarthy MJ (2018) Missing a beat: assessment of circadian rhythm abnormalities in bipolar disorder in the genomic era. Psychiatr Genet

McCarthy MJ, Wei H, Nievergelt CM, Stautland A, Maihofer AX, Welsh DK et al (2019) Chronotype and cellular circadian rhythms predict the clinical response to lithium maintenance treatment in patients with bipolar disorder. Neuropsychopharmacology 44(3):620–628

McClung CA (2007) Role for the clock gene in bipolar disorder. Cold Spring Harb Symp Quant Biol 72:637–644

Melo MC, Garcia RF, de Araujo CF, Luz JH, de Bruin PF, de Bruin VM (2019) Chronotype in bipolar disorder: an 18-month prospective study. Rev Bras Psiquiatr

Miklowitz DJ, Otto MW, Frank E, Reilly-Harrington NA, Wisniewski SR, Kogan JN et al (2007) Psychosocial treatments for bipolar depression: a 1-year randomized trial from the systematic treatment enhancement program. Arch Gen Psychiatry 64(4):419–426

Moon J-H, Cho C-H, Son GH, Geum D, Chung S, Kim H et al (2016 Sep) Advanced circadian phase in mania and delayed circadian phase in mixed mania and depression returned to normal after treatment of bipolar disorder. EBioMedicine 11:285–295

Mukherjee S, Coque L, Cao J-L, Kumar J, Chakravarty S, Asaithamby A et al (2010 Sep) Knockdown of clock in the ventral tegmental area through RNA interference results in a mixed state of mania and depression-like behavior. Biol Psychiatry 68(6):503–511

Ng TH, Chung K-F, Lee C-T, Yeung W-F, Ho FYY (2016) Eveningness and its associated impairments in remitted bipolar disorder. Behav Sleep Med 14(6):650–664

Novakova M, Prasko J, Latalova K, Sladek M, Sumova A (2015) The circadian system of patients with bipolar disorder differs in episodes of mania and depression. Bipolar Disord 17(3):303–314

Nurnberger JIJ, Adkins S, Lahiri DK, Mayeda A, Hu K, Lewy A et al (2000) Melatonin suppression by light in euthymic bipolar and unipolar patients. Arch Gen Psychiatry 57(6):572–579

Patke A, Murphy PJ, Onat OE, Krieger AC, Özçelik T, Campbell SS et al (2017) Mutation of the human circadian clock gene CRY1 in familial delayed sleep phase disorder. Cell 169(2):203–215.e13. http://www.sciencedirect.com/science/article/pii/S009286741730346X

Pawlak J, Dmitrzak-Weglarz M, Maciukiewicz M, Wilkosc M, Leszczynska-Rodziewicz A, Zaremba D et al (2015 Apr) Suicidal behavior in the context of disrupted rhythmicity in bipolar disorder--data from an association study of suicide attempts with clock genes. Psychiatry Res 226(2–3):517–520

Pawlak J, Szczepankiewicz A, Kapelski P, Rajewska-Rager A, Slopien A, Skibinska M et al (2017 Nov) Suicidal behavior in the context of disrupted rhythmicity in bipolar disorder-complementary research of clock genes with suicide risks factors and course of disease. Psychiatry Res 257:446–449

Randler C, Engelke J (2019) Gender differences in chronotype diminish with age: a meta-analysis based on morningness/chronotype questionnaires. Chronobiol Int 36(7):888–905

Randler C, Freyth-Weber K, Rahafar A, Florez Jurado A, Kriegs JO (2016) Morningness-eveningness in a large sample of German adolescents and adults. Heliyon 2(11):e00200

Reppert SM, Weaver DR (2001) Molecular analysis of mammalian circadian rhythms. Annu Rev Physiol 63:647–676

Robillard R, Naismith SL, Rogers NL, Scott EM, Ip TKC, Hermens DF et al (2013) Sleep-wake cycle and melatonin rhythms in adolescents and young adults with mood disorders: comparison of unipolar and bipolar phenotypes. Eur Psychiatry 28(7):412–416

Roybal K, Theobold D, Graham A, DiNieri JA, Russo SJ, Krishnan V et al (2007) Mania-like behavior induced by disruption of CLOCK. Proc Natl Acad Sci U S A 104(15):6406–6411

Schnell A, Albrecht U, Sandrelli F (2014) Rhythm and mood: relationships between the circadian clock and mood-related behavior. Behav Neurosci 128(3):326–343

Schroeder AM, Colwell CS (2013) How to fix a broken clock. Trends Pharmacol Sci 34 (11):605–619

Seleem MA, Merranko JA, Goldstein TR, Goldstein BI, Axelson DA, Brent DA et al (2015) The longitudinal course of sleep timing and circadian preferences in adults with bipolar disorder. Bipolar Disord 17(4):392–402

Serretti A, Benedetti F, Mandelli L, Lorenzi C, Pirovano A, Colombo C et al (2003) Genetic dissection of psychopathological symptoms: insomnia in mood disorders and CLOCK gene polymorphism. Am J Med Genet B Neuropsychiatr Genet 121B(1):35–38

Shen GHC, Alloy LB, Abramson LY, Sylvia LG (2008) Social rhythm regularity and the onset of affective episodes in bipolar spectrum individuals. Bipolar Disord 10(4):520–529

Steardo LJ, de Filippis R, Carbone EA, Segura-Garcia C, Verkhratsky A, De Fazio P (2019) Sleep disturbance in bipolar disorder: neuroglia and circadian rhythms. Front Psych 10:501

Swartz HA, Rucci P, Thase ME, Wallace M, Carretta E, Celedonia KL et al (2018) Psychotherapy alone and combined with medication as treatments for bipolar ii depression: a randomized controlled trial. J Clin Psychiatry 79(2)

Takaesu Y, Inoue Y, Murakoshi A, Komada Y, Otsuka A, Futenma K et al (2016) Prevalence of circadian rhythm sleep-wake disorders and associated factors in euthymic patients with bipolar disorder. PLoS One 11(7):e0159578

Takaesu Y, Inoue Y, Ono K, Murakoshi A, Futenma K, Komada Y et al (2017) Circadian rhythm sleep-wake disorders as predictors for bipolar disorder in patients with remitted mood disorders. J Affect Disord 220:57–61

Takaesu Y, Inoue Y, Ono K, Murakoshi A, Futenma K, Komada Y et al (2018) Circadian rhythm sleep-wake disorders predict shorter time to relapse of mood episodes in euthymic patients with bipolar disorder: a prospective 48-week study. J Clin Psychiatry 79(1)

Talih F, Gebara NY, Andary FS, Mondello S, Kobeissy F, Ferri R (2018 Oct) Delayed sleep phase syndrome and bipolar disorder: pathogenesis and available common biomarkers. Sleep Med Rev 41:133–140

Tordjman S, Chokron S, Delorme R, Charrier A, Bellissant E, Jaafari N et al (2017) Melatonin: pharmacology, functions and therapeutic benefits. Curr Neuropharmacol 15(3):434–443

Tseng P-T, Chen Y-W, Tu K-Y, Chung W, Wang H-Y, Wu C-K et al (2016) Light therapy in the treatment of patients with bipolar depression: a meta-analytic study. Eur Neuropsychopharmacol 26(6):1037–1047

Watson S, Gallagher P, Ritchie JC, Ferrier IN, Young AH (2004) Hypothalamic-pituitary-adrenal axis function in patients with bipolar disorder. Br J Psychiatry 184:496–502

Wehr TA, Wirz-Justice A, Goodwin FK, Duncan W, Gillin JC (1979) Phase advance of the circadian sleep-wake cycle as an antidepressant. Science 206(4419):710–713

Wehr TA, Turner EH, Shimada JM, Lowe CH, Barker C, Leibenluft E (1998) Treatment of rapidly cycling bipolar patient by using extended bed rest and darkness to stabilize the timing and duration of sleep. Biol Psychiatry 43(11):822–828

Wirz-Justice A, Quinto C, Cajochen C, Werth E, Hock C (1999) A rapid-cycling bipolar patient treated with long nights, bedrest, and light. Biol Psychiatry 45(8):1075–1077

World Health Organization (WHO). 1993. The ICD-10 Classification of Mental and Behavioural Disorders. Genève, Switzerland: World Health Organization.

Yatham LN, Kennedy SH, Parikh SV, Schaffer A, Bond DJ, Frey BN et al (2018) Canadian network for mood and anxiety treatments (CANMAT) and International Society for Bipolar Disorders (ISBD) 2018 guidelines for the management of patients with bipolar disorder. Bipolar Disord 20(2):97–170

Yoshikawa T, Honma S (2016) Lithium lengthens circadian period of cultured brain slices in area specific manner. Behav Brain Res 314:30–37

Zhang L, Hirano A, Hsu PK, Jones CR, Sakai N, Okuro M et al (2016) A PERIOD3 variant causes a circadian phenotype and is associated with a seasonal mood trait. Proc Natl Acad Sci U S A 113 (11):E1536–E1544. http://www.pnas.org/content/113/11/E1536.abstract

Neuroendocrine Stress System in Bipolar Disorder

Mario F. Juruena, Anthony J. Cleare, and Allan H. Young

Contents

Abstract Hormones have a crucial part in the progress and manifestation of a wide variety of different behaviors. The main influence of the neuroendocrine system on behavior is its action on the neurobiology of neuropsychiatric disorders and its relationship with the pharmacodynamics of medicines. Of all the neuroendocrine axes, the hypothalamic-pituitary-adrenal (HPA) axis has been the most extensively studied. There is evidence that disturbance in the HPA axis, the primary stress hormone system, could increase treatment resistance and relapse, worsen illness outcome, and cause cognitive deficits. Glucocorticoids mediate their actions in negative feedback binding in two different cytoplasmatic receptors described as mineralocorticoid receptors (MRs) and glucocorticoid receptors (GRs). Different psychopathologies underlying bipolar disorders are supposed to involve persistent dysfunctions in the expression and role of both MR and GR in the hippocampus. We review and analyze the evidence related to the correlation between bipolar disorders and the consequences and impact of stressful life events on the HPA axis, exploring

M. F. Juruena (✉), A. J. Cleare, and A. H. Young
Centre for Affective Disorders, Department of Psychological Medicine, Institute of Psychiatry, Psychology and Neuroscience, King's College London, London, UK
e-mail: mario.juruena@kcl.ac.uk

Curr Topics Behav Neurosci (2021) 48: 149–172
https://doi.org/10.1007/7854_2020_184
Published Online: 13 February 2021

the importance of these findings in bipolar disorders and as potential new targets for treatment.

Keywords Cortisol · Glucocorticoid receptors (GRs) · Hypothalamic-pituitary-adrenal (HPA) axis · Mineralocorticoid receptors (MRs)

1 Introduction

A central problem of diagnosis in psychiatry is that the current nosology systems for mental disorders are based almost entirely in descriptions of symptom-based categories (phenomenology). A comprehensive categorical classification described different clinical disorders, but there is no evidence of a biomarker that differentiates the different subtypes in the bipolar spectrum. Different psychiatric illness can manifest the same clinical symptoms, and the same mental illness can demonstrate itself differently in distinct or in the same patient. A future system of nosology criteria, in which etiology and neurobiology could be incorporated to diagnosis criteria would provide a more accurate pathway to precision psychiatry. The connection between stressful life events and mental disorders illness is a robust illustration of an area of study that can be better understood from such an integrative perspective (Ray 2004).

It is believed that disturbance of the stress hormone system (mainly, the hypothalamic-pituitary-adrenal axis) may cause cognitive impairment and make the depressive symptoms worse in bipolar disorder. Research has shown (Watson et al. 2004), that HPA axis dysfunction may predict an inadequate clinical response to drug treatment (Young et al. 2004; Juruena et al. 2009). Current studies demonstrate that the brain and its cognitive processes work in extraordinary synchrony with other bodily systems. Consequently, it is now possible to conceptualize a brain-body-mind complex when it is known that the three systems – neurological, endocrinological, and immunological – have receptors in crucial connections that can receive influences (by messenger molecules) from each of these systems. Body-mind interaction, an explicit functioning of the brain, is critical to maintaining homeostasis and well-being (Ray 2004). Today, it is widely accepted that psychological stress can alter an individual's internal homeostatic state.

In an acutely stressful event, several adaptive homeostasis and allostasis reactions may happen, including enlargement of the adrenal gland, increasing the release of hormones, mainly cortisol. Repeated stress may precipitate a psychiatric illness in genetically vulnerable individuals if this homeostasis is interrupted. Psychosocial stressful events can change neurophysiology and performance and influence the adaptation process (McEwen 2001). As experiences change our central nervous system, cognition can modify our brain. This neuropsychological process in the brain is an action in favor of neural and mental health. Genes, early stress, adulthood experiences, and stressful events contribute in the way all these may request the body to "pay the price" – or the "allostatic load" (McEwen 2001). These discoveries in current neuroscience and their consequences are essential for diagnosis and treatment of bipolar disorders (Juruena et al. 2007).

In a diagnosis point of view, the investigation has mainly concentrated in unipolar depression; just a few projects have studied biomarkers in well-defined and different subtypes of depressive and bipolar disorders. Several research projects and clinical trials have studied different subtypes of depression in the same group, in an explicit limitation of these projects that have not considered even phenomenology differences. It has been studied the differences between major depressive disorder and bipolar depression, mainly concerning their psychopathology, not correlating and stratifying biomarkers.

Cognitive impairment in bipolar patients is common (Robinson et al. 2006). Neuropsychological deficits on neurocognition (e.g., working memory, spatial memory and executive function) are also seen in family members of bipolar patients (Arts et al. 2008) and bipolar subjects in remission without residual symptoms (Thompson et al. 2005). Cognitive damages increased the severity significantly and decreased remission, recovery, and resilience (Burdick et al. 2010).

2 Endocrine Axis

The neuroendocrine system has a central importance in the evolution and manifestation of a wide variety of behaviors. Hormones have a probable influence on the neurobiology of neuropsychiatric illness and psychopharmacology medications, mainly in affective disorders.

Between all the neuroendocrine axes, the hypothalamic-pituitary-adrenal (HPA) axis is defined as the central stress axis and has been better studied. The HPA axis has a crucial function in responding to stressful events, but also any internal stimulation, including emotional stressors. Dysfunction of the stress axis has been described in patients with bipolar disorder. Also, the fundamental role of stressful events as a precipitant of manic and depressive episodes in vulnerable subjects is clear evidence. Dysfunction of the HPA axis is associated with fluctuations in the ability to circulate glucocorticoids (GCs) to exert their negative feedback on the release of hormones on the HPA axis by binding to mineralocorticoid receptors (MRs) and glucocorticoid receptors (GRs) in HPA axis (Juruena 2014), as shown in Fig. 1.

2.1 Regulation of the Hypothalamic-Pituitary-Adrenal Axis

The activity of the HPA axis is regulated by the secretion of corticotropin-releasing hormone (CRH) and arginine-vasopressin (AVP) by the hypothalamus, which activates the secretion of the adrenocorticotropic hormone (ACTH) by the pituitary,

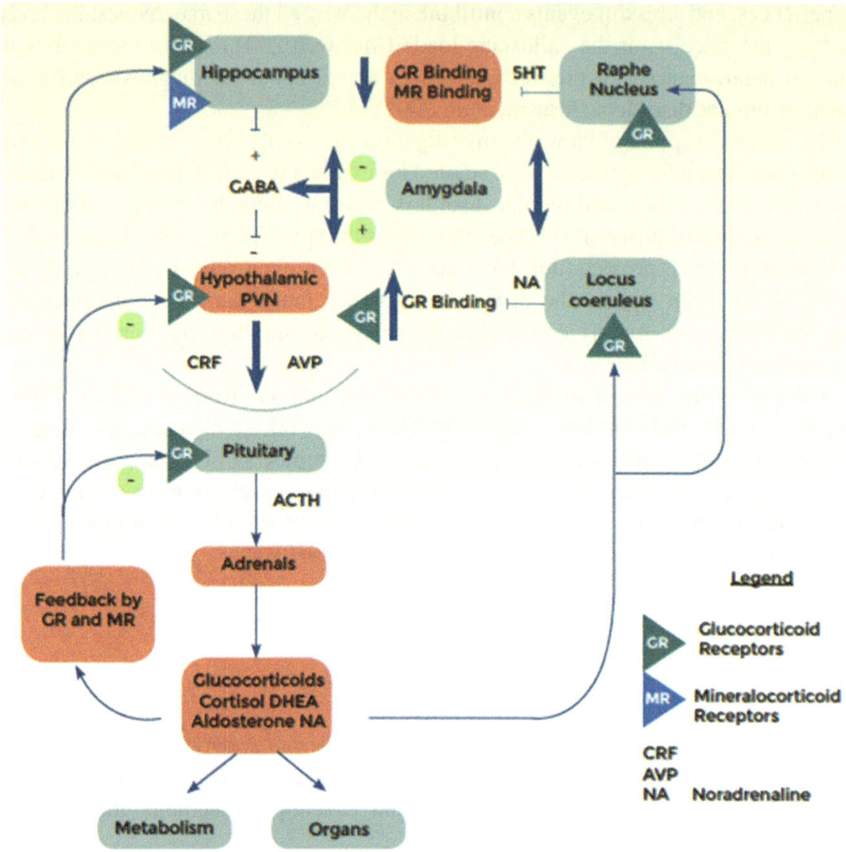

Fig. 1 Representation of the HPA axis. It summarizes the negative (−) feedback of hormones binding in GR and MR. The system includes essential areas of the limbic system, amygdala, hippocampus, locus coeruleus, raphe nucleus, and their relation via noradrenaline (NA) and serotonin (5HT), with adrenal hormones and GR/MR. It is adapted from Juruena et al. (2017). Illustrations are courtesy of Romayne Gadelrab

which ultimately stimulates the secretion of GCs by the adrenal gland (McEwen 2004). GCs then bind with their receptors in several target structures, mainly brain areas responsible for regulating the HPA axis, impacting on the negative feedback inhibition of ACTH (since the pituitary) and CRH (since the hypothalamus) secretion (Juruena 2014). Although GCs control the function of virtually all tissues in humans, the neurophysiological action of these neurohormones is mainly the regulation of metabolism. The GCs have significant immunosuppressive and/or anti-inflammatory actions (Rosenblat and McIntyre 2017). Several factors regulate the activity of the HPA axis. There is evidence of direct catecholaminergic, serotonergic, and dopaminergic innervation in CRH-producing neurons in the hypothalamus; these and other neurotransmitters appear to influence the release of CRH (McQuade and Young 2000). For example, serotonin exerts a stimulating influence on CRH through the 5-HT1A, 5-HT1B, 5-HT1C, and 5-HT2 receptors. The norepinephrine (NE) has an adjustable action; NE is stimulating at lower doses via alpha-1 receptors and has inhibition effect at higher doses via B-receptors (Yohn et al. 2017).

2.2 The Glucocorticoid Receptor (GR)

The steroid hormones (GCs, testosterone, mineralocorticoid, estrogen) are fat-soluble molecules that disseminated through the lipid-protein membrane. Different from the neurotransmitter's receptors, which are located on the lipid-protein membrane, receptors for steroid hormone are situated inside the cytoplasm. In response to coupling to these hormones, the stress hormone receptors migrate to the nucleus, where they induce some genes via binding to precise hormonal response elements (HREs) in respective controlling areas (Holsboer and Barden 1996; Juruena 2014).

The mineralocorticoid receptor (MR) has a higher affinity for corticosteroids and is considered to regulate circadian oscillations in these corticosteroids, especially in ACTH release in the progressive daytime fall in cortisol secretion. De Kloet et al. (1998) clarified that glucocorticoid receptor (GR) activation is crucial for regulating the HPA axis negative (−) feedback when GCs are higher in response to a stressful event and/or in a physiological higher activation point and also demonstrated that the MR has an essential function in the balance of MR/GR.

As said earlier, in the nucleus to cytoplasmic traffic model of receptor action, as detailed in Fig. 2, the GR – in its inactive configuration – primarily exists in the cytoplasm in association with molecular chaperone proteins, including numerous proteins (e.g., HSP56, HSP90).

After being coupled to the steroid, the receptor experiences a conformation alteration (initiation), itself since the molecular HSP protein composite, and migrate, then modulating negatively and/or positively to alter the gene transcription, coupling with glucocorticoid response elements (GREs). The GR then reprocesses to the cytoplasm. Subsequently, it works with a transcription factor regulated by the ligand by coupling to GREs (de Kloet et al. 1998; Juruena et al. 2003).

Evidence confirmed the GR has a lower affinity, but a with a stronger binding to glucocorticoids; GR is also significantly reactive to alteration in cortisol levels. Although it is considered that MR may be involved in the tonic inhibitory activity on the HPA axis, GRs seem to "shut down" cortisol production in times of stress (de Kloet and Reul 1987; de Kloet et al. 1998). Several research groups have suggested that the hyperreactivity of the stress axis in a depressive episode that could be associated with a dysfunction of GR at the amygdala, hippocampus and limbic system level. These dysfunctions result in a decrease in receptors and/or resistance to GCs. Several findings in depressive patients confirmed GR dysfunction. Most notable is that in melancholic depressive episodes, subjects do not exhibit many of the symptoms of an excess of GCs, despite the frequent incidence of higher

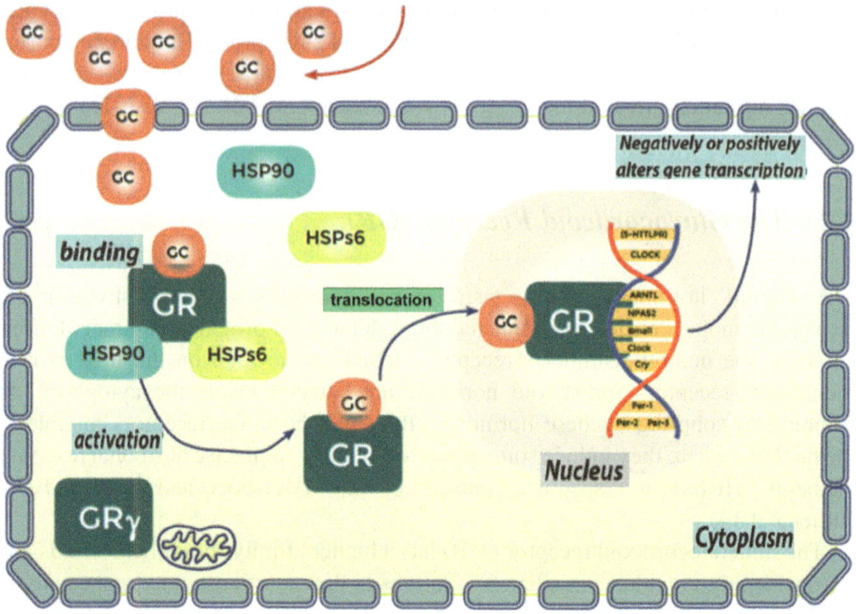

Fig. 2 Diagram of the glucocorticoid receptor (GR) stimulation; this receptor exists at the cytoplasm with chaperone proteins with numerous proteins (HSP56, HSP90). Glucocorticoid (GC) binds as GR ligands. Then the GR experiences an alteration disconnected from the HSP and migrated to the cell nucleus from the cytoplasm, modulating negatively or positively to alter the gene transcription. Adapted from Juruena 2014. Illustrations are courtesy of Romayne Gadelrab

cortisol, indicating that the peripheral receptors could be dysfunctional or downregulated in depression (Young et al. 2004; de Kloet et al. 1998).

The evidence confirmed that the GR is crucial to regulate the activity of the HPA axis, mainly in depressive subjects if glucocorticoids do are not leading to negative feedback in a scenario of hyperactive stress axis and hypercortisolism. Some data have concluded that the decreased GR function in depression could be reversed by the treatment that could be modified and improve the balance of GR/MR function (Holsboer and Barden 1996).

2.3 *Mineralocorticoid Receptors (MRs)*

MRs in the central nervous system are related to regulating the release of stress hormone and in the regulation of multifaceted behavior, such as sleep, emotion, and memory. In human, the function of MR in the neurobiology of stress and neuropsychiatric illness has not been sufficiently characterized. However, new studies found potential options for new psychopharmacotherapy via regulation of MR function (Geddes and Miklowitz 2013).

Significant research in animals concluded that MR at the hippocampus regulates the inhibition of limbic system over the HPA axis, in baseline or after induced stress (Reul et al. 2000).

Preclinical research using MR and/or GR antagonists have verified that MR stimulation is needed and enough to preserve lower basal corticosterone release in animals in a diurnal assessment. In contrast, GR stimulation is needed to limit corticosterone release during the daily peak and/or in stress. Nevertheless, during the diurnal peak or recent stress, MR stimulation induced an essential function in enabling the GR and MR activity of HPA axis function of glucocorticoids (Pace and Spencer 2005; Ratka et al. 1989; Spencer et al. 1998). Some preclinical studies antagonizing MR, with spironolactone, demonstrated anxiolytic properties in distinction to antagonists of GR (Smythe et al. 1997). Nevertheless, they demonstrated that both MR and GR antagonists stopped the impact of stress that induced anxiogenic properties in the high plus-maze paradigm (Calvo and Volosin 2001). The author determined that MR and GR are autonomously related to chronic glucocorticoid regulation of the anxiogenic reaction impact by restraining. Dysfunction of glucocorticoid receptor was described in disorders related to chronic stress. To clarify the function of different receptors, it should be studied, mainly related to memory function and stress-related events. A study described that the MR is upregulated and remains in this state after the treatment with antidepressant during 6 to 9 weeks, whereas GR proceeds to baseline levels (Mason and Pariante 2006). Spironolactone, an MR antagonist, was administrated 25 mg/kg, in the previous corticosteroid treated animals during 7 days. The authors described they decreased immobility in a forced swimming test and had a better performance in a new object recognition test. In summary, chronic glucocorticoid-induced and start several depressive related symptoms, and decrease MR expression in critical areas, as the hypothalamus and hippocampus. MR antagonist confers an antidepressant effect in chronic corticosteroid-pretreated individuals.

Up till now, just some studies described MR activity in severe mental illness and its relation to HPA axis and other brain structures. The first description was in postmortem research suicide subjects that confirmed lower MR messenger RNA in the hippocampus in relation to healthy controls (Lopez et al. 1998). To test the hypothesis that MR function is decreased in depression, a study used an antagonist (spironolactone) in a clinical trial to assess in depressed subjects with lower MR function. Paradoxically, a higher function was observed with a higher release of cortisol and ACTH after an MR antagonist is used in mildly depressed patients (Young et al. 2003). We can hypothesize that increased MR function may indicate resilient function to protect neurodegeneration in depression. Therefore, MR may prevent central nervous system from apoptosis and keep resilient mechanisms increasing serotonergic function (Crochemore et al. 2005; Cowen 2002; McAllister-Williams et al. 1999, 2001; Porter et al. 2004).

The spironolactone demonstrates a significant increase of cortisol release without circadian differences in the administration. Elizabeth Young et al.'s (1998) research concluded that the vital function of MR in HPA axis activity in human is at the peak and/or the nadir of the circadian rhythm (Young et al. 1998). Additionally, the neurophysiological alteration of HPA axis secretion in man during sleep onset in the non-REM phase was described to be blocked in the acute administration of MR antagonists (Born et al. 1997; Born and Fehm 1998). Several research projects have investigated the MR antagonists' impact on neuroendocrine hormone release and have generated different results (Dodt et al. 1993; Steiger et al. 1993, and Wiedemann et al. 1994).

The benzodiazepine receptor agonist alprazolam can block the stimulatory effect of MR antagonists upon HPA axis activity, probably on the GABA receptors on the hypothalamus and hippocampus (Grottoli et al. 2002).

Some research indicated that activation of MR increases the time of slow-wave sleep and canrenoate (an active metabolite of spironolactone) and decreases the time of slow-wave sleep (Born et al. 1991, 1997). However, other studies demonstrated that MR antagonism reduced REM sleep phase but did not change the time of the slow-wave sleep (Wiedemann et al. 1994). Kuningas et al. (2007), in elderly patients, found a higher prevalence of depression in patients carrying the MR-I180V polymorphism. Buckley and Schatzberg (2005) suggested a decrease in memory in healthy elderly could be improved, increasing the slow-wave sleep.

2.4 *Molecular Mechanisms for Resistance of Glucocorticoid Receptors*

Numerous studies have evaluated GR in depressive, bipolar, and schizophrenic patients. An autopsy of the central nervous system, in a sample of suicide patients with a history of depressive symptoms, found hippocampus decreased gene expression of GR/MR balance (Lopez et al. 1998). Therefore, the stress axis could be dysfunctional in bipolar and schizophrenic patients, and the impairment in the HPA axis could lead to different neuropsychiatric disorders.

Evidence shows that chronically high levels of cortisol can lead to receptor dysfunction and corticosteroid resistance, related to inflammation. Probably the chronic increase in cortisol released could develop GR resistance in bipolar disorder (Holsboer and Barden 1996). In patients with treatment response, high cortisol levels decreased, and lymphocyte resistant to dexamethasone returns to normal range. We can argue that treatment-resistant depression could also be related to glucocorticoid steroid resistance. However, high levels of corticosteroids and receptor resistance may not happen at the same time, and it could be that different subtypes of activation may lead to different courses in different types of bipolar disorder according to HPA axis activity (Holsboer and Barden 1996; Zobel et al. 2001).

Although the information presented here provides substantial evidence that there is resistance to glucocorticoids in depressive patients, and studies suggested that normal activity to cortisol in depression could be preserved peripherical. Mainly, it has been shown that depressed patients with severe metabolic syndrome, as in some patients with higher cortisol levels with Cushing's syndrome after prolonged treatment with a corticosteroid. Probably intra-abdominal receptors can maintain their affinity to cortisol while brain receptors are dysfunctional. Corroborating it, some data showed decreased bone mineral concentration in patients with depression, as well as association with bone loss (Checkley 1996; Young and Juruena 2018).

3 Abnormalities of the HPA Axis in Depression

In melancholic depression, there is evidence of increased activation of the HPA axis, increased release of ACTH and CRH, and increased cortisol levels, mainly post-challenge tests. On the other hand, the first studies with patients in a depressive episode with atypical features with inverted vegetative symptoms demonstrated reduced cortisol and lower HPA activity when compared with controls. However, most of the studies have not differentiated this depressive subtype from healthy controls (Juruena et al. 2018). The same dysfunction observed in one or recurrent melancholic depressive episodes may also be present in bipolar patients in mixed, manic, hypomanic, and/or depressive episodes (Juruena et al. 2011; Valiengo et al. 2012), see Fig. 3.

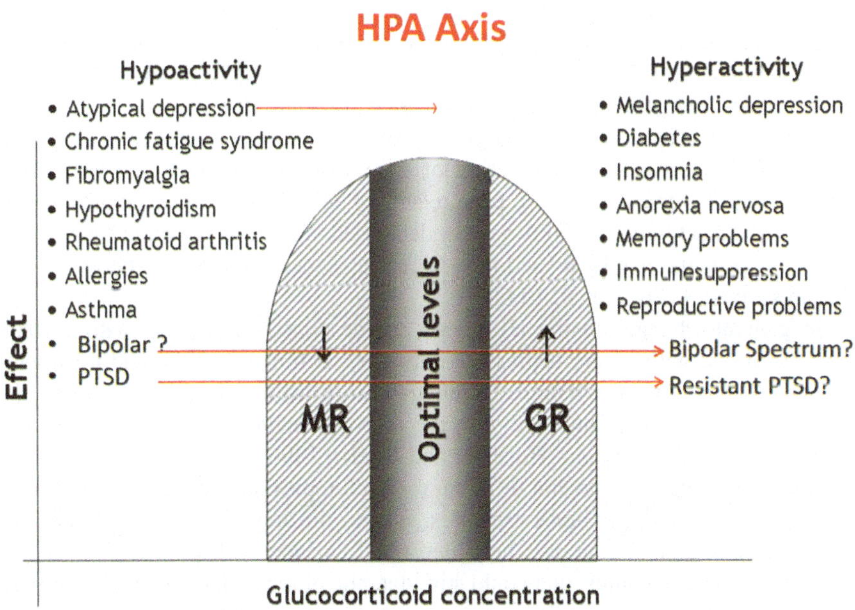

Fig. 3 The HPA axis dysfunction induces changes in the activity of GCs on the MR and GR and in the (−) feedback impacting on the release of HPA axis hormones. An impaired (−) feedback, leading to increased GCs, as in melancholic depression, diabetes (mainly type II), eating disorders (e.g., anorexia), and sleep disorders (e.g., insomnia). The hypoactivation of the HPA axis may increase the negative (−) feedback, in subjects with atypical depression (or normal activation), seasonal affective disorders, post-traumatic stress disorder (PTSD), fibromyalgia, and chronic fatigue syndrome. Bipolar and PTSD depending on the course could be related to lower or higher activity of the HPA axis. Adapted from de Kloet et al. (1998), Juruena et al. (2018), Palma et al. (2007), Gold and Chrousos (2002), and Young and Juruena (2018)

Most of the patients in these affective disorders' episode have the following abnormalities (Dallman 1984; Nemeroff et al. 1984):

- Augmented levels of cortisol: in the plasma, saliva, urine, and cerebrospinal fluid (CSF).
- Increased response of cortisol to ACTH.
- It is increased the pituitary and adrenal glands.

Several published data in the last years have confirmed that the increased HPA axis activity could be understood as an endophenotype of vulnerability that reacts to stressful events increasing the cortisol levels with an impairment in the negative feedback in depressive patients associated with different subtypes of depressive episodes (Juruena 2014):

- The subtype of the depressive episode, course, comorbidity, and severity of current episode
- The epigenetic and the genotype
- The history of childhood trauma (abuse and neglect)
- The presence or not of resilience

Adrenal gland hypertrophy has been found in depressed patients. The observation of adrenal hypertrophy in depressed episodes probably elucidates that the cortisol increase, stimulated by CRH, is analogous to the enlarged adrenal gland to compensate the blunted ACTH to compensate CRH levels, generally seen in depressive episodes (Dallman 1984; Nemeroff et al. 1984; Juruena 2014).

In depressive episodes, an enlargement of the pituitary volume was observed, which can be considered a biomarker of excessive stimulation of the HPA axis. Patients with psychosis have also been related to enlargement of the pituitary gland and increased activation of HPA axis. A decreased volume of the pituitary gland in psychotic patients may be related to recurrent episodes leading to hyperactivity of the stress axis (Axelson et al. 1992).

In general, changes in the HPA axis are described in patients with chronic depression and with severe depressive episodes, such as psychotic and treatment-resistant depression. Also, if these dysfunctions are related to the current episode, the tendency is for them to improve with the response and/or remission of the more recent episode of depression (Holsboer and Barden 1996).

Numerous data from several different research topics have described the impact of depressive episodes in brain structures, such as the amygdala, insula, hippocampus, and HPA axis. These data have described an increased release of CRH and AVP, at the hypothalamus, that could stimulate the pituitary and adrenal activity. In addition, CRH levels in the CSF are augmented in depression, mainly in untreated depressive episodes. In suicide victims, fewer CRH receptors were found in the frontal cortex, thought to be a down-regulatory response to higher CRH release. Several studies described that elevated CRH might influence depressive episode with inverted vegetative symptoms (Juruena and Cleare 2007; Juruena et al. 2018), such as:

- Diminished libido
- Diminished appetite
- Psychomotor changes
- Sleep disorders

Regarding neurobiological markers, cortisol is considered one of the primary mediators of allostatic load (McEwen 2001). Several studies in bipolar disorder – and major depressive disorder – demonstrate dysfunction of the HPA axis, with the maintenance in high levels of corticosteroids even in phases of remission of the disease, due to possible negative impaired feedback in the HPA axis (McEwen 2001). Also, abnormalities in axis regulation can predict depressive and manic relapses (Holsboer and Barden 1996; Varghese and Brown 2001; Vieta et al. 1999). It could be considered that the reduction of resilience, with the repetition of episodes and the increase of stressors in the progression of the disease, might be related to the dysfunction of the HPA axis (Juruena 2014).

Recently, several data described that the neurobiology of psychotic depression included significant and higher HPA axis activity compared with depression without psychotic symptoms (Schatzberg et al. 1983; Keller et al. 2017). These data describe the increased activity of the stress axis in psychotic depression, which is related to increased concentration of cortisol. Increased cortisol activity, particularly modifying the lowest point on the curve, is related to the severity of psychotic depression. Probably this comorbidity of psychosis and depression gravely harms the neuroendocrine system related to stress, comparable to Cushing neuroendocrine effects, and the imbalance in the activation of MR and GR (Schatzberg et al. 2014).

3.1 Impact of Stress on Bipolar Disorders

The progression of mood disorders, characterized by the recurrence of acute episodes, can be compared to models of sensitization to stress and models of electrophysiological kindling, as reviewed by Post (2007). This phenomenon of accelerating episodes was initially described by Kraepelin in 1899, suggesting that psychosocial stressors often initiate early episodes, while new recurrences may become autonomous and independent of environmental triggers (Post 2007, 2010).

Specifically, the progression of bipolar disorders has been linked to an increase in "allostatic load," which may help to explain the cumulative medical load associated with recurrent episodes of mood. It is believed that bipolar disorder patients are chronically exposed to stressful events and need to activate mechanisms to deal with them. Chronic activation of allostatic mechanisms (e.g., activation of the HPA axis and subsequent reduction in cortisol levels back to their basal levels) can lead to a reduction in resiliency mechanisms. This process can ultimately establish a vicious cycle progression, in which patients become more vulnerable to stress and triggers for new episodes as the disease progresses; see Table 1.

Table 1 Progression of bipolar disorder

A) Stressful events can act as triggers for acute mood episodes (mainly in the onset and early phases of the disorder), activating the stress axis and inducing the release of high levels of glucocorticoids in the circulation
B) High levels of cortisol can, in the long run, induce cell dysfunction, which may result in cell death (apoptosis) or reorganization of dendrites in the case of neurons
C) This reorganization can ultimately lead to significant neuroanatomical changes, such as an increase in the volume of the amygdala and a decrease in the volume of the hippocampus and the prefrontal cortex
D) These changes, consequently, lead to a decrease in the ability to deal with stressors (less resilience) and, therefore, greater vulnerability to the occurrence of new acute episodes of mood

Different studies have reported abnormalities in cortisol levels in depressive and bipolar episodes (Havermans et al. 2011; Juruena et al. 2003). Specifically, a large proportion of patients are resistant in the suppression of the HPA axis after a dexamethasone suppression test (DST). Also, they have increased corticosteroid release, irrespective of the stage of the disease (Watson et al. 2004; Sher et al. 2006). These observations indicate damage in the negative feedback loop of the HPA axis, which persists even after the remission of acute depression. Patients with abnormalities in the HPA axis have more vulnerability to recurrent episodes and relapses in unipolar depressive and bipolar episodes (Zobel et al. 2001; Vieta et al. 1997).

In addition to the already described HPA axis alterations in depressive episodes, studies suggest that such dysfunction also occurs during manic, hypomanic, and mixed episodes. Previous research using dexamethasone indicated a change in the release of glucocorticoids in manic episodes in mixed characteristics (Evans and Nemeroff 1983; Swann et al. 1992). A study that used the dexamethasone/CRH (DEX/CRH) test in patients with acute mania an augmented suppression to the DEX/CRH test was reported when compared to healthy subjects; these dysfunctions could also be observed after treatment in response and/or remission in manic patients (Schmider et al. 1995). Likewise, the CRH test in bipolar patients, during symptoms remission, could predict new episodes of mania. Studies with the release of cortisol from the HPA axis reported alterations in the stress axis flow during manic episodes. For example, one study found that plasma cortisol was significantly increased at night in manic patients compared to controls (Linkowski et al. 1994). Bipolar in different episodes – mania and/or hypomania, depression and/or euthymia – had high levels of cortisol when compared to controls, with no differences between patients (Cervantes et al. 2001). The deficiency in the negative feedback at different levels of the central neuroendocrine system in regulating circulating cortisol levels results in a mechanism that leads to an exaggerated increase in cortisol levels during stress and decreases the HPA axis' ability to resume these to baseline levels (Tatro et al. 2009). The increased cortisol, therefore, can have significant long-term consequences in patients with bipolar disorders, since glucocorticoids play essential roles in the process by which allostatic mediators interact with neurotransmitter systems and brain peptides (McEwen 2004; Juruena et al. 2017). In addition to altering neuroplasticity, the dysfunction of the stress axis can impact on different levels of the daily rhythm, e.g., cycle vigil/sleep. Most of these parameters have already been correlated with the neurophysiology of bipolar and depressive episodes (Murray and Harvey 2010).

In bipolar patients, there is evidence that points to high levels of cortisol regardless of the current mood state (Deshauer et al. 2003; Cassidy et al. 1998; Schmider et al. 1995; Linkowski et al. 1994). Patients in a first manic episode without treatment showed decreased plasma cortisol levels compared to controls, with high levels positively correlated with the presence of irritability (dysphoria) and elevated mood (euphoria) correlated with lower levels of cortisol (Valiengo et al. 2012). Also, the long-term assessment of cortisol levels did not show significant differences, being elevated only in patients with a later age of onset of the disease, after 30 years (Manenschijn et al. 2012). The rise in cortisol levels has also been reported as a biomarker for vulnerability to bipolar disorders in family members of bipolar patients (Ellenbogen et al. 2011).

3.2 Impact of Mediating Factors on the HPA Axis

The study of the functioning of the HPA axis in bipolar disorders presents sometimes inconsistent results. While some studies point to abnormalities comparable to those seen in depression, especially in patients in a current mixed episode, others do not indicate any change (Schlesser et al. 1980; Swann et al. 1992; Evans and Nemeroff 1983). There is evidence of increased activity of the stress axis in patients with a history of suicidal behavior. However, normal HPA axis function in bipolar patients without a past of suicide behavior, regardless of demographic factors, current episode of mood, severity, and course disease (Kamali et al. 2012), suggests that this could represent a marker for this subgroup of subjects.

The serotonergic system mediates the core symptoms of severe bipolar patients, e.g., aggressivity, impulsivity, and suicidal behavior. Also, significant evidence are connecting the serotonergic neurons (raphe nucleus), the HPA axis, the amygdala, and the hippocampus; see Fig. 1. Therefore, one hypothesis is that in the neurobiology of bipolar is based in the dysregulation of the HPA axis via disturbances in the serotonin neurons, in which the stress axis appears to demonstrate decreased negative feedback and consequently increased cortisol released, associated with reduced serotonergic neuron activity. Thus, the interaction between these systems, required in the regulation of the stress response, seems to be altered (Juruena et al. 2017; Yohn et al. 2017).

Early life stress (ELS), especially maltreatment, abuse, and neglect, seems to be an essential factor of vulnerability for bipolar patients. Other susceptibility factors, as unbalanced parental relations as epigenetic factors, are also important (Juruena et al. 2015; Aas et al. 2016; Etain et al. 2013). In bipolar disorders, several changes in the central nervous system are related to emotional control, response inhibition, and autobiographical memory, mainly the limbic system prefrontal regions, amygdala, and hippocampus. As described the HPA axis activity and neurotransmitters have a crucial function in these brain areas for emotional control and neuropsychological purposes (Watson et al. 2006; Wingenfeld 2010).

Bipolar and unipolar patients had been compared, and no significant differences between patients and healthy controls in the history of physical abuse were found. History of early life stress did not differ between bipolar and unipolar either (Watson et al. 2007).

Different subtypes of abuse and emotional neglect, although less studied, present significant evidence of influencing psychopathology and neuroendocrine systems. In studies carried out in animal models, it was observed that the sensitization of the stress system HPA axis occurred in reaction to different forms of stressful life events (Ladd et al. 2004). Also, data assessing the history of maltreatment, such different subtypes of abuse (e.g., physical, emotional, and sexual abuse) and neglect are significant predictors of severe affective psychopathology in adulthood (Brown et al. 2005). The history of emotional abuse demonstrated after a multiple logistic regression analysis as a significant factor for depression. Thus, these findings suggest that the history of emotional abuse is a significant risk factor involved in the pathogenesis of depression related to ELS (Martins-Monteverde et al. 2019).

The severity of early life stress may be another critical factor. Thus, physical neglect could play a role in activating the axis when present at lower levels, while its occurrence in greater severity may saturate the stress response, which can make it hypoactive. A study carried out with patients diagnosed with severe mood disorder showed a result suggestive of a similar relationship with emotional neglect, which correlated with the increased activation of the HPA axis when mild to moderate, but with no difference to healthy controls when this type of early stress was severe (Watson et al. 2007).

Several studies described patients where the relationship of ELS and stress axis in response to different types of stress impacts on the HPA axis, but results are somewhat controversial and varied, and not specific to bipolar disorders.

Studies with patients with a history of sexual abuse demonstrated increased activation of stress axis (Carpenter et al. 2007), and patients with a history of emotional abuse demonstrated hypoactivation and lower cortisol release (Carpenter et al. 2009; Feijo de Mello et al. 2007), and patients with history of emotional neglect and physical abuse also had decreased release of cortisol (Flory et al. 2009; Carpenter et al. 2007). Recently, a study suggests that flashbacks of traumatic memories as history of early life stress (e.g., abuse and/or neglect) are associated with female gender and increase the risk for affective disorders (e.g., depression, hypomanic and manic states; Haussleiter et al. 2020).

A study that compared bipolar and borderline personality disorder for influences associated with severity, early life stress, and cortisol found that borderline personality patients had a history of early life stress both overall and in subjects with emotional neglect, physical neglect, and emotional abuse than bipolar patients. The history of early life stress in patients with borderline personality and bipolar disorder was associated with decreased cortisol release. In this study the cortisol demonstrated contrary correlations in the history of sexual abuse, being a (−) correlation in bipolar patients and (+) correlation in borderline personality disorders (Mazer et al. 2019).

Overall, the risk factors of affective psychopathology are related to a complex interaction between several factors associated with the HPA axis and reflect the association of individual vulnerability with different subtypes of stressful life events for the progress of the neuropsychiatric illness. The assessment of possible biomarker related to HPA axis activity (e.g., cortisol) could be significant in precision psychiatry and diagnosis of bipolar disorders, such as the impact of a history of early stress and subsequent manifestations with late neuroendocrine changes. It remains unclear the extent to which such parameters may help with the difficulties of defining these disorders and whether they are different nosological entities or belong to the same continuum.

Although some drugs have been linked to changes in cortisol levels (Eroğlu et al. 1979; Venkatasubramanian et al. 2010; Heim and Nemeroff 1999; Watson et al. 2007), most of the clinical evaluation studies of the hormonal profile of the HPA axis are performed during the use of drugs usually prescribed for the condition studied (Valiengo et al. 2012; Juruena et al. 2006, 2010, 2013; Watson et al. 2007; Drevets et al. 2002).

Some studies, as mentioned earlier, found lowered cortisol levels in bipolar disorder contrasting with other evidence that described the increased activity of the HPA axis in bipolar disorder, regardless of the current affective state (Deshauer et al. 2003), during a mixed episode (Daban et al. 2005; Watson et al. 2004; Swann et al. 1992; Evans and Nemeroff 1983), in the presence of a history of suicidal behavior (Kamali et al. 2012), with the later age of occurrence of the first episode of the disorder (Manenschijn et al. 2012), or in the premorbid one evaluated in healthy family members. It has been described that these neuroendocrine changes may indicate a genetic vulnerability factor (endophenotype) for bipolar patients (Aydin et al. 2013; Duffy et al. 2012; Ellenbogen et al. 2011). However, in a clinical study evaluating patients in the first episode without previous treatment, decreased levels of cortisol were described, consistent with the finding of Mazer et al. (2019). In that study, cortisol measurements were correlated differently to mood presentation, according to Young Mania Rating Scale (YMRS): decreased cortisol in patients with higher scores of euphoria and increased cortisol in patients with higher irritability (Valiengo et al. 2012). Finally, subjects with normal cortisol release were described in another study (Schlesser et al. 1980) and also in bipolar patients without a history of suicide or younger age (Kamali et al. 2012; Manenschijn et al. 2012).

4 Factors Associated with an Endophenotype Increasing Vulnerability

We can understand better bipolar disorder if we can integrate an interaction of several influences, opening with the genotype, and including the environment experienced during childhood incorporating possible traumas, an individual temperament that confers on the specific skills to cope and deal with different stress. The neuroendocrine system is vital to understand the neurobiology of bipolar disorder (Evans and Nemeroff 1983).

The central stress system is the HPA axis that plays a fundamental role in acute and chronic stressful events (Kendler et al. 2003). Interestingly, the stress in children, such as abuse and neglect, could lead to permanent pathophysiological dysfunctions that are similar to the neurobiology of depressive and bipolar patients. Personality traits and temperament also have an essential influence on the sensibility of stressors and life events. Likewise, a predisposition to live a negative feeling or to be socially repressed could be a consequence of emotional or neuroendocrine stressful life event (Rijsdijk et al. 2001; Kagan et al. 1987).

Animal and human studies indicated that early life stress could develop permanent dysfunctions, and this impairment in the HPA axis may increase the vulnerability to bipolar disorders. Persistently increased activity of the HPA axis has also led to more frequent relapses and treatment resistance (Holsboer and Barden 1996).

A wide variety of stressors have been shown to stimulate the HPA axis. For this reason, glucocorticoids have been described as key molecules to develop the stress reaction and severe physiopathology situations. The persistent release of glucocorticoids is a combination of recent current stress or an epigenetic vulnerability to increase the HPA axis activity and severe damages in central nervous structures (mainly the hippocampus), which are crucial for controlling this system and cognition (Davis et al. 2018; Gallagher et al. 2014, 2015).

There is a hypothesis that this damage, in turn, leads to a feedforward circuit in which permanent stressors stimulate the overproduction of GCs indeterminately (GC cascade). Due to the ability of higher release of GCs to alter cell metabolism, which could trigger a large number of recurrences in bipolar, it is considered that the overproduction of these hormones directly contributes to several behaviors and psychological damages related to frequent chronically stressful events in bipolar disorders (McEwen 2001; Watson et al. 2007). Despite the popularity of the cascade hypothesis, several studies described the release of excess of GCs, on the GR/MR can develop stressful reaction and trigger treatment-resistant affective disorders (Juruena 2014).

Thus defined, insufficient signaling by the GCs implies the endpoint of the GC function, the key research query is if GC's message is arriving appropriately to HPA axis after stressful life events. In some subjects with hyperactivity of HPA axis and high levels of GCs, there may be an insufficiency of these hormones if the reduced sensitivity to GC, in relevant target tissues, overcomes the excess of circulating hormone (Raison and Miller 2003; Sapolsky et al. 2000).

5 Conclusion

The observations described in this chapter provide support describing the fundamental influence of the HPA axis in the neurobiology of bipolar disorders; but it is essential to be clear that bipolar disorder is considered a complex and heterogeneous clinical syndrome, as the current state of the evidence is insufficient to allow its characterization based on etiology or pathophysiology. To understand better the relationship between vulnerability and stressful environment in the genesis of bipolar disorder in adults, new research needs to study the several different influences in a comprehensive view of the interaction between genes and the environment. Such research will connect biological, epigenetic, and psychological approaches to complete the primary knowledge of neuropsychiatric disorders in general and bipolar disorder in particular.

Could the stress axis dysfunction be a fundamental dysfunction of bipolar disorder, or on the other hand, it is a tributary effect? While both directions of causality are possible, there are several indications that the influence of the HPA axis in the neurobiology of bipolar may play the underlying dysfunction in the vulnerability and development of bipolarity.

The primary system of the stress is HPA axis, during stressful life events (e.g., acute and chronic stress) may predispose and precipitate bipolar disorder. Intriguingly, several different early life stresses could develop permanent dysfunctions that look like some of the findings in the neurobiology of bipolar disorder. The relationship between bipolar, stressful events and the HPA axis are connected to and the impairment of the HPA axis feedback, observed in patients with bipolar disorders. This relationship could also develop in healthy first-degree relative with an affective disease. Therefore, we can conclude the importance of epigenetic and vulnerability to impair the HPA axis chronically.

The HPA axis is an essential bodily system to be clarified in the etiology of mental illnesses. However, this needs to be seen in the context that several other influences also require attention, such as genetics, the individual's relationship with the environment, early stress, and the resilience of some individuals, which may help explain interindividual differences in many of the aspects of HPA axis changes and stress responsiveness.

References

Aas M, Henry C, Andreassen OA, Bellivier F, Melle I, Etain B (2016) The role of childhood trauma in bipolar disorders. Int J Bipolar Dis 4(1):2

Arts B, Jabben N, Krabbendam L, van Os J (2008) Meta-analyses of cognitive functioning in euthymic bipolar patients and their first-degree relatives. Psychol Med 38:771–785

Axelson DA, Doraiswamy PM, Boyko OB, Escalona PR, McDonald WM, Ritchie JC, Patterson LJ, Ellinwood J, Nemeroff CB, Krishnan KRR (1992) In vivo assessment of pituitary volume with magnetic resonance imaging and systematic stereology: relationship to dexamethasone suppression test results in patients. Psychiatry Res 44:63–70

Aydin A, Selvi Y, Besiroglu L, Boysan M, Atli A, Ozdemir O, Kilic S, Balaharoglu R (2013) Mood and metabolic consequences of sleep deprivation as a potential endophenotype' in bipolar disorder. J Affect Disord 150(2):284–294

Born J, Fehm HL (1998) Hypothalamus-pituitary-adrenal activity during human sleep: a coordinating role for the limbic hippocampal system. Exp Clin Endocrinol Diabetes 106(3):153–163

Born J, DeKloet R, Wenz H, Kern W, Fehm HL (1991) Changes in slow wave sleep after glucocorticoids and antimineralocorticoids: a cue for central type I corticosteroid receptors in man. Am J Phys 260:E183–E188

Born J, Steinbach D, Wolff C, Fehm HL (1997) Blocking of central nervous mineralocorticoid receptors (MR) overcomes inhibition of pituitary-adrenal activity in human sleep. J Clin Endocrinol Metab 82:1106–1110

Brown J, Berenson K, Cohen P (2005) Documented and self-reported child abuse and adult pain in a community sample. Clin J Pain 21(5):374–377

Buckley TM, Schatzberg AF (2005) Aging and the role of the HPA axis and rhythm in sleep and memory-consolidation. Am J Geriatr Psychiatry 13(5):344–352

Burdick KE, Goldberg JF, Harrow M (2010) Neurocognitive dysfunction and psychosocial outcome in patients with bipolar I disorder at 15-year follow-up. Acta Psychiatr Scand 122:499–506

Calvo N, Volosin M (2001) Glucocorticoid and mineralocorticoid receptors are involved in the facilitation of anxiety-like response induced by restraint. Neuroendocrinology 73(4):261–271

Carpenter LL, Carvalho JP, Tyrka AR, Wier LM, Mello AF, Mello MF et al (2007) Decreased adrenocorticotropic hormone and cortisol responses to stress in healthy adults reporting significant childhood maltreatment. Biol Psychiatry 62(10):1080–1087

Carpenter LL, Tyrka AR, Ross NS, Khoury L, Anderson GM, Price L (2009) Effect of childhood emotional abuse and age on cortisol responsivity in adulthood. Biol Psychiatry 66:69–75

Cassidy F, Murry E, Forest K, Carroll BJ (1998) Signs and symptoms of mania in pure and mixed episodes. J Affect Disord 50:187–201

Cervantes P, Gelber S, Kin FN, Nair VN, Schwartz G (2001) Circadian secretion of cortisol in bipolar disorder. J Psychiatry Neurosci 26(5):411–416

Checkley S (1996) The neuroendocrinology of depression and chronic stress. Br Med Bull 52 (3):597–617

Cowen PJ (2002) Cortisol, serotonin and depression: all stressed out? Br J Psychiatry 180:99–100

Crochemore C, Lu J, Wu Y, Liposits Z, Sousa N, Holsboer F, Almeida OF (2005) Direct targeting of hippocampal neurons for apoptosis by glucocorticoids is reversible by mineralocorticoid receptor activation. Mol Psychiatry 10:790–798

Daban C, Vieta E, Mackin P, Young AH (2005) Hypothalamic-pituitary-adrenal axis and bipolar disorder. Psychiatr Clin North Am 28:469–480

Dallman MF (1984) Viewing the ventromedial hypothalamus from the adrenal gland. Am J Phys 246:R1–R12

Davis AC, Leppanen W, Mularczyk KP, Bedard T, Stroink ML (2018) Systems thinkers express an elevated capacity for the allocentric components of cognitive and affective empathy. Syst Res 35:216–229

de Kloet ER, Reul JM (1987) Feedback action and tonic influence of corticosteroids on brain function: a concept arising from the heterogeneity of brain receptor systems. Psychoneuroendocrinology 12:83–105

de Kloet ER, Vreugdenhil E, Oitzl MS, Joels M (1998) Brain corticosteroid receptor balance in health and disease. Endocr Rev 19:269–301

Deshauer D, Duffy A, Alda M, Grof E, Albuquerque J, Grof P (2003) The cortisol awakening response in bipolar illness: a pilot study. Can J Psychiatr 48:462–466

Dodt C, Kern W, Fehm HL, Born J (1993) Antimineralocorticoid canrenoate enhances secretory activity of the hypothalamus-pituitary-adrenocortical (HPA) axis in humans. Neuroendocrinology 58:570–574

Drevets WC, Price JL, Bardgett ME, Reich T, Todd RD, Raichle ME (2002) Glucose metabolism in the amygdala in depression: relationship to diagnostic subtype and plasma cortisol levels. Pharmacol Biochem Behav 71(3):431–447

Duffy A, Lewitzka U, Doucette S, Andreazza A, Grof P (2012) Biological indicators of illness risk in offspring of bipolar parents: targeting the hypothalamic-pituitary-adrenal axis and immune system. Early Interv Psychiatry 6(2):128–137

Ellenbogen MA, Hodgins S, Linnen AM, Ostiguy CS (2011) Elevated daytime cortisol levels: a biomarker of subsequent major affective disorder? J Affect Disord 132:265–269
Eroğlu L, Atamer-Simsek S, Yazici O, Keyer-Uysal M, Yüksel S (1979) A study of the relationship between serum lithium and plasma cortisol levels in manic depressive patients. Br J Clin Pharmacol 8(1):89–90
Etain B et al (2013) Childhood trauma is associated with severe clinical characteristics of bipolar disorders. J Clin Psychiatry 74(10):991–998
Evans DL, Nemeroff CB (1983) The dexamethasone suppression test in mixed bipolar disorder. Am J Psychiatr 140:615–617
Feijo de Mello A, Juruena MF, Pariante CM, Tyrka Audrey R, Price Lawrence H, Carpenter Linda L, Del Porto JA (2007) Depression and stress: is there an endophenotype. Braz J Psychiatr 29:13–18
Flory JD, Yehuda R, Grossman R, New AS, Mitropoulou V, Siever LJ (2009) Childhood trauma and basal cortisol in people with personality disorders. Compr Psychiatry 50(1):34–37
Gallagher P, Gray JM, Watson S, Young AH, Ferrier IN (2014) Neurocognitive functioning in bipolar depression: a component structure analysis. Psychol Med 44:961–974
Gallagher P, Nilsson J, Finkelmeyer AE, Goshawk M, Macritchie KA, Lloyd AJ, Thompson JM, Porter RJ, Young AH, Ferrier IN, McAllister-Williams RH, Watson S (2015) Neurocognitive intra-individual variability in mood disorders: effects on attentional response time distributions. Psychol Med 45:2985–2997
Geddes JR, Miklowitz DJ (2013) Treatment of bipolar disorder. Lancet 381(9878):1672–1682
Gold PW, Chrousos GP (2002) Organization of the stress system and its dysregulation in melancholic and atypical depression: high vs low CRH/NE states. Mol Psychiatry 7(3):254–275
Grottoli S, Giordano R, Maccagno B, Pellegrino M, Ghigo E, Arvat E (2002) The stimulatory effect of canrenoate, a mineralocorticoid antagonist, on the activity of the hypothalamus-pituitary-adrenal axis is abolished by alprazolam, a benzodiazepine, in humans. J Clin Endocrinol Metab 87:4616–4620
Haussleiter IS, Neumann E, Lorek S et al (2020) Role of child maltreatment and gender for bipolar symptoms in young adults. Int J Bipolar Dis 8:10
Havermans R, Nicolson NA, Berkhof J, de Vries MW (2011) Patterns of salivary cortisol secretion and responses to daily events in patients with remitted bipolar disorder. Psychoneuroendocrinology 36(2):258–265
Heim C, Nemeroff CB (1999) The impact of early adverse experiences on brain systems involved in the pathophysiology of anxiety and affective disorders. Biol Psychiat 46:1509–1522
Holsboer F, Barden N (1996) Antidepressants and hypothalamic-pituitary-adrenocortical regulation. Endocr Rev 17:187–205
Juruena MF (2014) Early-life stress and HPA axis trigger recurrent adulthood depression. Epilepsy Behav 38:148–159
Juruena MF, Cleare AJ (2007) Overlap between atypical depression, seasonal affective disorder and chronic fatigue syndrome. Braz J Psychiatr 29:20–27
Juruena MF, Cleare AJ, Bauer ME, Pariante CM (2003) Molecular mechanism of GR sensitivity and relevance for affective disorders. Acta Neuropsychiatr 15:354–367
Juruena MF, Cleare AJ, Papadopoulos AS, Poon L, Lightman S, Pariante CM (2006) Different responses to dexamethasone and prednisolone in the same depressed patients. Psychopharmacology 189:225–235
Juruena MF, Marques AH, Feijó de Mello A, Feijó de Mello M (2007) A paradigm for understanding and treating psychiatric illness. Braz J Psychiatr 29:1–2
Juruena MF, Pariante CM, Papadopoulos AS, Poon L, Lightman S, Cleare AJ (2009) Prednisolone suppression test in depression: prospective study of the role of HPA axis dysfunction in treatment resistance. Br J Psychiatry 194:342–349
Juruena MF, Cleare AJ, Papadopoulos AS, Poon L, Lightman S, Pariante CM (2010) The prednisolone suppression test in depression: dose response and changes with antidepressant treatment. Psychoneuroendocrinology 35:1486–1491

Juruena MF, Calil HM, Fleck MP, Del Porto JA (2011) Melancholia in Latin American studies: a distinct mood disorder for the ICD-11. Braz J Psychiatr 33:S48–S58

Juruena MF, Pariante CM, Papadopoulos AS, Poon L, Lightman S, Cleare AJ (2013) The role of mineralocorticoid receptor function in treatment-resistant depression. J Psychopharmacol 27 (12):1169–1179

Juruena MF, Baes CVW, Menezes IC, Graeff FG (2015) Depressive patients: role of glucocorticoid and mineralocorticoid receptors and of hypothalamic-pituitary-adrenal axis activity. Curr Pharm Des 21(11):1369–1378

Juruena MF, Agustini B, Cleare AJ, Young AH (2017) A translational approach to clinical practice via stress-responsive glucocorticoid receptor signalling. Stem Cell Invest 16(4):13

Juruena MF, Bocharova M, Agustini B, Young AH (2018) Atypical and non-atypical depression: is HPA axis function a biomarker? J Affect Disord 233:45–67

Kagan J, Reznick JS, Snidman N (1987) The physiology and psychology of behavioral inhibition in children. Child Dev 58(6):1459–1473

Kamali M, Saunders EF, Prossin AR, Brucksch CB, Harrington GJ, Langenecker SA et al (2012) Associations between suicide attempts and elevated bedtime salivary cortisol levels in bipolar disorder. J Affect Disord 136:350–358

Keller J, Gomez R, Williams G, Lembke A, Lazzeroni L, MurphyJr GM, Schatzberg AF (2017) HPA axis in major depression: cortisol, clinical symptomatology, and genetic variation predict cognition. Mol Psychiatr 22(4):527–536

Kendler KS, Hettema JM, Butera F, Gardner CO, Prescott CA (2003) Life event dimensions of loss, humiliation, entrapment, and danger in the prediction of onsets of major depression and generalized anxiety. Arch Gen Psychiatry 60(8):789–796

Kuningas M, de Rijk RH, Westendorp RG, Jolles J, Eline Slagboom P, van Heemst D (2007) Mental performance in old age dependent on cortisol and genetic variance in the mineralocorticoid and glucocorticoid receptors. Neuropsychopharmacology 32:1295–1301

Ladd CO, Huot RL, Thrivikraman KV, Nemeroff CB, Plotsky PM (2004) Long-term adaptations in glucocorticoid receptor and mineralocorticoid receptor mRNA and negative feedback on the hypothalamo-pituitary-adrenal axis following neonatal maternal separation. Biol Psychiatry 55:367–375

Linkowski P, Kerkhofs M, van Onderbergen A, Hubain P, Copinschi G, L'hermite-Baleriaux M et al (1994) The 24-hour profiles of cortisol, prolactin, and growth hormone secretion in mania. Arch General Psychiatr 51:616–624

Lopez JF, Chalmers DT, Little KY, Watson SJ, A.E. Bennett Research Award (1998) Regulation of serotonin1A, glucocorticoid, and mineralocorticoid receptor in rat and human hippocampus: implications for the neurobiology of depression. Biol Psychiatry 43:547–573

Manenschijn L, Spijker AT, Koper JW, Jetten AM, Giltay EJ, Haffmans J et al (2012) Long-term cortisol in bipolar disorder: associations with age of onset and psychiatric co-morbidity. Psychoneuroendocrinology 37(12):1960–1968

Martins-Monteverde CMS, Baes CVW, Reisdorfer E, Padovan T, Tofoli SMC, Juruena MF (2019) Relationship between depression and subtypes of early life stress in adult psychiatric patients. Front Psychiatry 10:19

Mason BL, Pariante CM (2006) The effects of antidepressants on the hypothalamic-pituitary-adrenal axis. Drug News Perspect 19(10):603–608

Mazer AK, Cleare AJ, Young AH, Juruena MF (2019) Bipolar affective disorder and borderline personality disorder: differentiation based on the history of early life stress and psychoneuroendocrine measures. Behav Brain Res 357:48–56

McAllister-Williams RH, Man MS, Young AH (1999) Effects of adrenalectomy on 8-OH-DPAT induced hypothermia in mice. Psychopharmacology 142:73–77

McAllister-Williams RH, Anderson AJ, Young AH (2001) Corticosterone selectively attenuates 8-OH-DPAT-mediated hypothermia in mice. Int J Neuropsychopharmacol 4:1–8

McEwen BS (2001) From molecules to mind. Stress, individual differences, and the social environment. Ann N Y Acad Sci 935:42–49

McEwen BS (2004) Protection and damage from acute and chronic stress: allostasis and allostatic overload and relevance to the pathophysiology of psychiatric disorders. Ann N Y Acad Sci 1032:1–7
McQuade R, Young AH (2000) Future therapeutic targets in mood disorders: the glucocorticoid receptor. Br J Psychiatry 177:390–395
Murray G, Harvey A (2010) Circadian rhythms and sleep in bipolar disorder. Bipolar Disord 12 (5):459–472
Nemeroff CB, Widerlov E, Bissette G, Walleus H, Karlsson I, Eklund K, Kilts CD, Loosen PT, Vale W (1984) Elevated concentrations of CSF corticotropin-releasing factor like immunoreactivity in depressed patients. Science 226:1342–1344
Pace TW, Spencer RL (2005) Disruption of mineralocorticoid receptor function increases corticosterone responding to a mild, but not moderate, psychological stressor. Am J Physiol Endocrinol Metab 288:E1082–E1088
Palma BD, Tiba P, Machado RB, Tufik S, Suchecki D (2007) Immune outcomes of sleep disorders: the hypothalamic-pituitary-adrenal axis as a modulatory factor. Braz J Psychiatr 29(Suppl 1): s33–s38
Porter RJ, Gallagher P, Watson S, Young AH (2004) Corticosteroid-serotonin interactions in depression: a review of the human evidence. Psychopharmacology 173:1–17
Post RM (2007) Kindling and sensitization as models for affective episode recurrence, cyclicity, and tolerance phenomena. Neurosci Biobehav Rev 31(6):858–873
Post RM (2010) Mechanisms of illness progression in the recurrent affective disorders. Neurotox Res 18:256–271
Raison CL, Miller AH (2003) When not enough is too much: the role of insufficient glucocorticoid signaling in the pathophysiology of stress-related disorders. Am J Psychiatry 160(9):1554–1565
Ratka A, Sutanto W, BLoemers M, de Kloet ER (1989) On the role of brain mineralocorticoid (type I) and glucocorticoid (type II) receptors in neuroendocrine regulation. Neuroendocrinology 50:117–123
Ray O (2004) The revolutionary health science of psychoendoneuroimmunology: a new paradigm for under- standing health and treating illness. Ann N Y Acad Sci 1032:35–51
Reul JM, Gesing A, Droste S, Stec IS, Weber A, Bachmann C, Bilang-Bleuel A, Holsboer F, Linthorst AC (2000) The brain mineralocorticoid receptor: greedy for ligand, mysterious in function. Eur J Pharmacol 405:235–249
Rijsdijk FV, Sham PC, Sterne A, Purcell S, McGuffin P, Farmer A, Goldberg D, Mann A, Cherny SS, Webster M, Ball D, Eley TC, Plomin R (2001) Life events and depression in a community sample of siblings. Psychol Med 31(3):401–410
Robinson LJ, Thompson JM, Gallagher P, Goswami U, Young AH, Ferrier IN et al (2006) A meta-analysis of cognitive deficits in euthymic patients with bipolar disorder. J Affect Disord 93:105–115
Rosenblat JD, McIntyre RS (2017) Bipolar disorder and immune dysfunction: epidemiological findings, proposed pathophysiology and clinical implications. Brain Sci 7(11)
Sapolsky RM, Romero M, Munck AU (2000) How do glucocorticoids influence stress responses? Integrating permissive, suppressive, stimulatory, and preparative actions. Endocr Rev 21 (1):55–89
Schatzberg AF, Rothschild AJ, Stahl JB, Bond TC, Rosenbaum AH, Lofgren SB, MacLaughlin RA, Sullivan MA, Cole JO (1983) The dexamethasone suppression test: identification of subtypes of depression. Am J Psychiatry 140(1):88–91
Schatzberg AF, Keller J, Tennakoon L, Lembke A, Williams G, Kraemer FB, Sarginson JE, Lazzeroni LC, Murphy GM Jr (2014) HPA axis genetic variation, cortisol, and psychosis in major depression. Mol Psychiatr 19(2):220–227
Schlesser MA, Winokur G, Sherman BM (1980) Hypothalamic-pituitary-adrenal axis activity in depressive illness. Its relationship to classification. Arch General Psychiatr 37:737–743

Schmider J, Lammers CH, Gotthardt U, Dettling M, Holsboer F, Heuser IJE (1995) Combined dexamethasone/corticotropin-releasing hormone test in acute and remitted manic patients, in acute depression, and in normal controls: I. Biol Psychiatry 38:797–802
Sher L, Cooper TB, Mann JJ, Oquendo MA (2006) Modified dexamethasone suppression-corticotropin-releasing hormone stimulation test: a pilot study of young healthy volunteers and implications for alcoholism research in adolescents and young adults. Int J Adolesc Med Health 18(1):133–137
Smythe JW, Murphy D, Timothy C, Costall B (1997) Hippocampal mineralocorticoid, but not glucocorticoid, receptors modulate anxiety-like behavior in rats. Pharmacol Biochem Behav 56 (3):507–513
Spencer RL, Kim PJ, Kalman BA, Cole MA (1998) Evidence for mineralocorticoid receptor facilitation of glucocorticoid receptor-dependent regulation of hypothalamic-pituitary-adrenal axis activity. Endocrinology 139:2718–2726
Steiger A, Rupprecht R, Spengler D, Guldner J, Hemmeter U, Rothe B, Damm K, Holsboer F (1993) Functional properties of deoxycorticosterone and spironolactone: molecular characterization and effects on sleep endocrine activity. J Psychiatr Res 27:275–284
Swann AC, Stokes PE, Casper R, Secunda SK, Bowden CL, Berman N et al (1992) Hypothalamic-pituitary-adrenocortical function in mixed and pure mania. Acta Psychiatr Scand 85:270–274
Tatro ET, Everall IP, Kaul M, Achim CL (2009) Modulation of glucocorticoid receptor nuclear translocation in neurons by immunophilins FKBP51 and FKBP52: implications for major depressive disorder. Brain Res 1286:1–12
Thompson JM, Gallagher P, Hughes JH, Watson S, Gray JM, Ferrier IN et al (2005) Neurocognitive impairment in euthymic patients with bipolar affective disorder. Br J Psychiatry 186:32–40
Valiengo LL, Soeiro-de-Souza MG, Andreazza AC, Juruena MF, Gattaz WF, Moreno DH, Marques AH, Machado-Vieira R (2012) Plasma cortisol in first episode drug-naïve mania: differential levels in euphoric versus irritable mood. J Affect Disord 138:149–152
Varghese P, Brown S (2001) The hypothalamic-pituitary adrenal Axis in major depressive disorder: a brief primer for primary care physicians Femina. Curr Gene Ther
Venkatasubramanian G, Chittiprol S, Neelakantachar N, Shetty T, Gangadhar BN (2010) Effect of antipsychotic treatment on insulin-like growth Factor-1 and cortisol in schizophrenia: a longitudinal study. Schizophr Res 119:131–137
Vieta E, Gasto C, Martinez de Osaba MJ et al (1997) Prediction of depressive relapse in remitted bipolar patients using corticotrophin-releasing hormone challenge test. Acta Psychiatr Scand 95 (3):205–211
Vieta E, Martínez-De-Osaba MJ, Colom F, Martínez-Arán A, Benabarre A, Gastó C (1999) Enhanced corticotropin response to corticotropin-releasing hormone as a predictor of mania in euthymic bipolar patients. Psychol Med
Watson S, Gallagher P, Ritchie JC, Ferrier IN, Young AH (2004) Hypothalamic-pituitary-adrenal axis function in patients with bipolar disorder. Br J Psychiatry 184:96–502
Watson S, Thompson JM, Ritchie JC, Nicol Ferrier I, Young AH (2006) Neuropsychological impairment in bipolar disorder: the relationship with glucocorticoid receptor function. Bipolar Disord 8:85–90
Watson S, Bruce M, Owen BM, Gallagher P, Hearn AJ, Young AH et al (2007) Family history, early adversity and the hypothalamic-pituitary-adrenal (HPA) axis: mediation of the vulnerability to mood disorders. Neuropsychiatr Dis Treat 3(5):647–653
Wiedemann K, Lauer C, Pollmächer T, Holsboer F (1994) Sleep endocrine effects of antigluco- and antimineralocorticoids in healthy males. Am J Phys 267:E109–E114
Wingenfeld K (2010) HPA axis and neuroimaging in BPD. Psychoneuroendocrinology 35:154–170
Yohn CN, Gergues MM, Samuels BA (2017) The role of 5-HT receptors in depression. Mol Brain 10(1):28
Young AH, Juruena MF (2018) Hypothalamic-pituitary-adrenal (HPA) axis. In: Ismail K, Barthel A, Bornstein SR, Licinio J (eds) Depression and type 2 diabetes, pp 59–72

Young EA, Lopez JF, Murphy-Weinberg V, Watson SJ, Akil H (1998) The role of mineralocorticoid receptors in hypothalamic-pituitary- adrenal axis regulation in humans. J Clin Endocrinol Metab 83:3339–3345

Young EA, Lopez JF, Murphy-Weinberg V, Watson SJ, Akil H (2003) Mineralocorticoid receptor function in major depression. Arch Gen Psychiatry 60:24–28

Young EA, Altemus M, Lopez JF, Kocsis JH, Schatzberg AF, DeBattista C et al (2004) HPA axis activation in major depression and response to fluoxetine: a pilot study. Psychoneuroendocrinology 29:1198–1204

Zobel AW, Nickel T, Sonntag A, Uhr M, Holsboer F, Ising M (2001) Cortisol response in the combined dexamethasone/CRH test as predictor of relapse in patients with remitted depression. A prospective study. J Psychiatr Res 35(2):83–94

Neuroanatomic and Functional Neuroimaging Findings

Alexandre Paim Diaz, Isabelle E. Bauer, Marsal Sanches, and Jair C. Soares

Contents

Abstract The search for brain morphology findings that could explain behavioral disorders has gone through a long path in the history of psychiatry. With the advance of brain imaging technology, studies have been able to identify brain morphology and neural circuits associated with the pathophysiology of mental illnesses, such as bipolar disorders (BD). Promising results have also shown the potential of neuroimaging findings in the identification of outcome predictors and response to treatment among patients with BD. In this chapter, we present brain imaging structural

A. P. Diaz (✉), I. E. Bauer, M. Sanches, and J. C. Soares
Louis A. Faillace, MD, Department of Psychiatry and Behavioral Sciences, University of Texas Health Science Center at Houston, Houston, TX, USA
e-mail: alexandre.paimdiaz@uth.tmc.edu

Curr Topics Behav Neurosci (2021) 48: 173–196
https://doi.org/10.1007/7854_2020_174
Published Online: 11 October 2020

and functional findings associated with BD, as well as their hypothesized relationship with the pathophysiological aspects of that condition and their potential clinical applications.

Keywords Bipolar disorder · Diffusion tensor imaging · Functional neuroimaging · Magnetic resonance imaging

1 Introduction

The search for brain morphology findings that could explain behavioral disorders has gone through a long path in the history of psychiatry. The German physician Theodor Meynert, one of the representatives of the so called "first biological psychiatry movement" in the nineteenth century, once said: "The study of human anatomy in its current form has passed from a solely descriptive science to something higher, to a form of knowledge that attempts to explain. . . The more psychiatry seeks, and finds, its scientific basis in a deep and finely grained understanding of the anatomical structure [of the brain], the more it elevates itself to the status of science that deals with causes" (Shorter 1997). Almost 130 years after his words, the psychiatric field has witnessed astonishing progress in the way we see the brain, both structurally and functionally, which has provided valuable insights on the neurophysiological processes that underlie the abnormalities in behavior, cognition, and emotion we observe in patients with psychiatric disorders. With the advances in brain imaging technology, studies have been able to identify patterns in brain circuitry associated with fundamental characteristics found in patients with bipolar disorders (BD), as alterations in emotional processing and regulation and reward processing (Phillips and Swartz 2014). Here, we critically analyze the brain imaging structural and functional findings associated with BD, not only regarding their correlation with pathophysiological aspects of the disease but also their potential clinical applications.

2 Structural Neuroimaging and Diffusion Tensor Imaging Studies

2.1 Structural Magnetic Resonance Imaging MRI (sMRI) Findings in BD

The use of sMRI in the study of brain abnormalities in individuals with BD has accumulated evidence over more than three decades of research, with countless studies. In one of the first of these studies, Johnstone et al. (1989) compared the temporal lobe structure in patients with BD, patients with schizophrenia, and healthy controls (HC). All participants with BD had a history of psychosis (Johnstone et al. 1989). With a scanning magnetic resonance imaging system that operated at 0.15 Tesla (T), the authors found a trend towards significance on the association between diagnosis and temporal lobe size, with lower measures among individuals with schizophrenia compared to the other two groups. Actually, in this study the authors did not find significant brain structural differences between participants with BD and HC. However, an association between increased ventricular size and poor outcome when evaluating participants with schizophrenia and BD together was reported, suggesting neuroprogression associated with severe psychiatric disorders (Johnstone et al. 1989; Gama et al. 2013). In a review of the literature comprising the first decade of sMRI studies in individuals with mood disorders, Soares and Mann (1997) reported a consistent finding of enlargement of the third ventricle in participants with BD. In addition, the majority of the studies that evaluated temporal lobe findings in individuals with BD identified abnormalities, especially decreased temporal lobe volumes. There were conflicting sMRI results related to amygdala, hippocampus, and basal ganglia (Soares and Mann 1997).

More recently, an analysis from the Enhancing Neuroimaging Genetics through Meta-Analysis (ENIGMA) Bipolar Disorder Working Group (2018) compared sMRI cortical gray matter thickness and surface area between adults with BD (n = 1,837) and HC (n = 2,582) (Hibar et al. 2018). The comparisons were controlled for age and sex (for the cortical thickness analysis) and also intracranial volume (for surface area analysis). The results showed a significant and diffuse pattern of reduced cortical thickness in individuals with BD compared with HC. The findings were more accentuated for the left pars opercularis, left fusiform gyrus, and left rostral middle frontal cortex regions. Regarding surface area comparisons, there were no significant differences between BD and controls. The authors also investigated associations between sMRI measures and clinical variables, such as illness duration and history of psychosis. The duration of illness was associated with a broader pattern of reduced cortical thickness. The surface area of regions of interest (ROIs) was not correlated with disease duration in this study. History of psychosis, however, was associated with reduced surface area in the right caudal anterior cingulate cortex and in the left inferior temporal gyrus (Hibar et al. 2018). The same working group also investigated subcortical volumetric brain abnormalities among individuals with BD compared with HC, including nucleus accumbens, amygdala, caudate, globus pallidus, putamen, thalamus, and lateral ventricles (Hibar et al. 2016). The results showed significantly lower mean volumes of the hippocampus, thalamus, and a trend towards significance regarding the lower mean volume of amygdala in participants with BD, bilaterally, as well as larger lateral ventricles. Decreased hippocampal volumes were also associated with older age (Hibar et al. 2016). DelBello et al. (2004) found volumetric differences related to subcortical brain regions in patients with BD already during adolescence. The authors compared adolescents with BD (mean age 16.3 years) and HC (mean age 17.2 years) regarding amygdala, globus pallidus, caudate, putamen, and thalamus volumes. In this study, in addition to lower total cerebral volume, adolescents with

BD presented significantly reduced volumes of amygdala and larger putamen (DelBello et al. 2004). Sanches et al. (2005) did not find striatal volumetric differences between adolescents with BD compared with controls. However, among the adolescents with BD, there was a significant inverse relationship between age and volumes of left caudate, right caudate, and left putamen, which were not found for the HC (Sanches et al. 2005). Studies with individuals at high genetic risk for BD also found smaller anterior cingulate cortex (ACC) and larger amygdala volumes in unaffected relatives of patients with BD compared with controls, suggesting potential brain imaging markers of disease vulnerability (Bauer et al. 2014; Sanches et al. 2019).

2.2 *Diffusion Tensor Imaging (DTI) Findings in BD*

DTI maps characterize the three-dimensional diffusion of water as a function of spatial location, which may be used to estimate the pattern of connectivity in the brain and to study the integrity of white matter tracts (Alexander et al. 2007; Benedetti et al. 2011). Adler et al. (2004) hypothesized that changes in prefrontal function found in individuals with BD are associated with a dysconnectivity syndrome due to impairment in the white matter tract integrity. The authors compared fractional anisotropy (FA) of ROIs, more specifically above anterior commissure, which was chosen a priori, according to previous findings from the group. Nine patients with bipolar disorder were compared with nine HC, controlling for age and education. The authors found significantly reduced FA in two of the four ROIs in patients with BD and medium-to-large effect sizes when all ROIs were combined (Adler et al. 2004). Mean FA and apparent diffusion coefficient (ADC) of four axonal pathways were compared between euthymic patients with BD type I (BDI), healthy first-degree relatives of these patients, and HC in the study of Mahapatra et al. (2017). The results showed a significantly lower FA for individuals with BD type I when compared with first degree relatives, who in their turn presented significantly lower FA compared with HC for the following regions: corpus callosum, dorsal part of the right cingulum bundle, hippocampal part of right cingulum bundle, hippocampal part of the left cingulum bundle, right uncinate fasciculus, and left uncinate fasciculus (Mahapatra et al. 2017). Duarte et al. (2016) reviewed studies of DTI with participants with BD focusing on FA in white matter tracts. Among the 18 studies included in this systematic review, decreased FA in commissural and association tracts, especially in the fronto-limbic tracts, was the most common finding (Duarte et al. 2016). Favre et al. (2019) compared FA from 43 ROIs between 1,482 participants with BD and 1,551 HC from 26 cohort studies. The authors found significant differences in most of the regions compared with higher effect sizes for corpus callosum and cingulum. In addition, there was a significant positive relationship between FA in the inferior fronto-occipital fasciculus and the age of onset within participants with BD. Moreover, similar to volumetric brain abnormalities, changes in white matter can be found in an early stage of the disease. Adler et al. (2006) reported that adolescents in their first episode of mania presented significantly decreased FA in the prefrontal white matter (more specifically in superior-frontal tracts) when compared with healthy participants.

2.3 Longitudinal sMRI and DTI Findings

Longitudinal neuroimaging studies are useful for the investigation of abnormalities in the brain maturation trajectories among individuals with BD, as well as to identify potential risk factors of neuroprogression and its clinical correlates. Abe et al. (2019) followed 90 patients with BD and 61 HC for 6 years to investigate putative alterations in cortical thickness. The results showed that patients with BD presented with increases in cortical thickness in visual/somatosensory areas and abnormal cortical thinning in bilateral middle temporal cortices compared with controls. While the decreases in cortical thickness suggest an expected neuroprogression of the disease, reflecting a gray matter loss, the increase in the cortical thickness of the medial occipital cortex and central sulcus was unexpected, especially considering that controls showed a decrease in the cortical thickness in these regions (Abe et al. 2019). Zak et al. (2019) also reported decreased cortical thickness in the left and right prefrontal and left temporal cortex of individuals with BD type II in a follow-up study. Both patients and controls displayed cortical thinning over the follow-up period, but patients with BD showed greater thinning in inferior temporal and left posterior cortices. Adolescents with BD, compared with HC, showed more reductions in the insula, orbitofrontal, and dorsolateral prefrontal cortices over a 2-year follow-up study (Najt et al. 2016). Bootsman et al. (2015) have investigated brain changes in subcortical regions in a twin study with individuals with BD. Despite significant differences between patients with BD and controls regarding thalamus, putamen, and nucleus accumbens at baseline, the authors did not find differences overtime after correction for multiple comparisons (Bootsman et al. 2015).

Furthermore, adolescents and young adults with BD (17.6 mean years old) and HC (16.6 mean years old) were part of a longitudinal study with DTI data at baseline and after 2.5 years of follow-up (Weathers et al. 2018). The authors investigated differences between the groups regarding uncinate fasciculus (UF), a tract that connects the amygdala and ventral prefrontal cortex, a brain network associated with emotion regulation (Weathers et al. 2018). In the patient group, there was no significant association between FA at the UF and age and no differences between the two time points, while in HC the authors did find positive associations between UF FA and age, as well as increases in UF over the follow-up period (Weathers et al. 2018). Finally, in another study, individuals at high risk for bipolar disorder (including a high-risk subgroup who later developed major depressive disorder) were compared with HC with respect to change in white matter integrity over a 2-year follow-up period (Ganzola et al. 2017). The authors hypothesized that patients, compared to controls, would show FA decreases in the corpus callosum and fronto-limbic connections. However, the results showed that, despite reductions in FA found in the whole sample, there were no statistically significant differences between groups (Ganzola et al. 2017).

3 Functional Neuroimaging

3.1 Positron Emission Tomography (PET)

PET uses radioactive tracers to investigate distinct neural functions, providing quantitative images of the spatial distribution of the compound. One highly used radioactive tracer, F-fluorodeoxyglucose (FDG), is a glucose analog and is, therefore, useful for the measurement of brain metabolism (Gonul et al. 2009). Mah et al. (2007) performed a PET study in individuals with BD type II and a current major depressive episode and HC. The authors did not find differences regarding whole-brain metabolism. However, for ROIs analysis, the results showed higher metabolism in the amygdala, left orbitofrontal cortex, right anterior cingulate cortex, accumbens area, left and right putamen, and left caudate regions in patients compared with controls (Mah et al. 2007). In another study, the FDG PET findings among BD patients with and without psychotic symptoms and HC were analyzed (Marotta et al. 2019). Patients with BD and no history of psychotic symptoms presented decreased FDG uptake in the middle occipital gyri bilaterally, as well as increased uptake in insula and temporal regions, compared with controls. However, in comparison with patients with BD and a history of psychotic symptoms, those without psychosis presented an increase in FDG uptake in the right fusiform gyrus (Marotta et al. 2019). The central serotonin transporter (5-HTT) system was also investigated in individuals with BD in comparison with HC by measuring a PET radioligand for the 5-HTT, the [^{11}C]DASB (Cannon et al. 2006). In this cross-sectional study, Cannon et al. (2006) found higher levels of the ligand in the insula, medial PFC, thalamus, caudate, and dorsal cingulate cortex and reduced levels in the brainstem of the patients. There were no significant correlations between 5-HTT binding and depression scores, illness duration, and age of onset (Cannon et al. 2006). Radioligand binding associated with microglia activation, which is suggestive of neuroinflammation, was reported as significantly increased in patients with BD compared with HC in the study of Haarman et al. (2014). Anand et al. (2011) investigated the dopaminergic system in individuals with BD by measuring the striatal binding potential of the dopamine transporter (DAT)-selective radiotracer [(11) C]CFT in individuals with BD (euthymic or depressed) and HC. The striatal binding potential was found to be significantly decreased among patients, particularly in the left and right dorsal caudate (Anand et al. 2011). These results are in agreement with previous reviews of the literature (such as the one by Berk et al. 2007), which highlight the evidence supporting the hypothesis that dysfunctions in the dopaminergic function seem to be involved in the pathophysiology of BD (Berk et al. 2007). Interestingly, a recent systematic review and meta-analysis indicated that a previous diagnosis of BD shows a more than threefold odds ratio associated with a subsequent diagnosis of idiopathic Parkinson's disease, an illness associated with degeneration of dopaminergic neurons in the substantia nigra (Kalia and Lang 2015; Faustino et al. 2019).

3.2 *Resting-State Functional MRI (rsfMRI)*

fMRI generates images using blood oxygen level-dependent (BOLD) contrast, which provides a spatial resolution that allows the localization and delimitation of activated brain areas. Thus, fMRI helps evaluate brain activity changes in specific areas, associated with the performance of specific tasks (Chow et al. 2017). Moreover, fMRI also allows the assessment of anatomically separated brain regions that are nevertheless related with respect to patterns of neuronal activity and are, therefore, functionally connected, even during "resting states" of the brain (van den Heuvel and Hulshoff Pol 2010). These complex set of neural networks mediate emotions, cognitions, and behaviors, whose dysfunction probably underlies the psychiatric manifestations are observed in the patients (Steinberg et al. 2015). Thus, brain imaging techniques that map brain regions functionally interconnected are fundamental tools to investigate neurobiological brain mechanisms of psychiatric illness.

For instance, Liu et al. (2012) investigated resting-state functional connectivity within the default mode network (DMN) in individuals with BD and HC. The DMN is related to self-oriented patterns of thought, which include rumination and introspective states, and comprise the medial, lateral, and inferior parietal cortices, the medial prefrontal cortex (mPFC), and the precuneus/posterior cingulate cortex (PCC) (Langenecker et al. 2014; Mak et al. 2017). The results showed that patients with BD displayed abnormal brain activity with significantly increased regional homogeneity (ReHo) in the left medial frontal gyrus and the left inferior parietal lobe, which could be interpreted as an enhancement of the local synchronization in those regions (Liu et al. 2012). The authors pointed out that hyperactivity of the medial frontal gyrus could be related to cognitive-emotional interference in patients with BD, as this region is implicated in cognitive regulation and emotion perception. Frontal-limbic connectivity dysfunction has been associated with the pathophysiology of BD, especially with respect to mood dysregulation due to impairment in the prefrontal cortex inhibitory control over subcortical structures as amygdala (Anticevic et al. 2013). Vizueta et al. (2012) showed that patients with BD type II and depression presented with significantly decreased activation in the left and right ventrolateral prefrontal cortices and the right amygdala, as well as reduced functional connectivity between the right amygdala and the orbitofrontal and dorsolateral prefrontal cortices in comparison with HC. In a systematic literature review, Vargas et al. (2013) reported that the most common findings from rsfMRI studies in individuals with BD were in the brain connectivity of the PFC areas and anterior cingulate cortex with meso-limbic areas as thalamus, insula, and amygdala in comparison with HC. Furthermore, altered connectivity between prefrontal and subcortical regions may already be present early in life. For instance, Singh et al. (2014) found that high-risk youth for BD were found to have decreased connectivity between pregenual cingulate and left amygdala and between left ventral lateral PFC and caudate compared with low-risk youth.

4 Summary of Main Findings

The advances in neuroimaging technology have allowed the noninvasive study of the brain, both structurally and functionally. Different study designs have helped to address distinct research questions, such as brain biomarkers related to BD, including information on cortical thickness, surface area, subcortical volumes, white matter integrity, and connectivity of neural circuits. They have also attempted to characterize longitudinal neuroprogression patterns that are potentially specific to individuals with BD in comparison with healthy individuals. Despite the high heterogeneity of the studies from a methodological standpoint, including sample size, sociodemographic factors, and clinical characteristics, some findings could be pointed out. Structural studies have shown associations between duration of illness and reduced cortical thickness, in addition to increased third ventricle size and reduced volume of hippocampus and amygdala. DTI studies show decreased connectivity in fronto-limbic tracts, corpus callosum, uncinate fasciculus, and cingulum. Longitudinal studies found reduced cortical thickness in prefrontal and temporal regions.

In contrast, PET studies show higher metabolism in the amygdala, orbitofrontal cortex, cingulate and accumbens with 18-FDG, increased 5-HTT binding in the insula, medial PFC, thalamus and cingulate cortex, as well as decreased striatal binding potential in caudate and putamen regions. Imaging findings also point to microglia activation, suggesting the involvement of neuroinflammation in patients with BD. Resting-state fMRI studies show enhancement of local synchronization within the default mode network, as well as reduced connectivity in fronto-limbic regions. Studies with early age individuals found a reduced volume of amygdala and increase putamen volume, as well as gray matter reductions in the insula, orbitofrontal and dorsolateral PFC regions and decrease FA in PFC white matter. Finally, individuals at high genetic risk for BD also show structural and functional abnormalities that suggest disease vulnerability. In summary, structural and functional studies indicate abnormalities in brain regions associated with emotion-processing, emotion-regulation, and reward-processing neural circuits in the pathophysiology of BD (Phillips and Swartz 2014).

5 Diagnostic Specificity of Neuroimaging Findings

Neuroimaging studies have been used to investigate the heterogeneity of BD, including potential differences between illness subtypes according to categorical classifications (for instance, Diagnostic and Statistical Manual of Mental Disorders [DSM] Bipolar disorders type I and II) (American Psychiatric Association 2000), and with other psychiatric disorders that could share similar clinical symptoms despite distinctions in terms of prognosis and responses to treatment (for example, unipolar depression).

5.1 Structural Neuroimaging and DTI Findings

A longitudinal study that evaluated cortical thickness among patients with BD and HC did not find differences between BD I and II subtypes (Abe et al. 2019). Similar results were found for the ENIGMA study, which compared 1,275 adults with BD type I and unrelated 345 adults with BD type II. The authors did not find differences regarding cortical thickness or the surface area between the BD subtypes I and II (Hibar et al. 2018). Subcortical volumetric regions, including lateral ventricles, thalamus, putamen, globus pallidus, hippocampus, caudate, amygdala, and nucleus accumbens, were similar between BD type I and II in the study of Hibar et al. (2016). In a systematic review, Hanford et al. (2016) reported that most of the studies comparing patients with BD type I and II regarding cortical thickness did not find significant differences (Hanford et al. 2016)

Foley et al. (2018) compared FA in the uncinate fasciculus in individuals with BD type I and II. In this study, patients with BD type I were found to have lower FA than those with BD type II, which, in turn, did not differ from HC. Similar results were found in the study of Caseras et al. (2015), in which FA in the left uncinate fasciculus of patients with BD type I was significantly reduced in comparison with patients with BD type II. Ambrosi et al. (2016) compared axial diffusivity (AD) and radial diffusivity (RD) besides FA in patients with subtypes of BD and HC. The authors found that patients with BD type II presented lower FA in the right inferior longitudinal fasciculus compared with both BD type I and HC (Ambrosi et al. 2016). Diffusion tensor images of patients with BD type I and II were also investigated by Liu et al. (2010), who reported that individuals with BD II had lower FA in the right inferior frontal gyrus, left inferior prefrontal area, and right precuneus in comparison with BD I.

Han et al. (2019) reviewed studies assessing structural and functional MRI findings in patients with unipolar depression (UP) and bipolar depression. BD were reported as showing greater cortical thickness in the right precuneus, left inferior parietal gyrus, and right dorsolateral prefrontal cortex, as well as a smaller hippocampus and amygdala volumes and increased anterior cingulate cortex volume, compared with individuals with UP (Han et al. 2019). Differences in cortical thickness were also found in the comparison between patients with schizophrenia (SCZ) and patients with BP in the study of Godwin et al. (2018), with respect to the frontal, parietal, and temporal cortices regions. In a study evaluating young individuals at high genetic risk for BD and SCZ, however, the authors did not find cortical thickness differences between groups, despite the high-risk SCZ group presenting significantly decreased surface area in the occipital lobe compared with those at high risk for BD (Sugranyes et al. 2017). Regarding volumetric structural neuroimaging findings, Ho et al. (2019), in a literature review, reported lower left, right, and total amygdala volumes in patients with schizophrenia compared with BD.

In the mentioned review of Han et al. (2019), the authors also reported differences regarding DTI measures, with patients with BD showing decreased FA in the posterior cingulum bundle and the genu of the corpus callosum compared with UP. Sexton et al. (2012) compared FA within the corpus callosum between participants with late-life BD and UP depression. The results showed that patients with BD had reductions in FA within the genu, body, and splenium of the corpus callosum, suggesting that altered inter-hemispheric connectivity might be a feature of late-life bipolar disorder (Sexton et al. 2012). Ho et al. (2019) reported no differences between patients with BD and SCZ regarding the FA of the uncinate fasciculus tract, a finding previously reported by studies comparing HC and BD (Ho et al. 2019; Mahapatra et al. 2017). On the other hand, comparisons on the corpus callosum white matter did not show differences between individuals with BD and SCZ diagnosis (Li et al. 2014). Negative findings were also found in the study of Skudlarski et al. (2013), which included patients with BD, SCZ, and their first-degree relatives. Positive findings were reported in the study of Tonnesen et al. (2018), in which patients with SCZ showed lower FA in the right inferior longitudinal fasciculus and right inferior fronto-occipital fasciculus compared with patients with BD.

5.2 *Functional Neuroimaging Findings (PET, rs-fMRI)*

Hosokawa et al. (2009) compared resting-state PET findings between patients with BD and UP and found a distinct pattern of decreased glucose metabolism related to HC, suggesting brain metabolism particularities in comparison with healthy individuals. However, the authors did not find differences between unipolar and bipolar depressed patients (Hosokawa et al. 2009). Altamura et al. (2017) used 18-FDG-PET scanning to compare brain metabolism differences between patients with BD type I and history of psychosis with or without substance use, and patients with substance-induced psychosis. The results showed that patients with substance-induced psychosis presented decreased glucose metabolism in the left posterior cingulate compared with patients with BD and history of psychosis with no substance use (Altamura et al. 2017). A study with fluorodihydroxyphenyl-l-alanine ([18F]-DOPA) PET did not show differences regarding dopamine synthesis capacity between patients with BD and history of psychosis and patients with SCZ. When both groups were combined, however, the findings showed a significant positive correlation between psychotic symptom severity and dopamine synthesis capacity, suggesting a transdiagnostic role for dopamine dysfunction in BD and SCZ (Jauhar et al. 2017). Glucose metabolism differences between psychiatric diagnosis were also investigated by Boen et al. (2019). In this study, patients with borderline personality disorder (BPD) and no comorbid BD were compared with patients with BD type II and HC. Both groups of patients (BPD and BD) showed decreased metabolism in similar regions, such as insula, brainstem, and frontal white matter compared to HC, which could, in part, explain the high comorbidity between these two psychiatric diagnoses (Frias et al. 2016). The results also showed that patients with BD presented higher metabolism in some cortical areas compared with those with BPD (Boen et al. 2019).

In a literature review with fMRI studies, McGrath et al. (2004) did not find evidence of differences between patients with BD type I and type II. When compared with patients with UP depression, however, patients with BD presented a distinct pattern of connectivity in the study of Goya-Maldonado et al. (2016). While the former showed increased connectivity in the DMN in the precuneus and hippocampus bilaterally, patients with BD showed increased functional connectivity in the frontoparietal network, especially in the dorsolateral and ventrolateral PFC compared with participants with UP depression (Goya-Maldonado et al. 2016). Karcher et al. (2019) investigated corticostriatal connectivity in individuals with SCZ and BD with a history of psychotic symptoms compared with HC. Both patient groups presented with reduced connectivity between the putamen and the medial prefrontal cortex and reduced salience network connectivity, suggesting a common pattern of corticostriatal dysconnectivity in patients with primary psychotic disorders (Karcher et al. 2019). Similar transdiagnostic findings were reported in the study of Ma et al. (2019), in which patients with SCZ, BD, and UP depression presented common network dysfunction.

5.3 *Neuroimaging and Pattern Classification Methods and the Diagnosis of BD*

Machine learning (ML) techniques are able to analyze multiple sources of data, which can provide information from an individual level, rather than between-groups average differences, with a potential role in terms of diagnostic prediction and accuracy (Nunes et al. 2018).

Nunes et al. (2018) analyzed clinical and neuroimaging data (including regional cortical thickness, surface area, and subcortical volumes) of 853 participants with a diagnosis of BD and 2,167 HC using support vector machines (SVM), aiming at discriminating patients from controls. In this study, despite the accuracy of 65.23% (95% CI = 63.47–67.00) was below the considered clinically relevant (80%), taking into account the high heterogeneity of the sample, from 13 cohort studies over the world, the results suggest ML approaches as a potential technique for improvement of diagnosis accuracy (Nunes et al. 2018). With DTI data, Mwangi et al. (2015) investigated the potential of SVM to accurately discriminate individuals with pediatric BD from HC (both groups with approximately 12 years old on average). The authors reported accuracy of 78.12%, with a sensitivity of 68.75% and specificity of 87.5%, with the most relevant regions discriminating patients from controls showing a consistent reduced pattern of FA (Mwangi et al. 2015).

The discrimination between UP depression and BD depression was also investigated by studies using SVM algorithms. Matsuo et al. (2019) found that gray matter volumes of the dorsolateral PFC bilaterally and anterior cingulate cortex contributed to the diagnosis classification of UP depression and BD depression with SVM models. Brain structural neuroimaging data was also used by Rubin-Falcone et al. (2018) to differentiate individuals with BD and UP depression, with results showing a combined accuracy of 75% using SVM for gray matter volume data. Li et al. (2017) used resting-state fMRI with SVM approach to test its accuracy to discriminate individuals with BD from unipolar depression. Results showed an accuracy of 86%, as well as a large proportion of disease-specific information, with a low overlap between individuals with UP depression and BD depression with respect to topographic abnormalities (Li et al. 2017).

Machine learning methods can help not only in terms of diagnostic discrimination but also in predicting specific outcomes and identifying specific clinical phenotypes in bipolar disorder, which can be useful for approaching the clinical heterogeneity of the illness. Sartori et al. (2018) used volumetric brain imaging data to predict functioning in patients with bipolar disorder and HC, utilizing an ML approach. Both groups displayed significantly different functional outcomes, which included the Functioning Assessment Short Test (FAST) scores and employment status (Reisberg 1988; Sartori et al. 2018). Left superior frontal cortex volume and left rostral middle frontal cortex were the central regions able to predict FAST scores in patients with BD (Sartori et al. 2018; Phillips and Swartz 2014). There were no significant findings in the HC group (Sartori et al. 2018). Neuroimaging and neurocognitive data were used to investigate clinical phenotypes in patients with BD in the study of Wu et al. (2017). The cognitive evaluation included measures associated with arousal, cognitive control, declarative memory, social communication, and valence systems according to the Research Domain Criteria (RDoC) Initiative (Wu et al. 2017; National Institute of Mental Health 2008). The authors found two phenotypes, which did not overlap with the DSM BD classifications. The ML algorithm, using FA, discriminated these phenotypes with 75.9% accuracy (the inferior frontal-occipital fasciculus and the minor forceps of the corpus callosum were the most relevant brain regions) (Wu et al. 2017).

Taken together, these studies show that ML techniques are very promising approaches not only for the improvement of diagnostic accuracy and prediction but also for the better characterization of the phenotypical heterogeneity in BD (Librenza-Garcia et al. 2017). In a position paper from the International Society for Bipolar Disorders Big Data Task Force, Passos et al. (2019) enumerated some of the challenges faced by ML, including model validation, computational power, multimodality, and lack of a uniform pipeline for ML studies. However, once overcome, these techniques, combined with big data analyze, could help with the aim of improving prognosis in the management of BD, with better prediction of clinical outcomes and response to treatment (Passos et al. 2019).

6 Perspectives on the Role of Neuroimaging in the Management of BD Patients

6.1 Neuroimaging Studies and Bipolar Disorders Mood States

In a longitudinal study, Zak et al. (2019) reported greater cortical thinning in the left temporal cortical among patients with BD with more than two depressives episodes from baseline to the follow-up, compared with patients with fewer depressive episodes. Another longitudinal study found a greater cortical thinning in the inferior frontal cortex of patients with mania compared with patients with non-manic BD type I. In addition, patients with BD type II that experienced hypomanic episodes during the follow-up showed more pronounced decreases in the inferior frontal cortex compared with patients with BD type II who did not have hypomanic episodes during the longitudinal evaluation (Abe et al. 2019). Using an FDG PET scan, Brooks et al. (2009) found a significant inverse correlation between global metabolic rates and scores in the Hamilton depression scale (HAM-D) (Brooks et al. 2009; Hamilton 1960). Brady et al. (2016) investigated rsfMRI activity in regions associated with affect perception, affect regulation, and reward-seeking behavior during different mood states in patients with BD type I, mania and euthymia, as well as in HC. The results showed that, compared with patients with BD in euthymia, those with mania presented with significantly increased connectivity between the right amygdala and the bilateral supplemental motor area in the frontal cortex, as well as decreased connectivity with the ACC, suggesting altered emotion regulation neural circuits associated with mania states. There were no significant differences regarding connectivity with the ventral striatum between patients with BD in mania or euthymia (Brady et al. 2016). In a longitudinal design, these authors were able to replicate these findings, suggesting that cortico-amygdala resting-state connectivity could be a biomarker of mood state in patients with BD (Brady et al. 2017).

6.2 Neuroimaging Studies, Bipolar Disorders, and Pharmacological Treatment

Hibar et al. (2018) found significant increases in cortical thickness associated with the use of lithium in patients with BD, mainly in the left paracentral gyrus and the left and right superior parietal gyrus, as well as increased surface area in the left paracentral lobe. The use of typical and atypical antipsychotic medications seemed to show different types of associations with the imaging findings. While the use of typical antipsychotics was associated with increased cortical surface area (especially in the left middle temporal gyrus, left inferior parietal gyrus, and right temporal pole), atypical antipsychotic use was associated with decreased cortical surface area in the right rostral middle frontal gyrus and right superior frontal gyrus (Hibar et al. 2018). Li et al. (2019) compared cortical thickness and subcortical volumes in HC with patients with BD taking valproate and patients with BD on lithium. Results showed that participants with BD on lithium had significantly increased cortical thickness in the right superior frontal cortex and in the left rostral middle frontal cortex compared with those taking valproate. The authors did not find differences in subcortical regions. In patients with pediatric bipolar disorder, a preliminary study showed decreases in the amygdala volume after a 6-week treatment period with sodium valproate (Cazala et al. 2018). Further, in the mentioned follow-up study by Abe et al. (2019), the authors reported that patients who used lithium presented with an increase in the cortical thickness of the medial occipital, which was not found for patients who were not on that medication. Brain structure changes with lithium use may also be related to duration of medication exposure, as showed by the study of Sani et al. (2018), in which short-term use was related to changes in amygdala volume and long-term use with changes in hippocampus and amygdala volumes.

Favre et al. (2019) reported associations between FA findings and use of medications in a cross-sectional study with 1,482 participants with BD. In this study, antipsychotics were associated with lower FA within the genu of the corpus callosum. The authors did not find differences related to antidepressant use. Regarding lithium, the authors reported higher FA in several ROIs among the patients who were on that medication, which may be related to the potential of lithium in promoting myelination (Favre et al. 2019; Brambilla et al. 2009). FA in the corpus callosum was significantly higher among patients on lithium in the study of Abramovic et al. (2018), with no significant differences in FA associated with antipsychotic medication.

Valproate treatment (in monotherapy or combination with lithium) was not associated with brain 5-HT2A receptor binding patterns in adult patients who met DSM criteria for a manic episode, in a study by Yatham et al. (2005). Using resting-state fMRI, Spielberg et al. (2019) investigated lithium effects on neural circuits related to mania and depression in BD. The results showed that treatment with lithium was associated with normalization of connectome indices observed during mania. In addition, changes in connectome indices associated with both mania and depression were correlated with symptom changes (Spielberg et al. 2019). Moreover, the velocity of normalization of neural circuits associated with pharmacological treatment seems to be distinct, depending on the medication considered. Dandash et al. (2018) reported that treatment with lithium showed a faster normalization of hyperconnectivity in the ventral striatum with the cerebellum compared to treatment with quetiapine.

6.3 *Neuroimaging to Predict Pharmacological Treatment Response*

Baseline FA connectivity of the cingulate and hippocampal regions significantly predicted 8-week global clinical impression (CGI) severity scores after treatment with lithium (4 weeks of treatment) in pediatric patients with BD, an effect found for both severity of mania and depression (Kafantaris et al. 2017). DTI was also used to investigate white matter connectivity as a possible predictor of response to antidepressant treatment in adult patients with BD type I. The results showed an inverse correlation between clinical improvements and white matter microstructure integrity of tracts, including corpus callosum, cingulum bundle, and inferior fronto-occipital fasciculus (Bollettini et al. 2015). Clinical response to ketamine infusion in patients with bipolar depression was associated with increased 18-FDG metabolism in the subgenual anterior cingulate cortex after placebo infusion in the study of Nugent et al. (2014). In addition, changes in metabolism in the right ventral striatum between placebo and ketamine infusions showed significant inverse correlations with changes in depression scores (Nugent et al. 2014). In another study, subgroups of patients defined by cluster analysis according to cortical thickness were associated with different responses to treatment in a randomized clinical trial, suggesting that neurobiological measures could address the clinical heterogeneity of bipolar disorder with respect to treatment response patterns (Zhang et al. 2018).

Furthermore, given the high complexity of bipolar disorder pathophysiology, improvements in the prediction of treatment response could be achieved by the inclusion of different clinical and neurobiological variables in addition to neuroimaging findings. The Response to Li Network (R-LiNK) initiative is a prospective multidisciplinary, international project which aims to identify individual predictors of clinical response to lithium. Data collection will include a combination of molecular, metabolic, structural, functional, and clinical biomarkers, with an ecological momentary assessment approach to monitor core BD symptoms, coupled with the investigation of moderators and mediators of response (Scott et al. 2019). Given the promising results on neuroimaging findings and treatment response prediction, such initiatives may provide precise and accurate information for the early and effective treatment of BD, which could potentially impact not only the burden associated with this illness but also psychiatric disorders in general.

6.4 *Neuroimaging and Psychotherapy in Bipolar Disorders*

Despite including principles of cognitive behavior therapy (CBT), such as extinction learning, identification and modification of maladaptive cognitions, and behavioral exposure (Ellard et al. 2010), transdiagnostic CBT focuses on maladaptive emotion processing. Ellard et al. used fMRI to investigate the association of brain connectivity of regions associated with emotion regulation and clinical outcomes of transdiagnostic CBT intervention in patients with BD (Ellard et al. 2018). The results showed that changes in affective control scores were predicted by weaker connectivity between the left anterior insula and the right ventrolateral PFC (salience network), and by stronger connectivity between the bilateral dorsal anterior insula and bilateral amygdala at baseline (Ellard et al. 2018).

According to Teasdale et al. (2000), mindfulness-based cognitive therapy (MBCT) aims at teaching individuals "to become more aware of thoughts and feelings and to relate to them in a wider, decentered perspective as 'mental events' rather than as aspects of the self or as necessarily accurate reflections of reality." (Teasdale et al. 2000). This therapy has shown promising results in the treatment of patients with BD and has a first-line indication for preventing depression relapse (Lovas and Schuman-Olivier 2018; Parikh et al. 2016). Ives-Deliperi et al. (2013) reported significant decreases in anxiety and emotion dysregulation, improvement in mindfulness, and executive performance associated with MBCT in patients with BD comparing with those in a waiting list group. In addition, there was a significant correlation between signal change in the medial PFC after the intervention, suggesting a potential action mechanism of MBCT (Ives-Deliperi et al. 2013).

7 Conclusions

Neuroimaging studies have contributed to a better understanding of the pathophysiology of BD as a brain disease, pointing to dysfunctions in neural circuits associated with emotion regulation and reward processing. Despite the limited validity of categorical diagnosis, as shown by genetic studies, scientific literature has consistently shown significant structural and functional findings among patients with BD (defined according to DSM criteria) and individuals at high genetic risk for BD. The heterogeneity of BD might be addressed with the association of different sources of biological and clinical information, in addition to neuroimaging techniques, allowing the better characterization of phenotypes and the identification of biomarkers, ultimately resulting in potentially important clinical implications.

References

Abe C, Liberg B, Song J, Bergen SE, Petrovic P, Ekman CJ, Sellgren CM, Ingvar M, Landen M (2019) Longitudinal cortical thickness changes in bipolar disorder and the relationship to genetic risk, mania, and lithium use. Biol Psychiatry 87(3):271–281. https://doi.org/10.1016/j.biopsych.2019.08.015

Abramovic L, Boks MPM, Vreeker A, Verkooijen S, van Bergen AH, Ophoff RA, Kahn RS, van Haren NEM (2018) White matter disruptions in patients with bipolar disorder. Eur Neuropsychopharmacol 28(6):743–751. https://doi.org/10.1016/j.euroneuro.2018.01.001

Adler CM, Holland SK, Schmithorst V, Wilke M, Weiss KL, Pan H, Strakowski SM (2004) Abnormal frontal white matter tracts in bipolar disorder: a diffusion tensor imaging study. Bipolar Disord 6(3):197–203. https://doi.org/10.1111/j.1399-5618.2004.00108.x

Adler CM, Adams J, DelBello MP, Holland SK, Schmithorst V, Levine A, Jarvis K, Strakowski SM (2006) Evidence of white matter pathology in bipolar disorder adolescents experiencing their first episode of mania: a diffusion tensor imaging study. Am J Psychiatry 163(2):322–324. https://doi.org/10.1176/appi.ajp.163.2.322

Alexander AL, Lee JE, Lazar M, Field AS (2007) Diffusion tensor imaging of the brain. Neurotherapeutics 4(3):316–329. https://doi.org/10.1016/j.nurt.2007.05.011

Altamura AC, Delvecchio G, Marotta G, Oldani L, Pigoni A, Ciappolino V, Caletti E, Rovera C, Dobrea C, Arici C, Benatti B, Camuri G, Prunas C, Paoli RA, Dell'osso B, Cinnante C, Triulzi FM, Brambilla P (2017) Structural and metabolic differentiation between bipolar disorder with psychosis and substance-induced psychosis: an integrated MRI/PET study. Eur Psychiatry 41:85–94. https://doi.org/10.1016/j.eurpsy.2016.09.009

Ambrosi E, Chiapponi C, Sani G, Manfredi G, Piras F, Caltagirone C, Spalletta G (2016) White matter microstructural characteristics in bipolar I and bipolar II disorder: a diffusion tensor imaging study. J Affect Disord 189:176–183. https://doi.org/10.1016/j.jad.2015.09.035

American Psychiatric Association (2000) Diagnostic and statistical manual of mental disorders, 4th edn, text revision. American Psychiatric Association, Washington, DC

Anand A, Barkay G, Dzemidzic M, Albrecht D, Karne H, Zheng QH, Hutchins GD, Normandin MD, Yoder KK (2011) Striatal dopamine transporter availability in unmedicated bipolar disorder. Bipolar Disord 13(4):406–413. https://doi.org/10.1111/j.1399-5618.2011.00936.x

Anticevic A, Brumbaugh MS, Winkler AM, Lombardo LE, Barrett J, Corlett PR, Kober H, Gruber J, Repovs G, Cole MW, Krystal JH, Pearlson GD, Glahn DC (2013) Global prefrontal and fronto-amygdala dysconnectivity in bipolar I disorder with psychosis history. Biol Psychiatry 73(6):565–573. https://doi.org/10.1016/j.biopsych.2012.07.031

Bauer IE, Sanches M, Suchting R, Green CE, El Fangary NM, Zunta-Soares GB, Soares JC (2014) Amygdala enlargement in unaffected offspring of bipolar parents. J Psychiatr Res 59:200–205. https://doi.org/10.1016/j.jpsychires.2014.08.023

Benedetti F, Yeh PH, Bellani M, Radaelli D, Nicoletti MA, Poletti S, Falini A, Dallaspezia S, Colombo C, Scotti G, Smeraldi E, Soares JC, Brambilla P (2011) Disruption of white matter integrity in bipolar depression as a possible structural marker of illness. Biol Psychiatry 69 (4):309–317. https://doi.org/10.1016/j.biopsych.2010.07.028

Berk M, Dodd S, Kauer-Sant'anna M, Malhi GS, Bourin M, Kapczinski F, Norman T (2007) Dopamine dysregulation syndrome: implications for a dopamine hypothesis of bipolar disorder. Acta Psychiatr Scand Suppl 434:41–49. https://doi.org/10.1111/j.1600-0447.2007.01058.x

Boen E, Hjornevik T, Hummelen B, Elvsashagen T, Moberget T, Holtedahl JE, Babovic A, Hol PK, Karterud S, Malt UF (2019) Patterns of altered regional brain glucose metabolism in borderline personality disorder and bipolar II disorder. Acta Psychiatr Scand 139(3):256–268. https://doi.org/10.1111/acps.12997

Bollettini I, Poletti S, Locatelli C, Vai B, Smeraldi E, Colombo C, Benedetti F (2015) Disruption of white matter integrity marks poor antidepressant response in bipolar disorder. J Affect Disord 174:233–240. https://doi.org/10.1016/j.jad.2014.11.010

Bootsman F, Brouwer RM, Kemner SM, Schnack HG, van der Schot AC, Vonk R, Hillegers MH, Boomsma DI, Hulshoff Pol HE, Nolen WA, Kahn RS, van Haren NE (2015) Contribution of genes and unique environment to cross-sectional and longitudinal measures of subcortical volumes in bipolar disorder. Eur Neuropsychopharmacol 25(12):2197–2209. https://doi.org/10.1016/j.euroneuro.2015.09.023

Brady RO Jr, Masters GA, Mathew IT, Margolis A, Cohen BM, Ongur D, Keshavan M (2016) State dependent cortico-amygdala circuit dysfunction in bipolar disorder. J Affect Disord 201:79–87. https://doi.org/10.1016/j.jad.2016.04.052

Brady RO Jr, Margolis A, Masters GA, Keshavan M, Ongur D (2017) Bipolar mood state reflected in cortico-amygdala resting state connectivity: a cohort and longitudinal study. J Affect Disord 217:205–209. https://doi.org/10.1016/j.jad.2017.03.043

Brambilla P, Bellani M, Yeh PH, Soares JC (2009) Myelination in bipolar patients and the effects of mood stabilizers on brain anatomy. Curr Pharm Des 15(22):2632–2636. https://doi.org/10.2174/138161209788957519

Brooks JO 3rd, Wang PW, Bonner JC, Rosen AC, Hoblyn JC, Hill SJ, Ketter TA (2009) Decreased prefrontal, anterior cingulate, insula, and ventral striatal metabolism in medication-free depressed outpatients with bipolar disorder. J Psychiatr Res 43(3):181–188. https://doi.org/10.1016/j.jpsychires.2008.04.015

Cannon DM, Ichise M, Fromm SJ, Nugent AC, Rollis D, Gandhi SK, Klaver JM, Charney DS, Manji HK, Drevets WC (2006) Serotonin transporter binding in bipolar disorder assessed using [11C]DASB and positron emission tomography. Biol Psychiatry 60(3):207–217. https://doi.org/10.1016/j.biopsych.2006.05.005

Caseras X, Murphy K, Lawrence NS, Fuentes-Claramonte P, Watts J, Jones DK, Phillips ML (2015) Emotion regulation deficits in euthymic bipolar I versus bipolar II disorder: a functional and diffusion-tensor imaging study. Bipolar Disord 17(5):461–470. https://doi.org/10.1111/bdi.12292

Cazala F, Suchting R, Zeni CP, Bauer IE, Mwangi B, Wu MJ, Passos IC, Spiker DE, Zunta-Soares GB, Soares JC (2018) Effects of valproate on brain volumes in pediatric bipolar disorder: a preliminary study. Psychiatry Res Neuroimaging 278:65–68. https://doi.org/10.1016/j.pscychresns.2018.05.006

Chow MS, Wu SL, Webb SE, Gluskin K, Yew DT (2017) Functional magnetic resonance imaging and the brain: a brief review. World J Radiol 9(1):5–9. https://doi.org/10.4329/wjr.v9.i1.5

Dandash O, Yucel M, Daglas R, Pantelis C, McGorry P, Berk M, Fornito A (2018) Differential effect of quetiapine and lithium on functional connectivity of the striatum in first episode mania. Transl Psychiatry 8(1):59. https://doi.org/10.1038/s41398-018-0108-8

DelBello MP, Zimmerman ME, Mills NP, Getz GE, Strakowski SM (2004) Magnetic resonance imaging analysis of amygdala and other subcortical brain regions in adolescents with bipolar disorder. Bipolar Disord 6(1):43–52. https://doi.org/10.1046/j.1399-5618.2003.00087.x

Duarte JA, de Araujo ESJQ, Goldani AA, Massuda R, Gama CS (2016) Neurobiological underpinnings of bipolar disorder focusing on findings of diffusion tensor imaging: a systematic review. Braz J Psychiatry 38(2):167–175. https://doi.org/10.1590/1516-4446-2015-1793

Ellard KK, Fairholme CP, Boisseau CL, Farchione TJ, Barlow DH (2010) Unified protocol for the transdiagnostic treatment of emotional disorders: protocol development and initial outcome data. Cogn Behav Pract 17(1):14

Ellard KK, Gosai AG, Bernstein EE, Kaur N, Sylvia LG, Camprodon JA, Dougherty DD, Nierenberg AA, Deckersbach T (2018) Intrinsic functional neurocircuitry associated with treatment response to transdiagnostic CBT in bipolar disorder with anxiety. J Affect Disord 238:383–391. https://doi.org/10.1016/j.jad.2018.06.002

Faustino PR, Duarte GS, Chendo I, Castro Caldas A, Reimao S, Fernandes RM, Vale J, Tinazzi M, Bhatia K, Ferreira JJ (2019) Risk of developing Parkinson disease in bipolar disorder: a systematic review and meta-analysis. JAMA Neurol 77(2):192–198. https://doi.org/10.1001/jamaneurol.2019.3446

Favre P, Pauling M, Stout J, Hozer F, Sarrazin S, Abe C, Alda M, Alloza C, Alonso-Lana S, Andreassen OA, Baune BT, Benedetti F, Busatto GF, Canales-Rodriguez EJ, Caseras X, Chaim-Avancini TM, Ching CRK, Dannlowski U, Deppe M, Eyler LT, Fatjo-Vilas M, Foley SF, Grotegerd D, Hajek T, Haukvik UK, Howells FM, Jahanshad N, Kugel H, Lagerberg TV, Lawrie SM, Linke JO, McIntosh A, Melloni EMT, Mitchell PB, Polosan M, Pomarol-Clotet E, Repple J, Roberts G, Roos A, Rosa PGP, Salvador R, Sarro S, Schofield PR, Serpa MH, Sim K, Stein DJ, Sussmann JE, Temmingh HS, Thompson PM, Verdolini N, Vieta E, Wessa M, Whalley HC, Zanetti MV, Leboyer M, Mangin JF, Henry C, Duchesnay E, Houenou J, Group EBDW (2019) Widespread white matter microstructural abnormalities in bipolar disorder: evidence from mega- and meta-analyses across 3033 individuals. Neuropsychopharmacology 44(13):2285–2293. https://doi.org/10.1038/s41386-019-0485-6

Foley SF, Bracher-Smith M, Tansey KE, Harrison JR, Parker GD, Caseras X (2018) Fractional anisotropy of the uncinate fasciculus and cingulum in bipolar disorder type I, type II, unaffected siblings and healthy controls. Br J Psychiatry 213(3):548–554. https://doi.org/10.1192/bjp.2018.101

Frias A, Baltasar I, Birmaher B (2016) Comorbidity between bipolar disorder and borderline personality disorder: prevalence, explanatory theories, and clinical impact. J Affect Disord 202:210–219. https://doi.org/10.1016/j.jad.2016.05.048

Gama CS, Kunz M, Magalhaes PV, Kapczinski F (2013) Staging and neuroprogression in bipolar disorder: a systematic review of the literature. Braz J Psychiatry 35(1):70–74

Ganzola R, Nickson T, Bastin ME, Giles S, Macdonald A, Sussmann J, McIntosh AM, Whalley HC, Duchesne S (2017) Longitudinal differences in white matter integrity in youth at high familial risk for bipolar disorder. Bipolar Disord 19(3):158–167. https://doi.org/10.1111/bdi.12489

Godwin D, Alpert KI, Wang L, Mamah D (2018) Regional cortical thinning in young adults with schizophrenia but not psychotic or non-psychotic bipolar I disorder. Int J Bipolar Disord 6 (1):16. https://doi.org/10.1186/s40345-018-0124-x

Gonul AS, Coburn K, Kula M (2009) Cerebral blood flow, metabolic, receptor, and transporter changes in bipolar disorder: the role of PET and SPECT studies. Int Rev Psychiatry 21 (4):323–335. https://doi.org/10.1080/09540260902962131

Goya-Maldonado R, Brodmann K, Keil M, Trost S, Dechent P, Gruber O (2016) Differentiating unipolar and bipolar depression by alterations in large-scale brain networks. Hum Brain Mapp 37(2):808–818. https://doi.org/10.1002/hbm.23070

Haarman BC, Riemersma-Van der Lek RF, de Groot JC, Ruhe HG, Klein HC, Zandstra TE, Burger H, Schoevers RA, de Vries EF, Drexhage HA, Nolen WA, Doorduin J (2014) Neuroinflammation in bipolar disorder – a [(11)C]-(R)-PK11195 positron emission tomography study. Brain Behav Immun 40:219–225. https://doi.org/10.1016/j.bbi.2014.03.016

Hamilton M (1960) A rating scale for depression. J Neurol Neurosurg Psychiatry 23:56–62

Han KM, De Berardis D, Fornaro M, Kim YK (2019) Differentiating between bipolar and unipolar depression in functional and structural MRI studies. Prog Neuro-Psychopharmacol Biol Psychiatry 91:20–27. https://doi.org/10.1016/j.pnpbp.2018.03.022

Hanford LC, Nazarov A, Hall GB, Sassi RB (2016) Cortical thickness in bipolar disorder: a systematic review. Bipolar Disord 18(1):4–18. https://doi.org/10.1111/bdi.12362

Hibar DP, Westlye LT, van Erp TG, Rasmussen J, Leonardo CD, Faskowitz J, Haukvik UK, Hartberg CB, Doan NT, Agartz I, Dale AM, Gruber O, Kramer B, Trost S, Liberg B, Abe C, Ekman CJ, Ingvar M, Landen M, Fears SC, Freimer NB, Bearden CE, Sprooten E, Glahn DC, Pearlson GD, Emsell L, Kenney J, Scanlon C, McDonald C, Cannon DM, Almeida J, Versace A, Caseras X, Lawrence NS, Phillips ML, Dima D, Delvecchio G, Frangou S, Satterthwaite TD, Wolf D, Houenou J, Henry C, Malt UF, Boen E, Elvsashagen T, Young AH, Lloyd AJ, Goodwin GM, Mackay CE, Bourne C, Bilderbeck A, Abramovic L, Boks MP, van Haren NE, Ophoff RA, Kahn RS, Bauer M, Pfennig A, Alda M, Hajek T, Mwangi B, Soares JC, Nickson T, Dimitrova R, Sussmann JE, Hagenaars S, Whalley HC, McIntosh AM, Thompson PM, Andreassen OA, Costa Rica/Colombia Consortium for Genetic Investigation of Bipolar E (2016) Subcortical volumetric abnormalities in bipolar disorder. Mol Psychiatry 21 (12):1710–1716. https://doi.org/10.1038/mp.2015.227

Hibar DP, Westlye LT, Doan NT, Jahanshad N, Cheung JW, Ching CRK, Versace A, Bilderbeck AC, Uhlmann A, Mwangi B, Kramer B, Overs B, Hartberg CB, Abe C, Dima D, Grotegerd D, Sprooten E, Boen E, Jimenez E, Howells FM, Delvecchio G, Temmingh H, Starke J, Almeida JRC, Goikolea JM, Houenou J, Beard LM, Rauer L, Abramovic L, Bonnin M, Ponteduro MF, Keil M, Rive MM, Yao N, Yalin N, Najt P, Rosa PG, Redlich R, Trost S, Hagenaars S, Fears SC, Alonso-Lana S, van Erp TGM, Nickson T, Chaim-Avancini TM, Meier TB, Elvsashagen T, Haukvik UK, Lee WH, Schene AH, Lloyd AJ, Young AH, Nugent A, Dale AM, Pfennig A,

McIntosh AM, Lafer B, Baune BT, Ekman CJ, Zarate CA, Bearden CE, Henry C, Simhandl C, McDonald C, Bourne C, Stein DJ, Wolf DH, Cannon DM, Glahn DC, Veltman DJ, Pomarol-Clotet E, Vieta E, Canales-Rodriguez EJ, Nery FG, Duran FLS, Busatto GF, Roberts G, Pearlson GD, Goodwin GM, Kugel H, Whalley HC, Ruhe HG, Soares JC, Fullerton JM, Rybakowski JK, Savitz J, Chaim KT, Fatjo-Vilas M, Soeiro-de-Souza MG, Boks MP, Zanetti MV, Otaduy MCG, Schaufelberger MS, Alda M, Ingvar M, Phillips ML, Kempton MJ, Bauer M, Landen M, Lawrence NS, van Haren NEM, Horn NR, Freimer NB, Gruber O, Schofield PR, Mitchell PB, Kahn RS, Lenroot R, Machado-Vieira R, Ophoff RA, Sarro S, Frangou S, Satterthwaite TD, Hajek T, Dannlowski U, Malt UF, Arolt V, Gattaz WF, Drevets WC, Caseras X, Agartz I, Thompson PM, Andreassen OA (2018) Cortical abnormalities in bipolar disorder: an MRI analysis of 6503 individuals from the ENIGMA Bipolar Disorder Working Group. Mol Psychiatry 23(4):932–942. https://doi.org/10.1038/mp.2017.73

Ho NF, Li Hui Chong P, Lee DR, Chew QH, Chen G, Sim K (2019) The amygdala in schizophrenia and bipolar disorder: a synthesis of structural MRI, diffusion tensor imaging, and resting-state functional connectivity findings. Harv Rev Psychiatry 27(3):150–164. https://doi.org/10.1097/HRP.0000000000000207

Hosokawa T, Momose T, Kasai K (2009) Brain glucose metabolism difference between bipolar and unipolar mood disorders in depressed and euthymic states. Prog Neuro-Psychopharmacol Biol Psychiatry 33(2):243–250. https://doi.org/10.1016/j.pnpbp.2008.11.014

Ives-Deliperi VL, Howells F, Stein DJ, Meintjes EM, Horn N (2013) The effects of mindfulness-based cognitive therapy in patients with bipolar disorder: a controlled functional MRI investigation. J Affect Disord 150(3):1152–1157. https://doi.org/10.1016/j.jad.2013.05.074

Jauhar S, Nour MM, Veronese M, Rogdaki M, Bonoldi I, Azis M, Turkheimer F, McGuire P, Young AH, Howes OD (2017) A test of the transdiagnostic dopamine hypothesis of psychosis using positron emission tomographic imaging in bipolar affective disorder and schizophrenia. JAMA Psychiat 74(12):1206–1213. https://doi.org/10.1001/jamapsychiatry.2017.2943

Johnstone EC, Owens DG, Crow TJ, Frith CD, Alexandropolis K, Bydder G, Colter N (1989) Temporal lobe structure as determined by nuclear magnetic resonance in schizophrenia and bipolar affective disorder. J Neurol Neurosurg Psychiatry 52(6):736–741. https://doi.org/10.1136/jnnp.52.6.736

Kafantaris V, Spritzer L, Doshi V, Saito E, Szeszko PR (2017) Changes in white matter microstructure predict lithium response in adolescents with bipolar disorder. Bipolar Disord 19(7):587–594. https://doi.org/10.1111/bdi.12544

Kalia LV, Lang AE (2015) Parkinson's disease. Lancet 386(9996):896–912. https://doi.org/10.1016/S0140-6736(14)61393-3. [pii] S0140-6736(14)61393-3

Karcher NR, Rogers BP, Woodward ND (2019) Functional connectivity of the striatum in schizophrenia and psychotic bipolar disorder. Biol Psychiatry Cogn Neurosci Neuroimaging 4(11):956–965. https://doi.org/10.1016/j.bpsc.2019.05.017

Langenecker SA, Jacobs RH, Passarotti AM (2014) Current neural and behavioral dimensional constructs across mood disorders. Curr Behav Neurosci Rep 1(3):144–153. https://doi.org/10.1007/s40473-014-0018-x

Li J, Kale Edmiston E, Chen K, Tang Y, Ouyang X, Jiang Y, Fan G, Ren L, Liu J, Zhou Y, Jiang W, Liu Z, Xu K, Wang F (2014) A comparative diffusion tensor imaging study of corpus callosum subregion integrity in bipolar disorder and schizophrenia. Psychiatry Res 221(1):58–62. https://doi.org/10.1016/j.pscychresns.2013.10.007

Li M, Das T, Deng W, Wang Q, Li Y, Zhao L, Ma X, Wang Y, Yu H, Li X, Meng Y, Palaniyappan L, Li T (2017) Clinical utility of a short resting-state MRI scan in differentiating bipolar from unipolar depression. Acta Psychiatr Scand 136(3):288–299. https://doi.org/10.1111/acps.12752

Li L, Ji E, Han X, Tang F, Bai Y, Peng D, Fang Y, Zhang S, Zhang Z, Yang H (2019) Cortical thickness and subcortical volumes alterations in euthymic bipolar I patients treated with different mood stabilizers. Brain Imaging Behav 13(5):1255–1264. https://doi.org/10.1007/s11682-018-9950-9

Librenza-Garcia D, Kotzian BJ, Yang J, Mwangi B, Cao B, Pereira Lima LN, Bermudez MB, Boeira MV, Kapczinski F, Passos IC (2017) The impact of machine learning techniques in the study of bipolar disorder: a systematic review. Neurosci Biobehav Rev 80:538–554. https://doi.org/10.1016/j.neubiorev.2017.07.004

Liu JX, Chen YS, Hsieh JC, Su TP, Yeh TC, Chen LF (2010) Differences in white matter abnormalities between bipolar I and II disorders. J Affect Disord 127(1–3):309–315. https://doi.org/10.1016/j.jad.2010.05.026

Liu CH, Ma X, Li F, Wang YJ, Tie CL, Li SF, Chen TL, Fan TT, Zhang Y, Dong J, Yao L, Wu X, Wang CY (2012) Regional homogeneity within the default mode network in bipolar depression: a resting-state functional magnetic resonance imaging study. PLoS One 7(11):e48181. https://doi.org/10.1371/journal.pone.0048181

Lovas DA, Schuman-Olivier Z (2018) Mindfulness-based cognitive therapy for bipolar disorder: a systematic review. J Affect Disord 240:247–261. https://doi.org/10.1016/j.jad.2018.06.017

Ma Q, Tang Y, Wang F, Liao X, Jiang X, Wei S, Mechelli A, He Y, Xia M (2019) Transdiagnostic dysfunctions in brain modules across patients with schizophrenia, bipolar disorder, and major depressive disorder: a connectome-based study. Schizophr Bull 46(3):699–712. https://doi.org/10.1093/schbul/sbz111

Mah L, Zarate CA Jr, Singh J, Duan YF, Luckenbaugh DA, Manji HK, Drevets WC (2007) Regional cerebral glucose metabolic abnormalities in bipolar II depression. Biol Psychiatry 61 (6):765–775. https://doi.org/10.1016/j.biopsych.2006.06.009

Mahapatra A, Khandelwal SK, Sharan P, Garg A, Mishra NK (2017) Diffusion tensor imaging tractography study in bipolar disorder patients compared to first-degree relatives and healthy controls. Psychiatry Clin Neurosci 71(10):706–715. https://doi.org/10.1111/pcn.12530

Mak LE, Minuzzi L, MacQueen G, Hall G, Kennedy SH, Milev R (2017) The default mode network in healthy individuals: a systematic review and meta-analysis. Brain Connect 7 (1):25–33. https://doi.org/10.1089/brain.2016.0438

Marotta G, Delvecchio G, Pigoni A, Mandolini G, Ciappolino V, Oldani L, Madonna D, Grottaroli M, Altamura AC, Brambilla P (2019) The metabolic basis of psychosis in bipolar disorder: a positron emission tomography study. Bipolar Disord 21(2):151–158. https://doi.org/10.1111/bdi.12710

Matsuo K, Harada K, Fujita Y, Okamoto Y, Ota M, Narita H, Mwangi B, Gutierrez CA, Okada G, Takamura M, Yamagata H, Kusumi I, Kunugi H, Inoue T, Soares JC, Yamawaki S, Watanabe Y (2019) Distinctive neuroanatomical substrates for depression in bipolar disorder versus major depressive disorder. Cereb Cortex 29(1):202–214. https://doi.org/10.1093/cercor/bhx319

McGrath BM, Wessels PH, Bell EC, Ulrich M, Silverstone PH (2004) Neurobiological findings in bipolar II disorder compared with findings in bipolar I disorder. Can J Psychiatry 49 (12):794–801. https://doi.org/10.1177/070674370404901202

Mwangi B, Wu MJ, Bauer IE, Modi H, Zeni CP, Zunta-Soares GB, Hasan KM, Soares JC (2015) Predictive classification of pediatric bipolar disorder using atlas-based diffusion weighted imaging and support vector machines. Psychiatry Res 234(2):265–271. https://doi.org/10.1016/j.pscychresns.2015.10.002

Najt P, Wang F, Spencer L, Johnston JA, Cox Lippard ET, Pittman BP, Lacadie C, Staib LH, Papademetris X, Blumberg HP (2016) Anterior cortical development during adolescence in bipolar disorder. Biol Psychiatry 79(4):303–310. https://doi.org/10.1016/j.biopsych.2015.03.026

National Institute of Mental Health (2008) Research domain criteria (RDoC). http://www.nimh.nih.gov/research-priorities/rdoc/index.shtml. Accessed 26 Oct 2013

Nugent AC, Diazgranados N, Carlson PJ, Ibrahim L, Luckenbaugh DA, Brutsche N, Herscovitch P, Drevets WC, Zarate CA Jr (2014) Neural correlates of rapid antidepressant response to ketamine in bipolar disorder. Bipolar Disord 16(2):119–128. https://doi.org/10.1111/bdi.12118

Nunes A, Schnack HG, Ching CRK, Agartz I, Akudjedu TN, Alda M, Alnaes D, Alonso-Lana S, Bauer J, Baune BT, Boen E, Bonnin CDM, Busatto GF, Canales-Rodriguez EJ, Cannon DM, Caseras X, Chaim-Avancini TM, Dannlowski U, Diaz-Zuluaga AM, Dietsche B, Doan NT,

Duchesnay E, Elvsashagen T, Emden D, Eyler LT, Fatjo-Vilas M, Favre P, Foley SF, Fullerton JM, Glahn DC, Goikolea JM, Grotegerd D, Hahn T, Henry C, Hibar DP, Houenou J, Howells FM, Jahanshad N, Kaufmann T, Kenney J, Kircher TTJ, Krug A, Lagerberg TV, Lenroot RK, Lopez-Jaramillo C, Machado-Vieira R, Malt UF, McDonald C, Mitchell PB, Mwangi B, Nabulsi L, Opel N, Overs BJ, Pineda-Zapata JA, Pomarol-Clotet E, Redlich R, Roberts G, Rosa PG, Salvador R, Satterthwaite TD, Soares JC, Stein DJ, Temmingh HS, Trappenberg T, Uhlmann A, van Haren NEM, Vieta E, Westlye LT, Wolf DH, Yuksel D, Zanetti MV, Andreassen OA, Thompson PM, Hajek T, Group EBDW (2018) Using structural MRI to identify bipolar disorders – 13 site machine learning study in 3020 individuals from the ENIGMA Bipolar Disorders Working Group. Mol Psychiatry. https://doi.org/10.1038/s41380-018-0228-9

Parikh SV, Quilty LC, Ravitz P, Rosenbluth M, Pavlova B, Grigoriadis S, Velyvis V, Kennedy SH, Lam RW, MacQueen GM, Milev RV, Ravindran AV, Uher R, Group CDW (2016) Canadian network for mood and anxiety treatments (CANMAT) 2016 clinical guidelines for the management of adults with major depressive disorder: section 2. Psychological treatments. Can J Psychiatry 61(9):524–539. https://doi.org/10.1177/0706743716659418

Passos IC, Ballester PL, Barros RC, Librenza-Garcia D, Mwangi B, Birmaher B, Brietzke E, Hajek T, Lopez Jaramillo C, Mansur RB, Alda M, Haarman BCM, Isometsa E, Lam RW, McIntyre RS, Minuzzi L, Kessing LV, Yatham LN, Duffy A, Kapczinski F (2019) Machine learning and big data analytics in bipolar disorder: a position paper from the International Society for Bipolar Disorders Big Data Task Force. Bipolar Disord 21(7):582–594. https://doi.org/10.1111/bdi.12828

Phillips ML, Swartz HA (2014) A critical appraisal of neuroimaging studies of bipolar disorder: toward a new conceptualization of underlying neural circuitry and a road map for future research. Am J Psychiatry 171(8):829–843. https://doi.org/10.1176/appi.ajp.2014.13081008

Reisberg B (1988) Functional assessment staging (FAST). Psychopharmacol Bull 24(4):653–659

Rubin-Falcone H, Zanderigo F, Thapa-Chhetry B, Lan M, Miller JM, Sublette ME, Oquendo MA, Hellerstein DJ, McGrath PJ, Stewart JW, Mann JJ (2018) Pattern recognition of magnetic resonance imaging-based gray matter volume measurements classifies bipolar disorder and major depressive disorder. J Affect Disord 227:498–505. https://doi.org/10.1016/j.jad.2017.11.043

Sanches M, Roberts RL, Sassi RB, Axelson D, Nicoletti M, Brambilla P, Hatch JP, Keshavan MS, Ryan ND, Birmaher B, Soares JC (2005) Developmental abnormalities in striatum in young bipolar patients: a preliminary study. Bipolar Disord 7(2):153–158. https://doi.org/10.1111/j.1399-5618.2004.00178.x

Sanches M, Amorim E, Mwangi B, Zunta-Soares GB, Soares JC (2019) Smaller left anterior cingulate cortex in non-bipolar relatives of patients with bipolar disorder. Braz J Psychiatry 41(3):254–256. https://doi.org/10.1590/1516-4446-2018-0051

Sani G, Simonetti A, Janiri D, Banaj N, Ambrosi E, De Rossi P, Ciullo V, Arciniegas DB, Piras F, Spalletta G (2018) Association between duration of lithium exposure and hippocampus/amygdala volumes in type I bipolar disorder. J Affect Disord 232:341–348. https://doi.org/10.1016/j.jad.2018.02.042

Sartori JM, Reckziegel R, Passos IC, Czepielewski LS, Fijtman A, Sodre LA, Massuda R, Goi PD, Vianna-Sulzbach M, Cardoso TA, Kapczinski F, Mwangi B, Gama CS (2018) Volumetric brain magnetic resonance imaging predicts functioning in bipolar disorder: a machine learning approach. J Psychiatr Res 103:237–243. https://doi.org/10.1016/j.jpsychires.2018.05.023

Scott J, Hidalgo-Mazzei D, Strawbridge R, Young A, Resche-Rigon M, Etain B, Andreassen OA, Bauer M, Bennabi D, Blamire AM, Boumezbeur F, Brambilla P, Cattane N, Cattaneo A, Chupin M, Coello K, Cointepas Y, Colom F, Cousins DA, Dubertret C, Duchesnay E, Ferro A, Garcia-Estela A, Goikolea J, Grigis A, Haffen E, Hoegh MC, Jakobsen P, Kalman JL, Kessing LV, Klohn-Saghatolislam F, Lagerberg TV, Landen M, Lewitzka U, Lutticke A, Mazer N, Mazzelli M, Mora C, Muller T, Mur-Mila E, Oedegaard KJ, Oltedal L, Palsson E, Papadopoulos Orfanos D, Papiol S, Perez-Sola V, Reif A, Ritter P, Rossi R, Schulze T,

Senner F, Smith FE, Squarcina L, Steen NE, Thelwall PE, Varo C, Vieta E, Vinberg M, Wessa M, Westlye LT, Bellivier F (2019) Prospective cohort study of early biosignatures of response to lithium in bipolar-I-disorders: overview of the H2020-funded R-LiNK initiative. Int J Bipolar Disord 7(1):20. https://doi.org/10.1186/s40345-019-0156-x

Sexton CE, Allan CL, Mackay CE, Ebmeier KP (2012) White matter integrity within the corpus callosum differentiates late-life bipolar and unipolar depression. Bipolar Disord 14(7):790–791. https://doi.org/10.1111/j.1399-5618.2012.01050.x

Shorter E (1997) A history of psychiatry: from the era of the asylum to the age of Prozac. Wiley, Lexington

Singh MK, Chang KD, Kelley RG, Saggar M, Reiss AL, Gotlib IH (2014) Early signs of anomalous neural functional connectivity in healthy offspring of parents with bipolar disorder. Bipolar Disord 16(7):678–689. https://doi.org/10.1111/bdi.12221

Skudlarski P, Schretlen DJ, Thaker GK, Stevens MC, Keshavan MS, Sweeney JA, Tamminga CA, Clementz BA, O'Neil K, Pearlson GD (2013) Diffusion tensor imaging white matter endophenotypes in patients with schizophrenia or psychotic bipolar disorder and their relatives. Am J Psychiatry 170(8):886–898. https://doi.org/10.1176/appi.ajp.2013.12111448

Soares JC, Mann JJ (1997) The anatomy of mood disorders--review of structural neuroimaging studies. Biol Psychiatry 41(1):86–106. https://doi.org/10.1016/s0006-3223(96)00006-6

Spielberg JM, Matyi MA, Karne H, Anand A (2019) Lithium monotherapy associated longitudinal effects on resting state brain networks in clinical treatment of bipolar disorder. Bipolar Disord 21 (4):361–371. https://doi.org/10.1111/bdi.12718

Steinberg EE, Christoffel DJ, Deisseroth K, Malenka RC (2015) Illuminating circuitry relevant to psychiatric disorders with optogenetics. Curr Opin Neurobiol 30:9–16. https://doi.org/10.1016/j.conb.2014.08.004

Sugranyes G, Sole-Padulles C, de la Serna E, Borras R, Romero S, Sanchez-Gistau V, Garcia-Rizo C, Goikolea JM, Bargallo N, Moreno D, Baeza I, Castro-Fornieles J (2017) Cortical morphology characteristics of young offspring of patients with schizophrenia or bipolar disorder. J Am Acad Child Adolesc Psychiatry 56(1):79–88. https://doi.org/10.1016/j.jaac.2016.09.516

Teasdale JD, Segal ZV, Williams JM, Ridgeway VA, Soulsby JM, Lau MA (2000) Prevention of relapse/recurrence in major depression by mindfulness-based cognitive therapy. J Consult Clin Psychol 68(4):615–623. https://doi.org/10.1037//0022-006x.68.4.615

Tonnesen S, Kaufmann T, Doan NT, Alnaes D, Cordova-Palomera A, Meer DV, Rokicki J, Moberget T, Gurholt TP, Haukvik UK, Ueland T, Lagerberg TV, Agartz I, Andreassen OA, Westlye LT (2018) White matter aberrations and age-related trajectories in patients with schizophrenia and bipolar disorder revealed by diffusion tensor imaging. Sci Rep 8(1):14129. https://doi.org/10.1038/s41598-018-32355-9

van den Heuvel MP, Hulshoff Pol HE (2010) Exploring the brain network: a review on resting-state fMRI functional connectivity. Eur Neuropsychopharmacol 20(8):519–534. https://doi.org/10.1016/j.euroneuro.2010.03.008

Vargas C, Lopez-Jaramillo C, Vieta E (2013) A systematic literature review of resting state network--functional MRI in bipolar disorder. J Affect Disord 150(3):727–735. https://doi.org/10.1016/j.jad.2013.05.083

Vizueta N, Rudie JD, Townsend JD, Torrisi S, Moody TD, Bookheimer SY, Altshuler LL (2012) Regional fMRI hypoactivation and altered functional connectivity during emotion processing in nonmedicated depressed patients with bipolar II disorder. Am J Psychiatry 169(8):831–840. https://doi.org/10.1176/appi.ajp.2012.11030349

Weathers J, Lippard ETC, Spencer L, Pittman B, Wang F, Blumberg HP (2018) Longitudinal diffusion tensor imaging study of adolescents and young adults with bipolar disorder. J Am Acad Child Adolesc Psychiatry 57(2):111–117. https://doi.org/10.1016/j.jaac.2017.11.014

Wu MJ, Mwangi B, Bauer IE, Passos IC, Sanches M, Zunta-Soares GB, Meyer TD, Hasan KM, Soares JC (2017) Identification and individualized prediction of clinical phenotypes in bipolar disorders using neurocognitive data, neuroimaging scans and machine learning. NeuroImage 145(Pt B):254–264. https://doi.org/10.1016/j.neuroimage.2016.02.016

Yatham LN, Liddle PF, Lam RW, Adam MJ, Solomons K, Chinnapalli M, Ruth TJ (2005) A positron emission tomography study of the effects of treatment with valproate on brain 5-HT2A receptors in acute mania. Bipolar Disord 7(Suppl 5):53–57. https://doi.org/10.1111/j.1399-5618.2005.00252.x

Zak N, Boen E, Boye B, Andreassen OA, Doan NT, Malt UF, Westlye LT, Elvsashagen T (2019) Mood episodes are associated with increased cortical thinning: a longitudinal study of bipolar disorder type II. Bipolar Disord 21(6):525–538. https://doi.org/10.1111/bdi.12771

Zhang W, Xiao Y, Sun H, Patino LR, Tallman MJ, Weber WA, Adler CM, Klein C, Strawn JR, Nery FG, Gong Q, Sweeney JA, Lui S, DelBello MP (2018) Discrete patterns of cortical thickness in youth with bipolar disorder differentially predict treatment response to quetiapine but not lithium. Neuropsychopharmacology 43(11):2256–2263. https://doi.org/10.1038/s41386-018-0120-y

Structural and Functional Brain Correlates of Neuroprogression in Bipolar Disorder

Diego Librenza-Garcia, Jee Su Suh, Devon Patrick Watts, Pedro Lemos Ballester, Luciano Minuzzi, Flavio Kapczinski, and Benicio N. Frey

Contents

Abstract Neuroprogression is associated with structural and functional brain changes that occur in parallel with cognitive and functioning impairments. There is substantial evidence showing early white matter changes, as well as trajectory-related gray matter alterations. Several structures, including prefrontal, parietal, temporal cortex, and limbic structures, seem to be altered over the course of bipolar

D. Librenza-Garcia
Mood Disorders Program, St. Joseph's Healthcare Hamilton, Hamilton, ON, Canada

J. S. Suh, D. P. Watts, and P. L. Ballester
Neuroscience Graduate Program, McMaster University, Hamilton, ON, Canada

L. Minuzzi and B. N. Frey (✉)
Neuroscience Graduate Program, McMaster University, Hamilton, ON, Canada

Department of Psychiatry and Behavioural Neurosciences, McMaster University, Hamilton, ON, Canada

Mood Disorders Program and Women's Health Concerns Clinic, St. Joseph's Healthcare Hamilton, Hamilton, ON, Canada
e-mail: freybn@mcmaster.ca

F. Kapczinski
Mood Disorders Program, St. Joseph's Healthcare Hamilton, Hamilton, ON, Canada

Neuroscience Graduate Program, McMaster University, Hamilton, ON, Canada

Department of Psychiatry and Behavioural Neurosciences, McMaster University, Hamilton, ON, Canada

Curr Topics Behav Neurosci (2021) 48: 197–214
https://doi.org/10.1007/7854_2020_177
Published Online: 11 October 2020

disorder, especially associated with the number of episodes and length of the disease. An important limitation is that most of the studies used either a cross-sectional design or a short follow-up period, which may be insufficient to identify all neuroprogressive changes over time. In addition, the heterogeneity of patients with bipolar disorder is another challenge to determine which subjects will have a more pernicious trajectory. Larger studies and the use of new techniques, such as machine learning, may help to enable more discoveries and evidence on the role of neuroprogression in BD.

Keywords Bipolar disorder · Functional MRI · Neuroprogression · Structural MRI

1 Introduction

Bipolar disorder (BD) has heterogeneous trajectories, with some patients experiencing a more severe course, marked by progressive features and impairments in both cognition and functioning. The term neuroprogression has been proposed as a pathological rewiring of the brain that occurs in parallel with clinical, neurocognitive, and functioning deterioration in BD patients (Passos et al. 2016). In these cases, there is cumulative evidence that several structures and systems seem to be affected. Neuroprogression has been associated with poor clinical outcomes such as worse response to treatment (Ketter et al. 2006; Swann et al. 1999), increased comorbidity (Matza et al. 2005), functional and cognitive impairments (Kessing 2004; Rosa et al. 2012; Torres et al. 2007), and higher suicide and hospitalization risk (Goldberg and Ernst 2002; Hawton et al. 2005). Refractory trajectories, which include difficulty to treat acute episodes, persistence of symptoms between episodes, or the emergence of mood episodes or cycling during optimal treatment, have been associated with an end-stage presentation of BD (Berk et al. 2007, 2014; Gitlin 2001; Poon et al. 2012). Among the factors contributing to these changes, the number of episodes seems to be the most robust aspect associated with clinical deterioration (Martino et al. 2016).

Although not all patients present a progressive course, it is paramount to understand factors that might be associated with neuroprogression. Among the distinct pathways that the clinical course of BD can assume, identifying those patients who will experience clinical and neurocognitive deterioration is a critical need. In addition, a better understanding of these processes will enable effective and targeted interventions with the aim of improving outcomes in patients with BD. Neuroimaging studies aim to shed light on some of the brain changes occurring over time, the factors associated with those changes, and its consequences for those suffering from BD.

In the current chapter, we will review findings from structural and functional neuroimaging studies that aimed to study brain changes in patients with BD associated with neuroprogression.

2 Structural Aspects of Neuroprogression in Neuroimaging

Changes in brain structures in bipolar disorder patients, when compared to controls, were shown by several cross-sectional and longitudinal studies, with patterns compatible with neuronal loss in cortical and subcortical tissues (Hibar et al. 2016). Most of the evidence comes from cross-sectional studies, which may limit definite conclusions about the directionality of the findings.

There is evidence pointing to a white matter pathology in earlier stages and gray matter loss in more advanced stages (López-Larson et al. 2002). A meta-analysis of eight studies assessing magnetic resonance imaging (MRI) in patients with first-episode BD found a reduction in total intracranial and white matter volumes, with no significant changes being observed for gray matter and whole brain volumes (Vita et al. 2009). Cortical thinning has been described in several brain regions, including the cingulate cortex, prefrontal cortex, fusiform cortex, and limbic structures, among others, with some studies finding an inverse correlation of duration of illness with cortical thickness of some regions, such as the left middle frontal cortex (Lyoo et al. 2010). These earlier changes in white matter may indicate that abnormal connectivity between regions may play a significant role in the onset of disorders, with more structural changes, predominantly in the gray matter, presenting over its clinical course and trajectory.

2.1 Evidence from Cross-Sectional Studies

2.1.1 Brain Volume and General Findings

A study comparing multiple-episode and first-episode BD found that the lateral ventricles were larger in those individuals with multiple episodes, which was associated with a higher number of manic episodes. Additionally, the authors also reported a smaller total cerebral volume (Strakowski et al. 2002). A voxel-based morphometry meta-analysis including 660 BD patients compared to 770 healthy controls showed gray matter reduction that was more pronounced in chronic illness, in the basal ganglia, subgenual anterior cingulate cortex (ACC), and amygdala, with lithium promoting an enlargement of these areas (Bora et al. 2010). Differences in the frontal, temporal and parietal cortex, amygdala, thalamus, globus pallidus, striatum, and hippocampus, among others, were also described comparing patients with healthy subjects (Hibar et al. 2018; Strakowski et al. 1999). An inverse correlation was also found between length of illness and gray matter volumes in

BD, but not in unipolar depression and healthy controls (Frey et al. 2008). There is also evidence pointing to a lower gyrification index (GI) in patients with late-stage BD and a correlation between the GI and the stage of the disorder (Cao et al. 2016). A study comparing 63 patients with first manic episode and healthy controls found no significant volumetric differences between the two groups (Arumugham et al. 2017), which reinforces the hypothesis of gray matter pathology being associated with the trajectory of the disorder. Longer duration of illness was associated with lower total gray matter volumes even when controlling for confounding factors, suggesting a cumulative loss over time (Gildengers et al. 2014).

2.1.2 Prefrontal Cerebral Cortex

Bipolar patients hospitalized for a manic episode showed smaller bilateral prefrontal gray matter volumes that were more noticeable in the middle and superior subregions when compared to healthy controls, with no significant differences in white matter content (López-Larson et al. 2002). A decreased gray matter density in fronto-limbic areas has been associated with poor clinical outcomes (Doris et al. 2004). Extensive prefrontal cortex thinning was also found by a meta-analysis of whole-brain voxel-based morphometry studies including 1720 subjects with BD, suggesting that this region plays a major role in the pathophysiology of BD (Lu et al. 2019). A small study ($n = 12$) with BD type I patients found a significant reduction of gray matter volume (GMV) in the right subgenual prefrontal cortex, with positive family history and sex being associated with the GMV (Sharma et al. 2003). A reduction of the subgenual prefrontal cortex was also observed in pediatric bipolar disorder patients with a history of first-degree relatives with a mood disorder (Baloch et al. 2010). A study with BD type II patients also showed significant thinning in the PFC, including the left perigenual ventromedial cortex, bilateral dorsomedial PFC, and bilateral dorsolateral PFC. However, no association was found with medication, mood state, illness duration, or family history (Elvsåshagen et al. 2013). The prefrontal gyrification index was reduced in ventral and dorsal regions of the PFC of BD patients when compared to controls, but not when compared to schizophrenia patients.

Moreover, those reductions were associated with cognitive impairments and reduced IQ (McIntosh et al. 2009). A study with patients experiencing a rapid-cycling course found reductions of the GMV in the bilateral medial orbital prefrontal cortex, ventromedial prefrontal cortex, and limbic regions when compared to controls and in the ventromedial prefrontal cortex when compared to BD patients without rapid cycling (Narita et al. 2011). Finally, a vertex-wise whole-brain analysis found reduced cortical thickness associated with executive function in BD type II patients in lateral prefrontal and occipital regions (Abé et al. 2018).

2.1.3 Cingulate Cortex

A decreased left ACC volume was found in untreated BD patients when compared to healthy controls and lithium-treated patients, while there was no difference between healthy controls and lithium-treated subjects (Sassi et al. 2004). Another study also found a decrease in the left ACC volume, with an inverse correlation of the ACC volume with the number of hospitalizations (Delvecchio et al. 2019). When comparing patients with a more severe presentation and poorer outcomes, another study found an abnormal gray matter density widespread on the cingulate cortex (Doris et al. 2004). In younger patients, a study reported smaller mean volumes in the left anterior, left posterior, and right posterior cingulate cortex when compared to controls (Kaur et al. 2005). Nevertheless, several studies, including a meta-analysis, pointed to an enlargement of the cingulate cortex when compared to controls, suggesting a possible compensatory mechanism (Lu et al. 2019).

2.1.4 Temporal-Limbic Structures

Temporal lobe clusters gray matter loss was also reported and significantly associated with reductions in full-scale IQ and performance IQ (Moorhead et al. 2007). A study with BD type II patients found significant thinning of the superior middle and inferior temporal gyrus but without association with illness duration, family history, medication, or mood state (Elvsåshagen et al. 2013). Cortical gray matter was found to be reduced in BD patients in the temporal study by another large study, finding associations with medication use (Hibar et al. 2018). BD patients treated with antipsychotics were found to have larger bilateral temporal lobe whiter matter volumes when compared to those not taking antipsychotics and healthy controls. Hyper perfusion of left frontal and temporal cerebral areas was found in BD patients, suggesting an over-activation of these regions (Agarwal et al. 2008).

A study stratifying patients in early, intermediate, and late stage as a function of the number of lifetime episodes found smaller hippocampus in late-stage patients when compared to healthy controls, as well as worse performances in the California Verbal Learning Test (CVLT) for intermediate- and late-stage patients (Cao et al. 2016). Anterior limbic regions also had the most robust decrease in gray matter in a meta-analysis comparing BD and HC, with a more robust decrease being associated with a longer duration of illness (Bora et al. 2010). Hippocampal volumes were decreased in a study with adolescents with BD, with negative correlations being found between the duration of the disorder and positive correlations between duration of medication use and levels of neurotrophins, such as NGF and BDNF (Inal-Emiroglu et al. 2015). In older adults with BD (mean age 57 years), a study showed smaller hippocampal and right amygdala volumes after controlling for intracranial volume, with the hippocampus but not the amygdala volume being negatively associated with the duration of depressive and manic episodes (Wijeratne et al. 2013). However, a study with males with BD, schizophrenia, and healthy controls found larger amygdala volumes of BD patients when compared to the two other groups (Altshuler et al. 2000), and a study of BD patients recently remitted from a manic episode found no difference in the amygdala volume when comparing to healthy controls (Arumugham et al. 2017). Another study, with euthymic patients with BD type I patients older than 50 years, found that the right and left hippocampal volume was negatively associated with inflammatory markers, including the tumor necrosis factor receptor-1 (sTNF-R1) and the soluble interleukin (IL)-2 receptor (sIL-2r) (Tsai et al. 2019). There is cumulative evidence that persistent inflammation plays a major role in bipolar disorder and neuroprogression (Kapczinski et al. 2008).

2.1.5 Other Brain Structures

A study comparing young individuals with bipolar disorder and healthy controls found a trend for smaller vermis V2 areas of the cerebellum in patients that was inversely correlated with the number of mood episodes in the male patient group (Monkul et al. 2008). Other studies also showed smaller cerebellum volumes for multiple-episode BD. Mills et al. compared first-episode and multiple-episode bipolar patients, and healthy controls, and found smaller vermis subregions V2 and V3 in patients with multiple episodes (Mills et al. 2005), while the V3 area was also found to be significantly smaller in multiple-episode patients with prior manic episodes and hospitalizations when compared to first-episode patients (DelBello 1999). Cerebellar loss in gray matter was also reported and associated with the number of manic and depressive episodes (Moorhead et al. 2007).

Altered myelination was also suggested in the development and course of BD. Lower signal intensity for the corpus callosum in all callosal subregions was reported for BD patients, although no effects of length of illness were found (Brambilla et al. 2004), and for first-episode BD with higher YMRS scores (Atmaca et al. 2007). The corpus callosum and total white matter volumes were also reduced in a study including patients in early and late stages, although gray matter was only altered in late stage, suggesting an early role for the demyelination process in BD (Duarte et al. 2018). Another study found a reduced posterior corpus callosum volume in late-stage women with BD type I when compared to early-stage type I BD and healthy controls, after controlling for confounding factors (Lavagnino et al. 2015). A PET-SCAN study found decreased cortical volume extending from the prefrontal cortex ventral to the genu of the corpus callosum, a region associated with the medication of emotional and autonomic responses to social stimuli and modulations of neurotransmitter system implicated in mood disorder treatment (Drevets et al. 1997). A DTI study also showed aberrations in several white matter structures in young patients, including the corpus callosum, in fibers connecting the fornix to the thalamus, and bilateral parietal and occipital corona radiata, in adolescents with BD (Barnea-Goraly et al. 2009). These findings are consistent with the hypothesis that white matter changes may occur early at the onset of the disease. In contrast, gray matter changes are more associated with the course of the disorder and, therefore, implicated in the clinical aspects of neuroprogression and deterioration.

Intracortical myelin (ICM) has recently been studied in the context of BD, with evidence for an association of verbal memory function with ICM and age-related deficits in adults with BD type I, including a correlation with age of onset and an inverse correlation with duration of illness and manic episodes (Sehmbi et al. 2018). Depressive episodes, however, were not associated with ICM signal in the study. This points to a role of ICM in cognitive dysfunction in BD since verbal memory impairment is one of the main findings in individuals with BD (Martínez-Arán et al. 2004) and suggests a role for ICM on neuroprogression.

Shape differences in the caudate heads and putamen were also observed for drug-naïve patients but not for treated subjects when compared to healthy subjects (Hwang et al. 2006). A systematic review including 508 subjects also analyzed hypoperfusion and showing widespread resting hypoperfusion in the cingulate gyrus, frontal, and anterior temporal regions when compared to healthy controls (Toma et al. 2018). A large study with 1837 patients found cortical thinning in bilateral frontal, temporal, and parietal regions, with longer duration of illness being associated with reduced cortical thickness in frontal, medial parietal, and occipital regions. Authors also found an association of several medications, such as lithium, anticonvulsants, and antipsychotics, with cortical thickness and brain surface area (Hibar et al. 2018). Finally, a pioneering study of blood-brain barrier imaging showed that patients with extensive BBB leakage had more severe depression and anxiety and a more chronic course of illness (Kamintsky et al. 2020).

2.2 *Evidence from Longitudinal Studies*

2.2.1 Brain Volume and General Findings

A study comparing remitted first manic episode patients, patients with recurrence of the episode, and healthy subjects, showed that the group with recurrence had a greater GMV loss when compared to those in remission, including left frontal and bilateral temporal regions (Kozicky et al. 2016). Loss of gray matter over time was also observed in longitudinal studies, with follow-ups up to 4 years that have been associated with deterioration in cognition and more pernicious illness courses (Moorhead et al. 2007).

2.2.2 Prefrontal Cerebral Cortex

Gray matter volume seems to be reduced at baseline in the frontal gyrus and right superior frontal gyrus in BD when compared to controls, but no reductions were found after a 2-year follow-up (Farrow et al. 2005). In adolescents diagnosed with BD and followed for 2 years, there seems to be a progressive bilateral reduction of GMV in both ventral and rostral portions of the PFC when compared to controls (Kalmar et al. 2009). Decreases of the GMV after 3–34 months of the initial scan were seen in the superior frontal gyrus in adults diagnosed with BD when compared to healthy controls (Lisy et al. 2011). Gene variants of the brain-derived neurotrophic factor (BDNF) such as the valine methionine were shown to be associated with a greater reduction in the gyrification index of the PFC that is more pronounced in the left hemisphere (Mirakhur et al. 2009). The role of BDNF as a neuroprotective factor with lower levels being associated with inflammation and neuroprogression is supported in the literature (Kapczinski et al. 2008).

Nevertheless, a study following first-episode psychosis found no difference in the PFC between BD patients and controls (Arango 2012). Some studies reported a progressive increase of GMV in the ventrolateral PFC in young BD patients (Gogtay et al. 2007; Lisy et al. 2011). Medial or orbital PFC seems to be decreased in patients with psychotic illnesses, including BD (Pantelis et al. 2003), although in young patients, there is evidence of increase in the GMV of the medial PFC in the left hemisphere (Gogtay et al. 2007). A study that followed bipolar disorder patients for 6 years found that those with a higher number of manic episodes had a significant decreased in frontal cortical volume in the dorsolateral and prefrontal inferior frontal cortex (Abé et al. 2015). The short period of follow-up of most studies may be an explanation for these contradictory results, since a longer follow-up may be needed to detect changes associated with neuroprogression.

2.2.3 Cingulate Cortex

A study of first-episode psychosis, including BD patients, found a gray matter deficit in the anterior cingulate cortex (Farrow et al. 2005). A reduced GMV in the ACC was also observed in young adults that developed psychotic illnesses (Pantelis et al. 2003), and bilateral GMV loss was shown in the ACC and the left posterior cingulate gyrus in young BD patients (Gogtay et al. 2007), and progressive losses of GMV when comparing adolescent and adult patients (Kalmar et al. 2009; Lyoo et al. 2010). These losses were shown to be reversible with lithium therapy (Lyoo et al. 2010; Moore et al. 2009).

2.2.4 Temporal-Limbic Structures

There are conflicting results regarding temporal lobe structures. Increased GMV was observed in medial and superior temporal gyri at 3–33 months (Lisy et al. 2011) and 4–8 years follow-up (Gogtay et al. 2007), while reduction of GMV in the temporal lobe was associated with reduced full-scale IQ and mainly verbal IQ (Moorhead et al. 2007). Reduced GMV was also observed in the left fusiform gyrus in high-risk subjects that went on to develop psychotic episodes, including BD patients (Pantelis et al. 2003).

The amygdala seems to be structurally altered in earlier phases of BD: a study showed reduced GMV in adolescents in a 2-year follow-up (Blumberg et al. 2005), with another study showing no difference over time in adult BD (Moorhead et al.

2007). In older patients, a 2-year follow-up also showed no difference between amygdala volume of BD and healthy control subjects (Delaloye et al. 2011). Changes found in the hippocampus include reductions of GMV in the left parahippocampal gyrus after the onset of psychosis (Pantelis et al. 2003); progressive loss of GMV in the left hippocampus and fusiform gyrus (Moorhead et al. 2007); increased GMV in the hippocampus of BD patients (Delaloye et al. 2011; Lisy et al. 2011); and an increased volume that was associated with lithium treatment and improved verbal memory (Yucel et al. 2007).

3 Functional Aspects of Neuroprogression in Neuroimaging

Currently, the knowledge of the functional changes of the brain as a consequence of neuroprogression in bipolar disorder remains limited. A major thrust of the work thus far has focused on biochemical and structural correlates, while fewer studies have examined the functional consequences of brain aging. Nonetheless, there have been some attempts to address this issue. For instance, a review comprising 12 longitudinal studies on functional neuroimaging found evidence for increased positive coupling with the insula and negative coupling with prefrontal regions as patients progressed from manic to depressed state. In brief, the insula is a region that separates the frontal and parietal lobes from the temporal lobe and plays a role in processing a number of basic emotions (Lim et al. 2013).

Moreover, in a motor task designed to assess long-term neural activity in frontal subcortical regions, strong ACC activation was found in those who were in remission from depression, suggesting that decreases in the activation of this region may be related to vulnerability to depression in BD. However, more than half of the studies within this review were gathered from pediatric samples, which limit the applicability of such findings in the context of abnormal brain aging. Apart from this, another study has compiled evidence on differences in the brain's response across distinct functional tasks and patient age groups (Schneider et al. 2012).

Currently, long-term longitudinal functional neuroimaging studies have yet to be conducted in the context of neuroprogression. Such studies will be important to assess the extent and probable mechanisms of brain changes as a result of bipolar disorder progression. Furthermore, it will be beneficial moving forward to control for clinical characteristics long thought to precipitate accelerated brain aging, such as the number of mood episodes, chronicity, severity, and medication use, to name a few.

Nonetheless, one way to extrapolate from existing data is to summarize findings from cross-sectional studies comparing BD cases to HC participants across varying age groups. For instance, a comprehensive meta-analysis from 2014 identified several significant differences between HC and BD in pediatric ($n = 21$) and adult samples ($n = 73$) using the activation likelihood estimation (ALE) technique (Wegbreit et al. 2014). Briefly, ALE is a robust quantitative method for pooling neuroimaging meta-analyses, as described elsewhere (Kirby and Robinson 2017).

Here, youth with BD exhibited significant hyper-activation in the right amygdala, left prefrontal cortex (PFC), and precuneus, as well as hypo-activation in the right pregenual ACC and caudate (Wegbreit et al. 2014). The authors indicated that this suggests potentially unique brain alterations between youth and adults with bipolar disorder.

Conversely, adults with BD generally exhibited hyper-activation in left inferior frontal gyrus (IFG), left pgACC, and right pallidus and significant hypo-activation in bilateral putamen, bilateral IFG, right lingual gyrus, and inferior parietal lobe compared to HC. Contradictory results such as both hyper- and hypo-activation in a region are a result of ALE methods, which measure convergence across studies. They indicate that a significant number of studies supported both directions of activation within a region. One salient detail from these findings is that BD youth exhibited hypo-activation in the pgACC while adults exhibited the opposite direction of activation compared to HC, a pattern that is directly converse to what is observed in HC (Weathers et al. 2012). This suggests that the development of the primary role of the ACC (error detection and conflict monitoring) is aberrant in childhood, resulting in increased compensation in adulthood reflected by hyper-activation of the region.

However, it is difficult to draw meaningful conclusions on the functional markers of neuroprogression by merely comparing BD and HC cases alone. Instead, the clinical characteristics of the patients should be closely considered, alongside the age range of the sample. For instance, Wegbreit et al. (2014) compared BD adults with BD youth in terms of regional activations for different types of functional tasks. For emotional face processing tasks, BD youth exhibited significant hyper-activation in the right amygdala compared to BD adults, replicating previous findings (Kim et al. 2012) that were excluded from this analysis due to data unavailability. For other emotional paradigms, BD youth exhibited significant hyper-activation in the left IFG and precuneus and significant hypo-activation in the pgACC for non-emotional paradigms compared to BD adults. This last result was retained when all studies were pooled, indicative of a trait deficit in BD youth when compared to both HC youth and BD adults. In a recent study of neural responses to emotional face presentations, BD youth ($n = 24$) exhibited hyper-activation in the PFC and lingual gyrus in comparison with BD adults ($n = 33$) as well as hypoconnectivity in the medial PFC and amygdala-temporoparietal junction (Kryza-Lacombe et al. 2019). Neither of these patterns were seen in comparisons of HC youth vs HC adults. These results seem indicative of a pattern of abnormal functional neurodevelopment in BD and may indicate "scarred" or compensatory mechanisms.

Notably, the aforementioned results do not appear to be explicitly affected by mood state, medications, or select comorbidities (ADHD and ODD). However, given the general limitation of limited sample size in these studies, these assurances have not been well-tested and seem to be provisional, at best.

More recently, new evidence on the disease burden of bipolar disorder and its relationship with functional changes has been measured with fMRI. Comparisons between first-episode and multi-episode bipolar disorder showed lower activation of several regions of the brain for multi-episode bipolar disorder, such as the ventrolateral prefrontal cortex and orbitofrontal cortex (Borgelt et al. 2019). Using spectroscopy, the authors also identified lower concentrations of glutamate and N-acetylaspartate in the anterior cingulate cortex of the multi-episode group. Individuals included in the study experienced one or more manic or mixed episodes, but the duration and severity of them were not included in the analysis. Another limitation of the study is the lack of information on the type of past episodes, as patients with more recurrent manic episodes than depressive ones might present a different progression of the disorder.

Studies on functional neuroprogression in BD are significantly limited in terms of adjusting for certain confounders. Although the above studies contend that their results are not significantly affected by mood state, a review on the topic suggests that functional activations across mood states vary by region (Townsend and Altshuler 2012), i.e., hypo-activation in PFC tends to be trait-dependent while functional changes in the amygdala are more state-dependent. To consider this more rigorously in the context of neuroprogression, functional differences across the life span should be considered within each mood state. Additionally, effects of medication are usually investigated in terms of current use, not lifetime, and are generally found to be non-significant. However, a recent review on the topic suggests this may be due to methodological problems, and when considering the data in aggregate, medication use may affect prefrontal activation in patients with BD (Laidi and Houenou 2016). A recent meta-analysis found that age of onset is associated with more severe depression, latency to treatment, substance use, and comorbid anxiety; however, there were none of the expected associations with greater severity of psychotic symptoms (Joslyn et al. 2016). Regardless, these are all factors that may influence brain activation and should be systematically considered.

The primary recommendation of this section for future studies is the conduction of large, longitudinal investigations of functional changes in BD across the life span. They would ideally follow individuals across significant life transitions (e.g., adolescence to adulthood, adulthood to older age) while taking into consideration mood state, lifetime and current medication use, and differing ages of onset which may correlate with overall illness severity across individuals. Furthermore, there is a growing need for prospective studies using state-of-the-art neuroimaging techniques to aid in our understanding of the neural correlates of neuroprogression.

4 Challenges (Limitations) and Perspectives

One of the major limitations is that most of the data we have available comes from cross-sectional studies, which hinders the establishment of causality. Nevertheless, a number of longitudinal studies also seem to support structural and functional changes to the brain that occur in parallel to distinct and unfavorable clinical trajectories, especially in gray matter. Another major challenge is to identify which patients will develop these changes and what are the distinct endophenotypes associated with neuroprogression. Studies with bipolar offspring, for example, may offer valuable insight about the trajectory of the disorder, including prodromal phases, baseline cognition and functioning, and its changes over time, as well as changes in the clinical presentation or development of treatment resistance. Nevertheless, one could still argue that these cohorts may capture specific subtypes of BD, but not all of them.

The reason for conflicting results in the field may go beyond the discussion of whether neuroprogression occurs. One reason for that may be that we are looking at BD through the lens of categorical diagnosis that is elaborated through symptom phenomenology alone. If several different endophenotypes of the disorder exist, they may be diluting the effect of one another. Moreover, we cannot be sure which of these phenotypes are being included in the studies and in which proportion, so multiple studies that are looking at the same bipolar disorder "label" may be looking at multiple different neurobiological entities that roughly manifest with the same symptomatology.

Group-level results, including patients based merely on the DSM-V criteria, may overlook these distinctions, given the heterogeneous presentation of BD. In this light, approaches that consider these changes at the individual level, such as machine learning, may help to pinpoint what subjects are presenting a neuroprogressive course. Therefore, a closer look at what these cases have in common may enable the discovery of endophenotypes and relevant environmental factors that interact and contribute to more severe trajectories. Predicting which cases may have a neuroprogressive trajectory will enable timely and tailored-made interventions that can prevent cognitive and functioning impairments, reduce the number of episodes, and avoid treatment refractoriness.

Finally, assuming that distinct presentations exist, the greatest challenge in the study of structural and functional neuroprogression changes in the brain of BD patients is the need for a sample large enough to capture a representative amount of patients, followed up by a long period of time and using multimodal assessments. This will enable us to detect progressive trajectories and/or subtypes associated with progression. In addition, assessing multiple neurobiological, cognitive, and functional levels, such as neuroimaging, cognitive testing, and blood biomarkers, may help to pinpoint which events have a greater contribution to neuroprogression in BD, how they interact with the biological blueprint of the patients, and which structural and functional changes follow, as well as its clinical consequences for patients.

References

Abé C, Ekman C-J, Sellgren C, Petrovic P, Ingvar M, Landén M (2015) Manic episodes are related to changes in frontal cortex: a longitudinal neuroimaging study of bipolar disorder 1. Brain 138 (11):3440–3448. https://academic.oup.com/brain/article-lookup/doi/10.1093/brain/awv266

Abé C, Rolstad S, Petrovic P, Ekman C-J, Sparding T, Ingvar M et al (2018) Bipolar disorder type I and II show distinct relationships between cortical thickness and executive function. Acta Psychiatr Scand 138(4):325–335. http://doi.wiley.com/10.1111/acps.12922

Agarwal N, Bellani M, Perlini C, Rambaldelli G, Atzori M, Cerini R et al (2008) Increased fronto-temporal perfusion in bipolar disorder. J Affect Disord 110(1–2):106–114. https://linkinghub.elsevier.com/retrieve/pii/S0165032708000347

Altshuler LL, Bartzokis G, Grieder T, Curran J, Jimenez T, Leight K et al (2000) An MRI study of temporal lobe structures in men with bipolar disorder or schizophrenia. Biol Psychiatry 48 (2):147–162. https://linkinghub.elsevier.com/retrieve/pii/S0006322300008362

Arumugham SS, Torres IJ, Lang DJ, Su W, Lam RW, Honer WG et al (2017) Subcortical structural volumes in recently remitted first episode mania. J Affect Disord 222:23–27. https://linkinghub.elsevier.com/retrieve/pii/S0165032717306390

Atmaca M, Ozdemir H, Yildirim H (2007) Corpus callosum areas in first-episode patients with bipolar disorder. Psychol Med 37(5):699. http://www.journals.cambridge.org/abstract_S0033291706009743

Baloch HA, Hatch JP, Olvera RL, Nicoletti M, Caetano SC, Zunta-Soares GB et al (2010) Morphology of the subgenual prefrontal cortex in pediatric bipolar disorder. J Psychiatr Res 44(15):1106–1110. https://linkinghub.elsevier.com/retrieve/pii/S0022395610001159

Barnea-Goraly N, Chang KD, Karchemskiy A, Howe ME, Reiss AL (2009) Limbic and corpus callosum aberrations in adolescents with bipolar disorder: a tract-based spatial statistics analysis. Biol Psychiatry 66(3):238–244. https://linkinghub.elsevier.com/retrieve/pii/S0006322309002923

Berk M, Conus P, Lucas N, Hallam K, Malhi GS, Dodd S et al (2007) Setting the stage: from prodrome to treatment resistance in bipolar disorder. Bipolar Disord 9(7):671–678. http://doi.wiley.com/10.1111/j.1399-5618.2007.00484.x

Berk M, Berk L, Dodd S, Cotton S, Macneil C, Daglas R et al (2014) Stage managing bipolar disorder. Bipolar Disord 16(5):471–477. http://doi.wiley.com/10.1111/bdi.12099

Blumberg HP, Fredericks C, Wang F, Kalmar JH, Spencer L, Papademetris X et al (2005) Preliminary evidence for persistent abnormalities in amygdala volumes in adolescents and young adults with bipolar disorder. Bipolar Disord 7(6):570–576. http://doi.wiley.com/10.1111/j.1399-5618.2005.00264.x

Bora E, Fornito A, Yücel M, Pantelis C (2010) Voxel-wise meta-analysis of gray matter abnormalities in bipolar disorder. Biol Psychiatry 67(11):1097–1105. https://linkinghub.elsevier.com/retrieve/pii/S0006322310000624

Borgelt L, Strakowski SM, DelBello MP, Weber W, Eliassen JC, Komoroski RA et al (2019) Neurophysiological effects of multiple mood episodes in bipolar disorder. Bipolar Disord 21 (6):503–513. https://onlinelibrary.wiley.com/doi/abs/10.1111/bdi.12782

Brambilla P, Nicoletti M, Sassi RB, Mallinger AG, Frank E, Keshavan MS et al (2004) Corpus callosum signal intensity in patients with bipolar and unipolar disorder. J Neurol Neurosurg Psychiatry 75(2):221–225. http://www.ncbi.nlm.nih.gov/pubmed/14742592

Cao B, Passos IC, Mwangi B, Bauer IE, Zunta-Soares GB, Kapczinski F et al (2016) Hippocampal volume and verbal memory performance in late-stage bipolar disorder. J Psychiatr Res 73:102–107. https://linkinghub.elsevier.com/retrieve/pii/S0022395615300261

Delaloye C, Moy G, de Bilbao F, Weber K, Baudois S, Haller S et al (2011) Longitudinal analysis of cognitive performances and structural brain changes in late-life bipolar disorder. Int J Geriatr Psychiatry 26(12):1309–1318. http://doi.wiley.com/10.1002/gps.2683

DelBello M (1999) MRI analysis of the cerebellum in bipolar disorder a pilot study. Neuropsychopharmacology 21(1):63–68. http://www.nature.com/doifinder/10.1016/S0893-133X(99)00026-3

Delvecchio G, Ciappolino V, Perlini C, Barillari M, Ruggeri M, Altamura AC et al (2019) Cingulate abnormalities in bipolar disorder relate to gender and outcome: a voxel-based morphometry study. Eur Arch Psychiatry Clin Neurosci 269(7):777–784. http://link.springer.com/10.1007/s00406-018-0887-1

Doris A, Belton E, Ebmeier KP, Glabus MF, Marshall I (2004) Reduction of cingulate gray matter density in poor outcome bipolar illness. Psychiatry Res Neuroimaging 130(2):153–159. https://linkinghub.elsevier.com/retrieve/pii/S0925492703001719

Drevets WC, Price JL, Simpson JR, Todd RD, Reich T, Vannier M et al (1997) Subgenual prefrontal cortex abnormalities in mood disorders. Nature 386(6627):824–827. http://www.nature.com/articles/386824a0

Duarte JA, Massuda R, Goi PD, Vianna-Sulzbach M, Colombo R, Kapczinski F et al (2018) White matter volume is decreased in bipolar disorder at early and late stages. Trends Psychiatry Psychother 40(4):277–284. http://www.scielo.br/scielo.php?script=sci_arttext&pid=S2237-60892018000400277&tlng=en

Elvsåshagen T, Westlye LT, Bøen E, Hol PK, Andreassen OA, Boye B et al (2013) Bipolar II disorder is associated with thinning of prefrontal and temporal cortices involved in affect regulation. Bipolar Disord 15(8):855–864. http://doi.wiley.com/10.1111/bdi.12117

Farrow TFD, Whitford TJ, Williams LM, Gomes L, Harris AWF (2005) Diagnosis-related regional gray matter loss over two years in first episode schizophrenia and bipolar disorder. Biol Psychiatry 58(9):713–723. https://linkinghub.elsevier.com/retrieve/pii/S0006322305004944

Frey BN, Zunta-Soares GB, Caetano SC, Nicoletti MA, Hatch JP, Brambilla P et al (2008) Illness duration and total brain gray matter in bipolar disorder: evidence for neurodegeneration? Eur Neuropsychopharmacol 18(10):717–722. https://linkinghub.elsevier.com/retrieve/pii/S0924977X08001132

Gildengers AG, Chung K-H, Huang S-H, Begley A, Aizenstein HJ, Tsai S-Y (2014) Neuroprogressive effects of lifetime illness duration in older adults with bipolar disorder. Bipolar Disord 16(6):617–623. http://doi.wiley.com/10.1111/bdi.12204

Gitlin MJ (2001) Treatment-resistant bipolar disorder. Bull Menn Clin 65(1):26–40. http://guilfordjournals.com/doi/10.1521/bumc.65.1.26.18709

Gogtay N, Ordonez A, Herman DH, Hayashi KM, Greenstein D, Vaituzis C et al (2007) Dynamic mapping of cortical development before and after the onset of pediatric bipolar illness. J Child Psychol Psychiatry 48(9):852–862. http://doi.wiley.com/10.1111/j.1469-7610.2007.01747.x

Goldberg JF, Ernst CL (2002) Features associated with the delayed initiation of mood stabilizers at illness onset in bipolar disorder. J Clin Psychiatry 63(11):985–991. http://article.psychiatrist.com/?ContentType=START&ID=10000451

Hawton K, Sutton L, Haw C, Sinclair J, Harriss L (2005) Suicide and attempted suicide in bipolar disorder. J Clin Psychiatry 66(6):693–704. http://article.psychiatrist.com/?ContentType=START&ID=10001333

Hibar DP, Westlye LT, van Erp TGM, Rasmussen J, Leonardo CD, Faskowitz J et al (2016) Subcortical volumetric abnormalities in bipolar disorder. Mol Psychiatry 21(12):1710–1716. http://www.nature.com/articles/mp2015227

Hibar DP, Westlye LT, Doan NT, Jahanshad N, Cheung JW, Ching CRK et al (2018) Cortical abnormalities in bipolar disorder: an MRI analysis of 6503 individuals from the ENIGMA bipolar disorder working group. Mol Psychiatry 23(4):932–942. http://www.nature.com/articles/mp201773

Hwang J, Lyoo IK, Dager SR, Friedman SD, Oh JS, Lee JY et al (2006) Basal ganglia shape alterations in bipolar disorder. Am J Psychiatry 163(2):276–285. http://psychiatryonline.org/doi/abs/10.1176/appi.ajp.163.2.276

Inal-Emiroglu FN, Resmi H, Karabay N, Guleryuz H, Baykara B, Cevher N et al (2015) Decreased right hippocampal volumes and neuroprogression markers in adolescents with bipolar disorder. Neuropsychobiology 71(3):140–148. https://www.karger.com/Article/FullText/375311

Joslyn C, Hawes DJ, Hunt C, Mitchell PB (2016) Is age of onset associated with severity, prognosis, and clinical features in bipolar disorder? A meta-analytic review. Bipolar Disord 18 (5):389–403. http://doi.wiley.com/10.1111/bdi.12419

Kalmar JH, Wang F, Spencer L, Edmiston E, Lacadie CM, Martin A et al (2009) Preliminary evidence for progressive prefrontal abnormalities in adolescents and young adults with bipolar disorder. J Int Neuropsychol Soc 15(3):476–481. https://www.cambridge.org/core/product/identifier/S1355617709090584/type/journal_article

Kamintsky L, Cairns KA, Veksler R, Bowen C, Beyea SD, Friedman A et al (2020) Blood-brain barrier imaging as a potential biomarker for bipolar disorder progression. NeuroImage Clin 26:102049. https://linkinghub.elsevier.com/retrieve/pii/S2213158219303961

Kapczinski F, Vieta E, Andreazza AC, Frey BN, Gomes FA, Tramontina J et al (2008) Allostatic load in bipolar disorder: implications for pathophysiology and treatment. Neurosci Biobehav Rev 32(4):675–692. https://linkinghub.elsevier.com/retrieve/pii/S0149763407001406

Kaur S, Sassi RB, Axelson D, Nicoletti M, Brambilla P, Monkul ES et al (2005) Cingulate cortex anatomical abnormalities in children and adolescents with bipolar disorder. Am J Psychiatry 162 (9):1637–1643. http://psychiatryonline.org/doi/abs/10.1176/appi.ajp.162.9.1637

Kessing LV (2004) Does the risk of developing dementia increase with the number of episodes in patients with depressive disorder and in patients with bipolar disorder? J Neurol Neurosurg Psychiatry 75(12):1662–1666. http://jnnp.bmj.com/cgi/doi/10.1136/jnnp.2003.031773

Ketter TA, Houston JP, Adams DH, Risser RC, Meyers AL, Williamson DJ et al (2006) Differential efficacy of olanzapine and lithium in preventing manic or mixed recurrence in patients with bipolar I disorder based on number of previous manic or mixed episodes. J Clin Psychiatry 67 (01):95–101. http://article.psychiatrist.com/?ContentType=START&ID=10002333

Kim P, Thomas LA, Rosen BH, Moscicki AM, Brotman MA, Zarate CA Jr et al (2012) Differing amygdala responses to facial expressions in children and adults with bipolar disorder. Am J Psychiatry 169(6):642–649. http://psychiatryonline.org/doi/abs/10.1176/appi.ajp.2012.11081245

Kirby LAJ, Robinson JL (2017) Affective mapping: an activation likelihood estimation (ALE) meta-analysis. Brain Cogn 118:137–148. https://linkinghub.elsevier.com/retrieve/pii/S0278262615000469

Kozicky J-M, McGirr A, Bond DJ, Gonzalez M, Silveira LE, Keramatian K et al (2016) Neuroprogression and episode recurrence in bipolar I disorder: a study of gray matter volume changes in first-episode mania and association with clinical outcome. Bipolar Disord 18 (6):511–519. http://doi.wiley.com/10.1111/bdi.12437

Kryza-Lacombe M, Brotman MA, Reynolds RC, Towbin K, Pine DS, Leibenluft E et al (2019) Neural mechanisms of face emotion processing in youths and adults with bipolar disorder. Bipolar Disord 21(4):309–320. https://onlinelibrary.wiley.com/doi/abs/10.1111/bdi.12768

Laidi C, Houenou J (2016) Brain functional effects of psychopharmacological treatments in bipolar disorder. Eur Neuropsychopharmacol 26(11):1695–1740. https://linkinghub.elsevier.com/retrieve/pii/S0924977X16300906

Lavagnino L, Cao B, Mwangi B, Wu M-J, Sanches M, Zunta-Soares GB et al (2015) Changes in the corpus callosum in women with late-stage bipolar disorder. Acta Psychiatr Scand 131 (6):458–464. http://doi.wiley.com/10.1111/acps.12397

Lim CS, Baldessarini RJ, Vieta E, Yucel M, Bora E, Sim K (2013) Longitudinal neuroimaging and neuropsychological changes in bipolar disorder patients: review of the evidence. Neurosci Biobehav Rev 37(3):418–435. https://linkinghub.elsevier.com/retrieve/pii/S0149763413000043

Lisy ME, Jarvis KB, DelBello MP, Mills NP, Weber WA, Fleck D et al (2011) Progressive neurostructural changes in adolescent and adult patients with bipolar disorder. Bipolar Disord 13(4):396–405. http://doi.wiley.com/10.1111/j.1399-5618.2011.00927.x

López-Larson MP, DelBello MP, Zimmerman ME, Schwiers ML, Strakowski SM (2002) Regional prefrontal gray and white matter abnormalities in bipolar disorder. Biol Psychiatry 52 (2):93–100. https://linkinghub.elsevier.com/retrieve/pii/S0006322302013501

Lu X, Zhong Y, Ma Z, Wu Y, Fox PT, Zhang N et al (2019) Structural imaging biomarkers for bipolar disorder: meta-analyses of whole-brain voxel-based morphometry studies. Depress Anxiety 36(4):353–364. https://onlinelibrary.wiley.com/doi/abs/10.1002/da.22866

Lyoo IK, Dager SR, Kim JE, Yoon SJ, Friedman SD, Dunner DL et al (2010) Lithium-induced gray matter volume increase as a neural correlate of treatment response in bipolar disorder: a longitudinal brain imaging study. Neuropsychopharmacology 35(8):1743–1750. http://www.nature.com/articles/npp201041

Martínez-Arán A, Vieta E, Reinares M, Colom F, Torrent C, Sánchez-Moreno J et al (2004) Cognitive function across manic or hypomanic, depressed, and euthymic states in bipolar disorder. Am J Psychiatry 161(2):262–270. http://psychiatryonline.org/doi/abs/10.1176/appi.ajp.161.2.262

Martino DJ, Samamé C, Marengo E, Igoa A, Strejilevich SA (2016) A critical overview of the clinical evidence supporting the concept of neuroprogression in bipolar disorder. Psychiatry Res 235:1–6. https://linkinghub.elsevier.com/retrieve/pii/S0165178115303395

Matza LS, Rajagopalan KS, Thompson CL, de Lissovoy G (2005) Misdiagnosed patients with bipolar disorder. J Clin Psychiatry 66(11):1432–1440. http://article.psychiatrist.com/?ContentType=START&ID=10001911

McIntosh AM, Moorhead TWJ, McKirdy J, Hall J, Sussmann JED, Stanfield AC et al (2009) Prefrontal gyral folding and its cognitive correlates in bipolar disorder and schizophrenia. Acta Psychiatr Scand 119(3):192–198. http://doi.wiley.com/10.1111/j.1600-0447.2008.01286.x

Mills NP, DelBello MP, Adler CM, Strakowski SM (2005) MRI analysis of cerebellar vermal abnormalities in bipolar disorder. Am J Psychiatry 162(8):1530–1533. http://psychiatryonline.org/doi/abs/10.1176/appi.ajp.162.8.1530

Mirakhur A, Moorhead TWJ, Stanfield AC, McKirdy J, Sussmann JED, Hall J et al (2009) Changes in gyrification over 4 years in bipolar disorder and their association with the brain-derived neurotrophic factor valine66 methionine variant. Biol Psychiatry 66(3):293–297. https://linkinghub.elsevier.com/retrieve/pii/S0006322308015680

Monkul ES, Hatch JP, Sassi RB, Axelson D, Brambilla P, Nicoletti MA et al (2008) MRI study of the cerebellum in young bipolar patients. Prog Neuro-Psychopharmacol Biol Psychiatry 32 (3):613–619. https://linkinghub.elsevier.com/retrieve/pii/S0278584607003454

Moore GJ, Cortese BM, Glitz DA, Zajac-Benitez C, Quiroz JA, Uhde TW et al (2009) A longitudinal study of the effects of lithium treatment on prefrontal and subgenual prefrontal gray matter volume in treatment-responsive bipolar disorder patients. J Clin Psychiatry 70 (5):699–705. http://www.psychiatrist.com/abstracts/abstracts.asp?abstract=200905/050908.htm

Moorhead TWJ, McKirdy J, Sussmann JED, Hall J, Lawrie SM, Johnstone EC et al (2007) Progressive gray matter loss in patients with bipolar disorder. Biol Psychiatry 62(8):894–900. https://linkinghub.elsevier.com/retrieve/pii/S0006322307002338

Narita K, Suda M, Takei Y, Aoyama Y, Majima T, Kameyama M et al (2011) Volume reduction of ventromedial prefrontal cortex in bipolar II patients with rapid cycling: a voxel-based morphometric study. Prog Neuro Psychopharmacol Biol Psychiatry 35(2):439–445. https://linkinghub.elsevier.com/retrieve/pii/S0278584610004616

Pantelis C, Velakoulis D, McGorry PD, Wood SJ, Suckling J, Phillips LJ et al (2003) Neuroanatomical abnormalities before and after onset of psychosis: a cross-sectional and longitudinal MRI comparison. Lancet 361(9354):281–288. https://linkinghub.elsevier.com/retrieve/pii/S0140673603123239

Passos IC, Mwangi B, Vieta E, Berk M, Kapczinski F (2016) Areas of controversy in neuroprogression in bipolar disorder. Acta Psychiatr Scand 134(2):91–103. http://doi.wiley.com/10.1111/acps.12581

Poon SH, Sim K, Sum MY, Kuswanto CN, Baldessarini RJ (2012) Evidence-based options for treatment-resistant adult bipolar disorder patients. Bipolar Disord 14(6):573–584. http://doi.wiley.com/10.1111/j.1399-5618.2012.01042.x

Arango C (2012) Progressive brain changes in children and adolescents with first-episode psychosis. Arch Gen Psychiatry 69(1):16. http://archpsyc.jamanetwork.com/article.aspx?doi=10.1001/archgenpsychiatry.2011.150

Rosa AR, González-Ortega I, González-Pinto A, Echeburúa E, Comes M, Martínez-Àran A et al (2012) One-year psychosocial functioning in patients in the early vs. late stage of bipolar disorder. Acta Psychiatr Scand 125(4):335–341. http://doi.wiley.com/10.1111/j.1600-0447.2011.01830.x

Sassi RB, Brambilla P, Hatch JP, Nicoletti MA, Mallinger AG, Frank E et al (2004) Reduced left anterior cingulate volumes in untreated bipolar patients. Biol Psychiatry 56(7):467–475. https://linkinghub.elsevier.com/retrieve/pii/S000632230400767X

Schneider MR, DelBello MP, McNamara RK, Strakowski SM, Adler CM (2012) Neuroprogression in bipolar disorder. Bipolar Disord 14(4):356–374. http://doi.wiley.com/10.1111/j.1399-5618.2012.01024.x

Sehmbi M, Rowley CD, Minuzzi L, Kapczinski F, Steiner M, Sassi RB et al (2018) Association of intracortical myelin and cognitive function in bipolar I disorder. Acta Psychiatr Scand 138 (1):62–72. http://doi.wiley.com/10.1111/acps.12875

Sharma V, Menon R, Carr TJ, Densmore M, Mazmanian D, Williamson PC (2003) An MRI study of subgenual prefrontal cortex in patients with familial and non-familial bipolar I disorder. J Affect Disord 77(2):167–171. https://linkinghub.elsevier.com/retrieve/pii/S016503270200109X

Strakowski SM, DelBello MP, Sax KW, Zimmerman ME, Shear PK, Hawkins JM et al (1999) Brain magnetic resonance imaging of structural abnormalities in bipolar disorder. Arch Gen Psychiatry 56(3):254. http://archpsyc.jamanetwork.com/article.aspx?doi=10.1001/archpsyc.56.3.254

Strakowski SM, DelBello MP, Zimmerman ME, Getz GE, Mills NP, Ret J et al (2002) Ventricular and periventricular structural volumes in first- versus multiple-episode bipolar disorder. Am J Psychiatry 159(11):1841–1847. http://psychiatryonline.org/doi/abs/10.1176/appi.ajp.159.11.1841

Swann AC, Bowden CL, Calabrese JR, Dilsaver SC, Morris DD (1999) Differential effect of number of previous episodes of affective disorder on response to lithium or divalproex in acute mania. Am J Psychiatry 156(8):1264–1266. http://ajp.psychiatryonline.org/cgi/content/full/156/8/1264

Toma S, MacIntosh BJ, Swardfager W, Goldstein BI (2018) Cerebral blood flow in bipolar disorder: a systematic review. J Affect Disord 241:505–513. https://linkinghub.elsevier.com/retrieve/pii/S0165032718309698

Torres IJ, Boudreau VG, Yatham LN (2007) Neuropsychological functioning in euthymic bipolar disorder: a meta-analysis. Acta Psychiatr Scand 116(s434):17–26. http://doi.wiley.com/10.1111/j.1600-0447.2007.01055.x

Townsend J, Altshuler LL (2012) Emotion processing and regulation in bipolar disorder: a review. Bipolar Disord 14(4):326–339. http://doi.wiley.com/10.1111/j.1399-5618.2012.01021.x

Tsai S-Y, Gildengers AG, Hsu J-L, Chung K-H, Chen P-H, Huang Y-J (2019) Inflammation associated with volume reduction in the gray matter and hippocampus of older patients with bipolar disorder. J Affect Disord 244:60–66. https://linkinghub.elsevier.com/retrieve/pii/S0165032718312370

Vita A, De Peri L, Sacchetti E (2009) Gray matter, white matter, brain, and intracranial volumes in first-episode bipolar disorder: a meta-analysis of magnetic resonance imaging studies. Bipolar Disord 11(8):807–814. http://doi.wiley.com/10.1111/j.1399-5618.2009.00759.x

Weathers JD, Stringaris A, Deveney CM, Brotman MA, Zarate CA Jr, Connolly ME et al (2012) A developmental study of the neural circuitry mediating motor inhibition in bipolar disorder. Am J Psychiatry 169(6):633–641. http://psychiatryonline.org/doi/abs/10.1176/appi.ajp.2012.11081244

Wegbreit E, Cushman GK, Puzia ME, Weissman AB, Kim KL, Laird AR et al (2014) Developmental meta-analyses of the functional neural correlates of bipolar disorder. JAMA Psychiat 71 (8):926. http://archpsyc.jamanetwork.com/article.aspx?doi=10.1001/jamapsychiatry.2014.660

Wijeratne C, Sachdev S, Wen W, Piguet O, Lipnicki DM, Malhi GS et al (2013) Hippocampal and amygdala volumes in an older bipolar disorder sample. Int Psychogeriatrics 25(1):54–60. https://www.cambridge.org/core/product/identifier/S1041610212001469/type/journal_article

Yucel K, McKinnon MC, Taylor VH, Macdonald K, Alda M, Young LT et al (2007) Bilateral hippocampal volume increases after long-term lithium treatment in patients with bipolar disorder: a longitudinal MRI study. Psychopharmacology 195(3):357–367. http://link.springer.com/10.1007/s00213-007-0906-9

Inflammation as a Mechanism of Bipolar Disorder Neuroprogression

Tatiana Barichello, Vijayasree Vayalanellore Giridharan, Gursimrat Bhatti, Pavani Sayana, Tejaswini Doifode, Danielle Macedo, and Joao Quevedo

The original version of this chapter was revised. A correction to this chapter can be found at https://doi.org/10.1007/7854_2020_214

T. Barichello (✉)
Translational Psychiatry Program, Faillace Department of Psychiatry and Behavioral Sciences, McGovern Medical School, The University of Texas Health Science Center at Houston (UTHealth), Houston, TX, USA

Laboratory of Experimental Pathophysiology, Graduate Program in Health Sciences, Health Sciences Unit, University of Southern Santa Catarina (UNESC), Criciúma, SC, Brazil

Neuroscience Graduate Program, The University of Texas MD Anderson Cancer Center UTHealth Graduate School of Biomedical Sciences, Houston, TX, USA
e-mail: Tatiana.Barichello@uth.tmc.edu

V. V. Giridharan, G. Bhatti, P. Sayana, and T. Doifode
Translational Psychiatry Program, Faillace Department of Psychiatry and Behavioral Sciences, McGovern Medical School, The University of Texas Health Science Center at Houston (UTHealth), Houston, TX, USA

D. Macedo
Neuropsychopharmacology Laboratory, Drug Research, and Development Center, Faculty of Medicine, Universidade Federal do Ceará, Fortaleza, CE, Brazil

National Institute for Translational Medicine (INCT-TM, CNPq), Ribeirao Preto, Brazil

J. Quevedo
Translational Psychiatry Program, Faillace Department of Psychiatry and Behavioral Sciences, McGovern Medical School, The University of Texas Health Science Center at Houston (UTHealth), Houston, TX, USA

Neuroscience Graduate Program, The University of Texas MD Anderson Cancer Center UTHealth Graduate School of Biomedical Sciences, Houston, TX, USA

Laboratory of Neurosciences, Graduate Program in Health Sciences, University of Southern Santa Catarina-UNESC, Criciúma, SC, Brazil

Center of Excellence on Mood Disorders, Department of Psychiatry and Behavioral Sciences, McGovern Medical School, The University of Texas Health Science Center at Houston (UTHealth), Houston, TX, USA

Curr Topics Behav Neurosci (2021) 48: 215–238
https://doi.org/10.1007/7854_2020_173
Published Online: 11 October 2020

Contents

Abstract Bipolar disorder (BD) is a severe, debilitating psychiatric condition with onset in adolescence or young adulthood and often follows a relapsing and remitting course throughout life. The concept of neuroprogression in BD refers to the progressive path with an identifiable trajectory that takes place with recurrent mood episodes, which eventually leads to cognitive, functional, and clinical deterioration in the course of BD. Understanding the biological basis of neuroprogression helps to explain the subset of BD patients who experience worsening of their disorder over time. Additionally, the study of the neurobiological mechanisms underpinning neuroprogression will help BD staging based on systems biology. Replicated epidemiological studies have suggested inflammatory mechanisms as primary contributors to the neuroprogression of mood disorders. It is known that dysregulated inflammatory/immune pathways are often associated with BD pathophysiology. Hence, in this chapter, we focus on the evidence for the involvement of inflammation and immune regulated pathways in the neurobiological consequences of BD neuroprogression. Herein we put forth the evidence of immune markers from autoimmune disorders, chronic infections, and gut-brain axis that lead to BD neuroprogression. Further, we highlighted the peripheral and central inflammatory components measured along with BD progression.

Keywords Bipolar disorder · Host immune response · Inflammation · Neuroprogression

1 Introduction

Bipolar disorder (BD) is a severe episodic mental illness with chronic and accelerating course, composed of mood swings ranging from extreme elation, called mania, to extreme lows, called depression, which may sometimes occur together (Association AP 2013). BD is classified as BD-1 (which represents classic manic depressive disorder), BD-2 (represented by the manifestation of hypomanic and depressive symptoms), and cyclothymic disorder. In BD, the obligatory symptoms (gate A criteria) are hypomanic or manic episodes. Lifetime prevalence is about 1.0% for BD-1, 1.1% for BD-2, and 2.4% for sub-threshold BD (Merikangas et al. 2007). Among mental disorders, the absolute risk of suicide was found to be the highest for men with BD (7.77%), while for women, it was 4.78% (Nordentoft et al. 2011). There was a reported increase in the suicide risk due to comorbid occurrence of substance abuse and unipolar affective disorders, while the co-occurrence of deliberate self-harm generally doubled the suicide risk (Nordentoft et al. 2011). Growing evidence has suggested that BD, like many other chronic illnesses, may have a progressive course with functional impairment and neuroanatomical changes.

The current clinical focus is mainly on stabilizing the acute mood episodes and preventing their recurrence in BD patients leading to negligence of the need to promote functional recovery (Kapczinski et al. 2017). Clinical, neuroimaging, and neurocognitive studies further support the progressive nature of BD. In a large cross-sectional study, it was found that 42% of the euthymic BD patients had poor overall functioning (Samalin et al. 2016). Also, it was found that episode density, level of residual depressive symptoms, estimated verbal intelligence, and inhibitory control are the risk factors predictive of poor functional outcome of BD (Reinares et al. 2013). Clinical staging model, which is used in oncology and medicine for a long time and recently introduced in psychiatry, proposes that there is a stepwise progression through a series of identifiable steps, which have characteristic features and potential treatment implications (Berk et al. 2007). A large study has reported progressive functional impairment from stage I to stage IV of BD on clinical staging (Rosa et al. 2014). More specifically, stage I includes individuals who exhibit the same status in the interepisodic period as they did before the onset of BD (i.e., premorbid status); stage II includes individuals whose interepisodic period is characterized by psychiatric comorbidities or residual symptoms that require changes in pharmacological treatment, but who are able to maintain daily activities; stage III includes individuals who require occupational and social rehabilitation and face difficulties in their daily activities; and stage IV includes individuals who are unable to maintain personal self-care and to live autonomously.

Various studies have described that the presence of widespread structural brain abnormalities in BD is associated with the incidence of manic episodes and higher illness burden – which points to neuroprogression (Abe et al. 2015; Mwangi et al. 2016). A study found that reduced hippocampal volume and severe cognitive impairment were associated with the increased number of manic episodes and hospitalizations in BD patients (Cao et al. 2016), while another study reported decreased posterior corpus callosum volume in women with late-stage BD (Lavagnino et al. 2015). Reduced hippocampal volume has also been seen in particular viral infections pointing towards an association between inflammation and structural changes seen in severe BD (Almanzar et al. 2005). Besides, a longitudinal study found reduced frontal cortex volume (dorsolateral prefrontal and inferior frontal cortex) in patients who had at least one manic episode (Abe et al. 2015). Similar findings were described in a study that having larger lateral ventricles was associated with a higher number of prior manic episodes in patients with BD (Strakowski et al. 2002). Given these findings, the concept of neuroprogression was postulated to encompass the progressive functional impairment and neuroanatomical changes in BD presentation (Grande et al. 2016).

Neuroprogression has thus been proposed as the pathological alterations in the brain that take place simultaneously with the clinical and neurocognitive deterioration in the course of BD (Berk et al. 2011). However, the clinical implications and molecular foundations of neuroprogression remain incompletely understood. It seems that changes in some peripheral biomarkers from oxidative, inflammatory, and neurotrophic pathways are associated with neuroprogression. Hence, in this book chapter, we will highlight the association of various immune pathways and also the gut microbiota with neuroprogression in a subset of more severe patients with BD. Light will also be shed on the association of neuroprogression and the peripheral biomarker changes in BD.

2 Evidence of Inflammatory and Infectious Diseases in BD Neuroprogression

Even though BD is thought to be a neuroprogressive disorder (Berk et al. 2011), evidence suggests that disruption in neurodevelopmental pathways may play a pivotal role in the etiopathology of this illness (Savitz et al. 2014). Neurodevelopment can be disrupted by several pathways, causing the mood dysregulations seen in BD (Harrison 2016). A multiple-hit model has been postulated as a series of three factors, with hit 1 being a genetic predisposition to BD, hit 2 being the perinatal environment, which gives rise to phenotypes of vulnerability, and hit 3 is the later life experiences and exposures (Daskalakis et al. 2013). Although the mechanism is still unclear, this multiple-hit model has been suggested to dysregulate the homeostasis chronically in a process that is thought to involve immune dysfunction (Leboyer et al. 2016).

2.1 *Maternal Immune Activation as a Risk Factor for BD Development and Neuroprogression*

Epidemiologic evidence has repeatedly suggested that prenatal environmental influences, such as maternal immune activation (MIA), are involved in the pathophysiology of neuropsychiatric disorders (Brown and Derkits 2010; Estes and McAllister 2016). Prenatal infections and inflammation are potential risk factors associated with schizophrenia (Brown and Derkits 2010; Estes and McAllister 2016), autism spectrum disorders (Canetta et al. 2014), and BD (Brown 2015; Canetta et al. 2014). Animal models of MIA have also demonstrated behavioral, chemical, anatomical, and physiologic disturbances in the CNS of the progeny (Meyer 2014; Meyer et al. 2009). While the role of prenatal and postnatal infections as a risk factor for schizophrenia has been studied extensively (Debnath et al. 2015; Khandaker et al. 2012), not many studies have evaluated the role of infection during the prenatal period as a risk factor for BD (Canetta et al. 2014). For instance, multiple studies have reported an association between serologically documented maternal influenza infection and increased risk of BD in the offspring (Canetta et al. 2014; Parboosing et al. 2013). A case–control study has demonstrated a nearly four times increased risk of BD in adult offspring after maternal influenza infection at any time during pregnancy (Parboosing et al. 2013). Another study of T2-weighted magnetic resonance imaging (MRI) findings described that in BD patients three times more periventricular white matter hyperintensities were seen as compared to controls (Altshuler et al. 1995). Increased number of BD patients showing deep subcortical white matter lesions were born during the winter (Moore et al. 2001) when influenza incidence is high (Kilbourne 1987). These results suggest that structural and functional abnormalities are induced in the CNS of the offspring by maternal infections or inflammation, which might be responsible for neuroprogression of BD in later life.

Although animal models of MIA have not explicitly been explored for their validity as a BD model, some of the experimentally induced phenotypes may be considered for this disorder. Deficits in sensorimotor gating, as present in various rodent MIA models (Estes and McAllister 2016; Meyer 2014; Meyer et al. 2009), are also seen in acute mania (Perry et al. 2001) and euthymic BD patients (Giakoumaki et al. 2007). Besides, several animal studies have reported depression-like behaviors in offspring exposed to MIA (Khan et al. 2014; Ronovsky et al. 2016). The latter phenotypes may not only be seen in unipolar depression, but also depressive episodes in BD. Evaluation of other core behavioral symptoms of BD, such as poor decision-making, altered risk-taking behavior, impulsivity, and loss of inhibitory control, remain unexplored in MIA models.

2.2 *Role of Infectious Disease as a Trigger to Develop BD Neuroprogression*

Infections with several pathogenic agents have been studied as risk factors for BD (Stich et al. 2015; Yolken and Torrey 2008). Herpes virus can lead to latent and lytic infection in the brain, and it has been associated with memory impairment (Kapur et al. 1994), mania (Koehler and Guth 1979), and psychosis (Schlitt et al. 1985). A study reported that serologic evidence of herpes simplex virus type 1 (HSV-1) infection was an independent predictor of low cognitive functioning in BD patients (Dickerson et al. 2004). Similar results were reported in another study showing a negative association between HSV-1 infection and cognitive performance in both BD patients and controls (Yolken 2011), suggesting that HSV-1 infection is associated with worse functioning in BD patients. Rising evidence also suggests a possible association between cytomegalovirus (CMV) and BD. A study described that CMV IgG levels are associated with a reduction in hippocampal volume and worse episodic verbal memory in BD patients (Houenou et al. 2014). The association between CMV latent infection and lower cognitive functioning is thought to be mediated by a chronic inflammatory response and subsequent reduced hippocampal volume (Almanzar et al. 2005). On the other hand, the hippocampal volume is usually normal in BD and reported to be reduced only in the most severe forms of BD (Strasser et al. 2005). Hence, this evidence suggests that exposure to CMV infection may contribute to the pathological rewiring of neurons and neuroprogression of BD.

A case–control study showed that BD patients had an increased seroprevalence for *Toxoplasma gondii* compared to controls (Tedla et al. 2011). Indeed, *T. gondii* is a neurotropic protozoan with high seropositivity rates globally. Two separate studies have reported that BD patients with manic episodes had elevated *T. gondii* IgM antibody levels as compared to healthy controls. Also, these studies reported a significant negative correlation between *T. gondii* IgM antibody levels and cognitive scores in both controls and BD patients (Dickerson et al. 2014a, b). The negative association between the IgM antibody load and cognitive functioning points towards the worsening of BD with exposure to *T. gondii* infection. Evidence also shows that the cognitive deterioration index (DI) in BD patients correlated to high IL-6 mRNA expression only among *T. gondii* seropositive group (Hamdani et al. 2015), asserting that inflammatory pathways may be involved in the cognitive dysfunction caused by the infection.

Various infections activate a common inflammatory pathway within the cell. Infectious agents are recognized by pattern-recognition receptors (PRRs), which are critical components of the innate immune system (Mook-Kanamori et al. 2011; Sellner et al. 2010). Activation of PRRs causes the release of mediators, such as pro-inflammatory cytokines and chemokines, that propagate and regulate the immune response necessary to remove invaded microorganisms (Iwasaki and Medzhitov 2010). Some of the pro-inflammatory cytokines that are produced during infections, like TNF-α, IL-1β, and IL-6, are increased in BD patients compared to healthy controls (Dong and Zhen 2015; Munkholm et al. 2013; Soderlund et al. 2011), providing a potential pathophysiological mechanism linking infections to neuroprogression of BD.

2.3 *Autoimmune Disorders and Their Association with BD Neuroprogression*

Growing evidence has described an association between BD and various autoimmune disorders. Multiple epidemiologic studies have pointed towards an association between autoimmune diseases, autoantibodies, and BD (Rosenblat and McIntyre 2015). The most common associations established over the years have been those with autoimmune thyroiditis (strongest association), multiple sclerosis, systemic lupus erythematosus (SLE), rheumatoid arthritis (RA), autoimmune hepatitis, inflammatory bowel disease, and Guillain barre syndrome (Hillegers et al. 2007; Hsu et al. 2014; Marrie et al. 2018a, b; Tiosano et al. 2017).

Several mechanisms have been suggested through which the autoimmune diseases have been hypothesized to affect the neuroprogression of BD. Benros et al. proposed a correlation between autoantibodies and psychiatric symptoms (Benros et al. 2013). The most well-accepted hypothesis has been synthesized from the observation of psychiatric symptoms in the autoimmune encephalitis. In autoimmune encephalitis, IgG autoantibodies are formed against the NMDA (N-methyl-D-aspartate receptor) receptors (NR1 subunit) (Vitaliani et al. 2005). As the NMDA receptor is a primary receptor for the excitatory neurotransmitter glutamate, the activation of the pathway will lead to an increase in excitatory neurotransmission. When the NMDA receptor is activated, it will lead to a cascade of activation of various components such as protein kinases. One of the protein kinases (PKC) that is significant in the presentation of a manic episode is also activated via this pathway (Traynelis et al. 2010). In a similar study, it was observed that the levels of autoantibodies against the NR2 subunit of NMDA were high during a manic episode in BD and schizoaffective disorder (Dickerson et al. 2012). Astrocytes are responsible for the clearance of glutamate, by converting it into glutamine. However, due to the interaction of astrocytes with pro-inflammatory cytokines, the glutamate clearance is hampered (Zou et al. 2010). This increased glutamate receptor activation causes an increase in the levels of calcium in the mitochondria. These changes cause neuroplastic alterations and excitotoxicity (Berk et al. 2000; Kato 2007).

Another mechanism that can affect the neuroprogression of BD involves the activation of the tryptophan–kynurenine pathway by the systemic pro-inflammatory cytokines. NMDAR and tryptophan–kynurenine pathways are responsible for the regulation of serotonin (directly) and dopamine levels (indirectly) (Dantzer et al. 2008). Some of the pro-inflammatory cytokines that are observed to be increased in BD patients are C-reactive protein (CRP), interleukin (IL)-1 beta, soluble IL-2 receptor, IL-4, IL-6, tumor necrosis factor-alpha (TNF-α), and soluble receptor of TNF-α type 1 (sTNFR1) (Barbosa et al. 2014a; Brietzke et al. 2009a, b; Modabbernia et al. 2013). Serum levels of these cytokines have been seen to be mood dependent. In this regard, high serum concentrations of IL-4, IL-6, IL-RA, TNF-α, sTNFR1, CXCL10, and CXCL11 are seen in the manic phases. Similarly, elevated serum concentrations of IL-6, IL-1β, CRP, TNF-α, sTNFR1, and CXCL10 are also seen in the depressed period (Barbosa et al. 2014a, b; Rosenblat et al. 2014). IL-6 and TNF-α levels correlate directly with the severity of the disease (Kauer-Sant'Anna et al. 2009). Increased levels of these cytokines lead to neuroplastic changes in the brain. According to one hypothesis, high levels of TNF-α could lead to reduced expression of muscarinic acetylcholine receptor (M2 receptors) in the cortex (Gibbons et al. 2009). The M2 receptors are observed to be reduced in the cortex of patients with major depressive disorder and BD.

Autoimmune diseases have a baseline inflammatory condition. Inflammation causes an increase in the permeability of the blood–brain barrier (BBB). In this instance, the pro-inflammatory cytokines and other components of the inflammatory pathways, such as autoantibodies, can enter the CSF directly and cause various psychiatric symptoms (Modabbernia et al. 2013). The baseline chronic inflammatory state in autoimmune disorders can lead to excessive microglial activation (Frick et al. 2013). The microglial activation can cause detrimental changes in some of the neuronal circuits associated with mood and cognitive functions. Another effect of the microglial overactivation would be a surge in the levels of reactive oxygen species (ROS), which can lead to oxidative stress and further damage to the neuronal circuits, and hence, the neuroprogression of the BD (Frick et al. 2013; Stertz et al. 2013).

2.4 Role of the Gut–Brain Axis on BD Neuroprogression

In the last decade, there has been mounting evidence suggesting that gut microbiota makes a substantial contribution to mental health and, subsequently, to the neuroprogression of various neuropsychiatric disorders such as BD, depression, and anxiety (Cryan and Dinan 2012; Forsythe et al. 2010; Painold et al. 2019). The gut–brain axis refers to a bidirectional communication between the intestine (enteric nervous system, gut microbiota, and metabolites of gut microbiome) and the brain (Carabotti et al. 2015). Multiple studies have shown that the diversity of the gut bacteria is inversely linked to the illness duration of BD (Carabotti et al. 2015; Painold et al. 2019). However, the diversity of the gut bacteria is not solely responsible for the severity of the BD. It is one of the contributing factors to the overall disease state. Patients suffering from BD have decreased levels of *Faecalibacterium* sp. and of an unknown bacterium of the *Ruminococcaceae* family (Bengesser et al. 2019; Carabotti et al. 2015; Evans et al. 2017; Huang et al. 2019). Notably, *Ruminococcus* species are related to the synthesis of butyrate, which presents anti-inflammatory and mood regulatory activities (Hwang et al. 2017). The *Enterobacteriaceae* family was found in higher fractions in BD patients suffering from depressive symptoms (Carabotti et al. 2015; Evans et al. 2017; Huang et al. 2019). A high proportion of genus *Lactobacillus* and genus *Streptococcus* in the gut was associated with higher levels of pro-inflammatory cytokines such as IL-6 (Painold et al. 2019) suggesting activation of inflammatory pathways by these bacteria. A low-grade inflammatory state is seen in a subgroup of BD patients (Bechter 2013; Fillman et al. 2014; Miller et al. 2011). Over the years, various studies have suggested that the cause of this inflammation might be related to the dysbiosis of the gut microbiome. The concept of microbial translocation (gut bacteria leaking into the circulation due to changes in the permeability of the intestinal lumen) plays a central role in this hypothesis. Microbial translocation has been measured by markers such as soluble CD14 (sCD14) and fungal antibodies. The inflammatory reaction mounted in response to the gut bacteria releases mediators in circulation that, in turn, have effects on the behavioral and cognitive patterns (Dickerson et al. 2017). It can be postulated that the chronic inflammatory state produced by long-term changes in the gut microbiota diversity might cause activation of microglia. This may cause a detrimental effect on the neuronal networks directly and indirectly (via the formation of ROS damaging DNA, and proteins). In animal models, it has been shown that the gut flora is responsible for regulating serotonin levels in plasma (Collins and Bercik 2009). The inflammatory state produced by the gut bacteria upregulates the indoleamine 2,3 deoxygenase (IDO) enzyme. The IDO enzyme upregulation causes increased degradation of serotonin. This pathway plays a vital role in the acute manic episodes in BD (Myint et al. 2007).

According to the monoamine hypothesis, the decreased levels of monoamines play a role in the pathophysiology of depressive symptoms. Building on this hypothesis, Bengessar et al. showed that the change in gut microbiota diversity affects the CpG methylation status of the clock gene of aryl hydrocarbon receptor nuclear translocator-like protein 1 (ARNTL), which plays a vital role in the neuroprogression of BD (Bengesser et al. 2018, 2019). The ARNTL gene codes for one of the transcription factors of the monoamine oxidase A (MAO-A) enzyme (responsible for the degradation of monoamine neurotransmitters). The gut bacteria diversity is negatively correlated to the methylation status of the ARNTL gene. Hence, a decrease in gut bacteria diversity would lead to more methylation of the ARNTL gene and decreased expression of the MAO-A transcription factor. This cascade will lead to a reduced breakdown of the monoamine neurotransmitters. The increased levels of monoamine neurotransmitters would have a pro-manic effect, according to this model. However, this hypothesis has loopholes as well since the mechanism of action of mood stabilizers does not support this hypothesis as the MAO-A levels in the brain do not correlate with the mechanisms of these drugs.

3 Inflammatory and Oxidative Mechanisms in BD Neuroprogression

3.1 Mechanisms of Inflammation and Their Contribution to BD Neuroprogression

Many experimental facts solidify the link between BD and inflammation such as an increase in inflammatory biomarker levels seen in both manic and depressive phases of BD, pro-inflammatory cytokine infusion being the best experimental model of depression, epidemiologic evidence of increased rates of inflammatory medical comorbidities in BD associated with an upsurge in the levels of inflammatory mediators, and most importantly anti-inflammatory agents being considered as the novel therapeutic agents in BD trials (Ayorech et al. 2015; Barbosa et al. 2014b; Goldstein et al. 2009; Sayana et al. 2017; Wadee et al. 2002).

Neuroprogression in BD is evident in the form of brain structural changes, neuronal and glial cell abnormalities, biochemical alterations that comprise of inflammatory cytokines, neurotrophins including brain-derived neurotrophic factor (BDNF), oxidative stress, autoimmunity, mitochondrial and endoplasmic reticulum stress, dopaminergic and glutamatergic system alterations, kynurenine pathway imbalance, and involvement of epigenetic changes such as histone and DNA methylation leading to gene expression variations (Berk et al. 2008, 2011; Grayson et al. 2010; Post 2007; Wadee et al. 2002). BD has also been linked to changes in neuroplasticity and neuronal survival that are determined by neurotransmitters, hormones, neurotrophins, and inflammatory biomarkers such as cytokines and chemokines; and acute phase proteins such as immunoglobulins, complement proteins, factor B and high-sensitivity CRP (Brietzke et al. 2009b; Cunha et al. 2008).

The cytokines influence the BD course via their direct actions on the immune system as well as their effect on neurotransmitter and neuropeptide systems. Cytokine production is controlled by phosphoinositides, the arachidonic acid (AA) cascade, adenyl cyclase, tyrosine phosphorylation, and the protein kinase C systems. Similar to neurotransmitters and hormones, cytokines act through the hypothalamus-pituitary-adrenal (HPA) axis to maintain stress response, alter serotonin–catecholamine associated pathways in the brain, and cause mood changes (Ortiz-Dominguez et al. 2007). Studies have demonstrated increased levels of circulating pro-inflammatory cytokines in various phases of BD that lead to activation of neutrophils, the proliferation of B cells, synthesis of acute-phase proteins, and an increase in vascular permeability (Bai et al. 2014).

Pro-inflammatory cytokines, such as IL-1, IL-6, and TNF-α, are involved in neural processes like the induction of long-term potentiation, neurogenesis, survival regulation of neurons, and the development of astrocytes, impacting on several cognitive functions in normal states. However, cytokines may contribute to the neurodegenerative process in neurotoxic and stress-induced states of BD (Eyre and Baune 2012; Khairova et al. 2009). For example, enhanced TNF-α levels seem to be

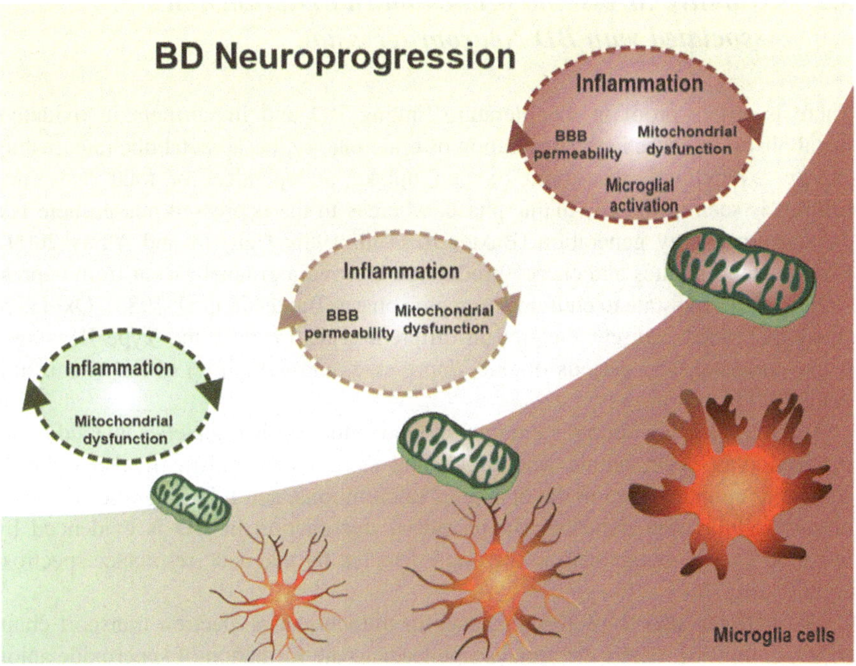

Fig. 1 Inflammation as a mechanism of bipolar disorder neuroprogression. Infection, maternal immune activation, dysbiosis, and autoimmune diseases can activate the host immune response increasing the inflammatory mediators in the bloodstream, subsequently triggering cell damage and mitochondrial dysfunction. The inflammatory mediators and damage-associated molecular patterns (DAMPs) released from the mitochondrial dysfunction can increase the blood-brain barrier (BBB) permeability inducing microglial activation leading to bipolar disorder (BD) neuroprogression

involved in neuronal death via the activation of caspases and apoptotic machinery in BD cases (Cacci et al. 2005).

Cyclo-oxygenase 2 inhibitor (celecoxib), an anti-inflammatory drug, showed better improvement of depressive symptoms compared to the standard treatment. Other immune-based treatments in vogue for BD are anti-inflammatory drugs (aspirin, statins), immune-based drugs (minocycline), and an anti-TNF-α monoclonal antibody, infliximab (Austin and Tan 2012; Berk et al. 2013; Elisa and Beny 2010; Nery et al. 2008; Savitz et al. 2012). Thus, the immune basis in BD seems to be a promising novel therapeutic target for BD patients, and the direct and indirect effects of inflammation help us to understand the behavioral modification by the immunological system. The Fig. 1 illustrates the details of inflammation mediated bipolar neuroprogression.

3.2 Oxidative Stress and Mitochondrial Dysfunction Associated with BD Neuroprogression

There is ample proof in the literature linking BD and impairment in oxidative metabolism. Increase in the generation of brain energy, basal metabolic rate, resting energy expenditure, maximum oxygen uptake, independent of total consumed calories is seen in the BD-manic phase, whereas in the depressed phase, there is a decrease in energy generation (Baxter Jr. et al. 1985; Caliyurt and Altiay 2009). Brain metabolic rates and energy production showed a gradual ascent from depression to the mixed state to euthymia to manic phase (Baxter Jr. et al. 1985). Oxidative stress biomarkers are encountered in different types (Type I and Type II), stages (early and late), and periods (manic, depressive, and euthymic) of BD (Panizzutti et al. 2015).

Mitochondria are double membrane-bound cytoplasmic organelles that aid in the production of ATP, amino acid, lipid, and steroid metabolism. It also involves activating apoptosis and in the uptake of calcium ions, and it is a significant source of intracellular free radicals. Mitochondrial dysfunction in BD is evidenced by impaired energy metabolism in the brain detected by magnetic resonance spectroscopy (Kato 2007).

BD patients also show discrepancies in mitochondrial electron transport chain (ETC) complex I, where electron escape leads to the formation of superoxide anion from molecular oxygen, which is the precursor for most ROS (Guo et al. 2018). Studies on BD patients showed a reduction in ETC complex 1 activity in prefrontal postmortem brain tissue and administration of lithium increased the activity of mitochondrial complexes I/II and II/III in the brain, altered expression of ETC complex 1 subunits, linking BD to chromosome 19p13 that has multiple ETC complex 1 subunit genes (Cheng et al. 2006; Konradi et al. 2004; Maurer et al. 2009; Sun et al. 2006). In the literature, there are alterations in antioxidant enzymes in BD, such as increased activity of SOD during manic and depressive phases, reduced activity of catalase during the euthymic period, and increased activity of both enzymes in unmedicated manic patients (Andreazza et al. 2007; Machado-Vieira et al. 2007). N-acetyl cysteine (NAC), which is a free radical scavenger and glutathione precursor, when used in BD reduced the depressive symptoms and presented overall functional improvement (Berk et al. 2011). Experimental evidence has proven that oxidative parameters have shown a stage-dependent pattern, where glutathione reductase and glutathione s-transferase (GST) are increased during late-stage, suggesting a failure of compensatory mechanisms during BD progression (Andreazza et al. 2009). Overall oxidative stress, mitochondrial dysfunction, and antioxidant enzyme alterations can cause neuronal cell death via apoptosis or aggregated antioxidants that may result in impairment of mood-stabilizing mechanisms.

3.3 *Peripheral Inflammatory Mediators as a Trigger or Accelerator of BD Neuroprogression*

In chronic BD, mood relapses lead to neuroprogression with higher frequencies and rapid cycling, resulting in worst outcomes (Berk et al. 2011). One of the crucial mechanisms accountable for neuroprogression in BD is the aberrant exacerbation of the inflammatory mediators during mood episodes (mania and depression) in BD (Sayana et al. 2017).

Inflammation appears to be phase dependent on BD. On marker analysis, manic patients experienced an elevation in peripheral TNF-α and IL-4 along with a reduction in IL-1β and IL-2 levels, while depressed patients showed elevated levels of IL-6 and TNF-α and decreased IL-2 levels (Ortiz-Dominguez et al. 2007). Another study showed an increase in IL-2, IL-4, and IL-6 in mania and IL-6 in depression (Brietzke et al. 2009b). Kim and his team showed an increase in IL-4, IFN-γ, TNF-α, IL-6, IFN-γ/transforming growth factor (TGF)β1, IL-4/TGF-β1, IL-6/IL-4, TNF-α/IL-4, IL-2/IL-4, and IFN-γ/IL-4 ratios and low levels of TGF-β1 in manic episodes (Kim et al. 2004, 2007). Data from the available literature suggest that hsCRP levels were significantly elevated in mania compared to euthymic, depressed phase or controls, and they positively correlated with Bech Rafaelson Manic Rating Scale (BRMRS) and Young Mania Rating Scale (YMRS). The levels of acute-phase reactants, notably CRP, demonstrated an elevation in depressive phase compared to controls and positively correlated with Hamilton Rating Scale for Depression (HAM-D) scores in one study, whereas in another study they did not show any association with bipolar depression or HAM-D scores (Cunha et al. 2008; De Berardis et al. 2008; Dickerson et al. 2007).

When the early and late stages of BD were compared, IL-6 and TNF-α were elevated in both groups, while IL-10 levels were higher in the early stages. However, TNF-α was more elevated in late stages than in early (Kauer-Sant'Anna et al. 2009). In addition to these cytokines, eotaxin/CCL11, a chemokine, was also shown to increase in late-stage euthymic BD compared to controls, suggesting a link between pathological aging, eosinophil function marker CCL11, and BD neuroprogression (Panizzutti et al. 2015). Also, damage-associated molecular patterns (DAMPs) such as circulating nuclear DNA, HSP70, and HSP90α that bind to Toll-like receptors (TLRs) cause systemic toxicity via immune activation in BD patients, as TLRs activate the signaling pathways of immune system triggering inflammatory pathways. The DAMPs activation of TLR signaling cascades can explain how initial insults such as drugs, stress, and relapses can cause systemic inflammation (Kapczinski et al. 2017).

The inflammatory pathways play a crucial role in progressive cognitive impairment in BD. Peripherally measured markers such as TNF-α, hsCRP, sCD40L, IL-1Ra, and sTNFR1 seem to influence cognitive performance in BD (Barbosa et al. 2012; Chung et al. 2013; Hope et al. 2015; Hoseth et al. 2016). The circulating levels of TNF-α correlated with inhibitory control part of executive dysfunction in BD patients, impairment of which is regarded as a cognitive endophenotype of BD

(Barbosa et al. 2012). Increased CRP levels are associated with cognitive decline, as evidenced by the low Repeatable Battery for the Assessment of the Neuropsychological Status (RBANS) (Dickerson et al. 2013). Also raised serum hsCRP levels are negatively correlated with the volume of the orbitofrontal cortex, which is, in turn, associated with the poor cognitive performance (Chung et al. 2013). IL-1 receptor antagonist (IL-1Ra) and sTNFR1 levels were associated with the worst performance on the Global Assessment of Functioning (GAF) scale (Hope et al. 2015; Hoseth et al. 2016). The pro-inflammatory profile characterized by activation of cell-mediated immunity, systemic inflammation, phase, and stage related changes in BD with deleterious clinical, cognitive, and neurological consequences, seem to act as a significant player in disease neuroprogression.

3.4 Cerebrospinal Fluid (CSF) System Inflammatory Markers in BD Neuroprogression

BD patients present with higher CSF concentrations of markers of neuroinflammation, glial activation, and neuronal injury compared to controls (Isgren et al. 2017). In this regard, CSF studies demonstrated an increased CSF/serum albumin ratio indicating increased BBB permeability, elevated CSF cell count, IgG index, oligoclonal bands, IL-1β, IL-6, and IL-8 suggesting inflammation and intrathecal immunoglobulin production (Orlovska-Waast et al. 2019).

Studies revealed elevation of inflammatory CSF markers such as IL-8, monocyte activation marker, monocyte chemoattractant protein 1 (MCP-1; also known as CCL-2), glial activation marker, chitinase 3 like protein 1 (CHI3L1; also known as YKL-40) and axonal damage marker, neurofilament light chain (NFL) in BD patients compared to controls. The IL-8 showed a positive association with lithium and antipsychotics, whereas NFL to atypical antipsychotic drugs (Isgren et al. 2015; Jakobsson et al. 2014, 2015). On the evaluation of cognitive decline in BD patients, CSF biomarkers, especially microglial marker, YKL-40 showed a significant impairment in executive functions for euthymic BD patients compared to controls, which is independent of patient age, medication, disease status, and type of BD (Rolstad et al. 2015a). Moreover, the marker, NFL concentrations showed a negative association with verbal function and working memory (Rolstad et al. 2015b). Previously assessed CSF proteins may be involved in adaptive immune processes or may reflect immune aberrations or a state of vulnerability for BD rather than being of predictive value for disease progression.

3.5 *Postmortem Inflammatory Markers in BD Neuroprogression*

The role of neuroinflammation in BD and alterations in microglial, astrocyte, and oligodendrocyte markers are evident in postmortem BD studies (Giridharan et al. 2019). The innate immune cells that contribute to the neuroinflammation are microglia, astrocytes, macrophages, natural killer (NK) cells, mast cells, as well as oligodendrocytes and neurons (Stephenson et al. 2018). Postmortem BD studies also revealed increased neuroinflammation with decreased anti-inflammatory marker levels in the frontal cortex (Bezchlibnyk et al. 2001; Rao et al. 2010). Accurately, increased protein and mRNA levels of IL-1β, IL-1R, and myeloid differentiation primary response 88 (MyD88) were described, as well as upregulation of nuclear factor kappa B (NF-kB) transcription factor and its subunits (p50 and p65), and astroglial and microglial markers (GFAP, inducible nitric oxide synthase (iNos), c-fos and CD11b) in the pre frontal cortex of BD (Rao et al. 2010).

The neurodegenerative process mediated by TNF-α may result in the volumetric reduction and hypoactivation of frontal lobes in BD patients, along with the disinhibition of limbic structures (Brooks 3rd et al. 2009; Kupferschmidt and Zakzanis 2011). A study examining TNF parameters in BA 24 and BA 46 demonstrated that BD patients presented increased transmembrane TNF-α (tmTNF-α) protein level in the anterior cingulate cortex (ACC; BA 24), and decreased TNFR2 protein levels in the dorso lateral PFC (BA 46). Peripheral tissue inflammation notably increased TNF-α levels, leads to reduced expression of muscarinic M2 receptors in the cortex of MDD and BD, and ultimately results in cognitive deficits in BD (Gibbons et al. 2009; Haddad et al. 1996; Jones et al. 2004).

On the evaluation of kynurenine pathway metabolites, an increase in quinolinic acid (QUIN) expression and QUIN-immunopositive microglia have been observed in the subgenual and supracallosal anterior cingulate cortex (ACC) in depressed patients (Steiner et al. 2011). Overall, the changes in QUIN levels signify the importance of NMDA-R signaling, glutamate transmission, and mononuclear phagocyte system in BD depression. The ratio of kynurenic acid (KA) to kynurenine was lower in the BD group than in the control group, and KA levels were unchanged. There was also an elevation in the density and intensity of both TDO (Tryptophan-2,3-dioxygenase) 2-positive white matter glia and TDO2-positive gray matter glia in the BD group (Miller et al. 2006). Overall, postmortem studies showed that BD patients presented increased markers of neuroinflammation and decreased anti-inflammatory markers that lead to neuroprogression in the form of neuroanatomical changes, neurotransmitter imbalance, cognitive decline, and progressive deterioration of mental health.

4 Conclusions and Future Directions

In summary, we are now beginning to understand the underlying processes of neuroprogression in BD that include the involvement of inflammatory cytokines, neurotrophins, and epigenetic effects. Infection and other inflammatory processes have clearly shown to be associated not only with increased risk of developing BD but also with worsening cognitive impairment and structural changes indicative of neuroprogression in BD. Various cytokines and chemokines activate multitudes of immune pathways leading to mitochondrial dysfunction, oxidative stress generation, and activation of microglial cells in the CNS, leading to worsening of this illness. Cross-sectional studies suggest that inflammatory markers can be used as both a biomarker of illness activity and stage of the disease. Longitudinal studies are needed to clarify the exact role of inflammation and neuroinflammation in BD. Thus prevention of infections with neurotropic pathogens in pregnant women as well as later in life can be one of the strategies implemented to prevent the development and progression of BD in genetically predisposed individuals. Also, specific steps in the molecular pathways of neuroprogression in BD patients may provide new targets for further research to develop new therapeutic drugs for this chronic mental disorder.

References

Abe C, Ekman CJ, Sellgren C, Petrovic P, Ingvar M, Landen M (2015) Manic episodes are related to changes in frontal cortex: a longitudinal neuroimaging study of bipolar disorder 1. Brain 138 (Pt 11):3440–3448

Almanzar G, Schwaiger S, Jenewein B, Keller M, Herndler-Brandstetter D, Wurzner R et al (2005) Long-term cytomegalovirus infection leads to significant changes in the composition of the CD8+ T-cell repertoire, which may be the basis for an imbalance in the cytokine production profile in elderly persons. J Virol 79(6):3675–3683

Altshuler LL, Curran JG, Hauser P, Mintz J, Denicoff K, Post R (1995) T2 hyperintensities in bipolar disorder: magnetic resonance imaging comparison and literature meta-analysis. Am J Psychiatry 152(8):1139–1144

Andreazza AC, Cassini C, Rosa AR, Leite MC, de Almeida LM, Nardin P et al (2007) Serum S100B and antioxidant enzymes in bipolar patients. J Psychiatr Res 41(6):523–529

Andreazza AC, Kapczinski F, Kauer-Sant'Anna M, Walz JC, Bond DJ, Goncalves CA et al (2009) 3-Nitrotyrosine and glutathione antioxidant system in patients in the early and late stages of bipolar disorder. J Psychiatry Neurosci 34(4):263–271

Association AP (2013) Diagnostic and statistical manual of mental disorders, 5th edn. American Psychiatric Publishing, Arlington, pp 123–154

Austin M, Tan YC (2012) Mania associated with infliximab. Aust N Z J Psychiatry 46(7):684–685

Ayorech Z, Tracy DK, Baumeister D, Giaroli G (2015) Taking the fuel out of the fire: evidence for the use of anti-inflammatory agents in the treatment of bipolar disorders. J Affect Disord 174:467–478

Bai YM, Su TP, Tsai SJ, Wen-Fei C, Li CT, Pei-Chi T et al (2014) Comparison of inflammatory cytokine levels among type I/type II and manic/hypomanic/euthymic/depressive states of bipolar disorder. J Affect Disord 166:187–192

Barbosa IG, Rocha NP, Huguet RB, Ferreira RA, Salgado JV, Carvalho LA et al (2012) Executive dysfunction in euthymic bipolar disorder patients and its association with plasma biomarkers. J Affect Disord 137(1–3):151–155

Barbosa IG, Bauer ME, Machado-Vieira R, Teixeira AL (2014a) Cytokines in bipolar disorder: paving the way for neuroprogression. Neural Plast 2014:360481

Barbosa IG, Machado-Vieira R, Soares JC, Teixeira AL (2014b) The immunology of bipolar disorder. Neuroimmunomodulation 21(2–3):117–122

Baxter LR Jr, Phelps ME, Mazziotta JC, Schwartz JM, Gerner RH, Selin CE et al (1985) Cerebral metabolic rates for glucose in mood disorders. Studies with positron emission tomography and fluorodeoxyglucose F 18. Arch Gen Psychiatry 42(5):441–447

Bechter K (2013) Updating the mild encephalitis hypothesis of schizophrenia. Prog Neuro-Psychopharmacol Biol Psychiatry 42:71–91

Bengesser SA, Reininghaus EZ, Lackner N, Birner A, Fellendorf FT, Platzer M et al (2018) Is the molecular clock ticking differently in bipolar disorder? Methylation analysis of the clock gene ARNTL. World J Biol Psychiatry 19(Supp 2):S21–Ss9

Bengesser SA, Morkl S, Painold A, Dalkner N, Birner A, Fellendorf FT et al (2019) Epigenetics of the molecular clock and bacterial diversity in bipolar disorder. Psychoneuroendocrinology 101:160–166

Benros ME, Waltoft BL, Nordentoft M, Ostergaard SD, Eaton WW, Krogh J et al (2013) Autoimmune diseases and severe infections as risk factors for mood disorders: a nationwide study. JAMA Psychiat 70(8):812–820

Berk M, Plein H, Belsham B (2000) The specificity of platelet glutamate receptor supersensitivity in psychotic disorders. Life Sci 66(25):2427–2432

Berk M, Hallam KT, McGorry PD (2007) The potential utility of a staging model as a course specifier: a bipolar disorder perspective. J Affect Disord 100(1–3):279–281

Berk M, Ng F, Dean O, Dodd S, Bush AI (2008) Glutathione: a novel treatment target in psychiatry. Trends Pharmacol Sci 29(7):346–351

Berk M, Kapczinski F, Andreazza AC, Dean OM, Giorlando F, Maes M et al (2011) Pathways underlying neuroprogression in bipolar disorder: focus on inflammation, oxidative stress and neurotrophic factors. Neurosci Biobehav Rev 35(3):804–817

Berk M, Dean O, Drexhage H, McNeil JJ, Moylan S, O'Neil A et al (2013) Aspirin: a review of its neurobiological properties and therapeutic potential for mental illness. BMC Med 11:74

Bezchlibnyk YB, Wang JF, McQueen GM, Young LT (2001) Gene expression differences in bipolar disorder revealed by cDNA array analysis of post-mortem frontal cortex. J Neurochem 79(4):826–834

Brietzke E, Kauer-Sant'Anna M, Teixeira AL, Kapczinski F (2009a) Abnormalities in serum chemokine levels in euthymic patients with bipolar disorder. Brain Behav Immun 23 (8):1079–1082

Brietzke E, Stertz L, Fernandes BS, Kauer-Sant'anna M, Mascarenhas M, Escosteguy Vargas A et al (2009b) Comparison of cytokine levels in depressed, manic and euthymic patients with bipolar disorder. J Affect Disord 116(3):214–217

Brooks JO 3rd, Wang PW, Bonner JC, Rosen AC, Hoblyn JC, Hill SJ et al (2009) Decreased prefrontal, anterior cingulate, insula, and ventral striatal metabolism in medication-free depressed outpatients with bipolar disorder. J Psychiatr Res 43(3):181–188

Brown AS (2015) The Kraepelinian dichotomy from the perspective of prenatal infectious and immunologic insults. Schizophr Bull 41(4):786–791

Brown AS, Derkits EJ (2010) Prenatal infection and schizophrenia: a review of epidemiologic and translational studies. Am J Psychiatry 167(3):261–280

Cacci E, Claasen JH, Kokaia Z (2005) Microglia-derived tumor necrosis factor-alpha exaggerates death of newborn hippocampal progenitor cells in vitro. J Neurosci Res 80(6):789–797

Caliyurt O, Altiay G (2009) Resting energy expenditure in manic episode. Bipolar Disord 11 (1):102–106

Canetta SE, Bao Y, Co MD, Ennis FA, Cruz J, Terajima M et al (2014) Serological documentation of maternal influenza exposure and bipolar disorder in adult offspring. Am J Psychiatry 171 (5):557–563

Cao B, Passos IC, Mwangi B, Bauer IE, Zunta-Soares GB, Kapczinski F et al (2016) Hippocampal volume and verbal memory performance in late-stage bipolar disorder. J Psychiatr Res 73:102–107

Carabotti M, Scirocco A, Maselli MA, Severi C (2015) The gut-brain axis: interactions between enteric microbiota, central and enteric nervous systems. Ann Gastroenterol 28(2):203–209

Cheng R, Juo SH, Loth JE, Nee J, Iossifov I, Blumenthal R et al (2006) Genome-wide linkage scan in a large bipolar disorder sample from the National Institute of Mental Health genetics initiative suggests putative loci for bipolar disorder, psychosis, suicide, and panic disorder. Mol Psychiatry 11(3):252–260

Chung KH, Huang SH, Wu JY, Chen PH, Hsu JL, Tsai SY (2013) The link between high-sensitivity C-reactive protein and orbitofrontal cortex in euthymic bipolar disorder. Neuropsychobiology 68(3):168–173

Collins SM, Bercik P (2009) The relationship between intestinal microbiota and the central nervous system in normal gastrointestinal function and disease. Gastroenterology 136(6):2003–2014

Cryan JF, Dinan TG (2012) Mind-altering microorganisms: the impact of the gut microbiota on brain and behaviour. Nat Rev Neurosci 13(10):701–712

Cunha AB, Andreazza AC, Gomes FA, Frey BN, da Silveira LE, Goncalves CA et al (2008) Investigation of serum high-sensitive C-reactive protein levels across all mood states in bipolar disorder. Eur Arch Psychiatry Clin Neurosci 258(5):300–304

Dantzer R, O'Connor JC, Freund GG, Johnson RW, Kelley KW (2008) From inflammation to sickness and depression: when the immune system subjugates the brain. Nat Rev Neurosci 9 (1):46–56

Daskalakis NP, Bagot RC, Parker KJ, Vinkers CH, de Kloet ER (2013) The three-hit concept of vulnerability and resilience: toward understanding adaptation to early-life adversity outcome. Psychoneuroendocrinology 38(9):1858–1873

De Berardis D, Conti CM, Campanella D, Carano A, Scali M, Valchera A et al (2008) Evaluation of C-reactive protein and total serum cholesterol in adult patients with bipolar disorder. Int J Immunopathol Pharmacol 21(2):319–324

Debnath M, Venkatasubramanian G, Berk M (2015) Fetal programming of schizophrenia: select mechanisms. Neurosci Biobehav Rev 49:90–104

Dickerson FB, Boronow JJ, Stallings C, Origoni AE, Cole S, Krivogorsky B et al (2004) Infection with herpes simplex virus type 1 is associated with cognitive deficits in bipolar disorder. Biol Psychiatry 55(6):588–593

Dickerson F, Stallings C, Origoni A, Boronow J, Yolken R (2007) Elevated serum levels of C-reactive protein are associated with mania symptoms in outpatients with bipolar disorder. Prog Neuro-Psychopharmacol Biol Psychiatry 31(4):952–955

Dickerson F, Stallings C, Vaughan C, Origoni A, Khushalani S, Yolken R (2012) Antibodies to the glutamate receptor in mania. Bipolar Disord 14(5):547–553

Dickerson F, Stallings C, Origoni A, Vaughan C, Khushalani S, Yolken R (2013) Elevated C-reactive protein and cognitive deficits in individuals with bipolar disorder. J Affect Disord 150(2):456–459

Dickerson F, Stallings C, Origoni A, Katsafanas E, Schweinfurth L, Savage C et al (2014a) Antibodies to Toxoplasma gondii and cognitive functioning in schizophrenia, bipolar disorder, and nonpsychiatric controls. J Nerv Ment Dis 202(8):589–593

Dickerson F, Stallings C, Origoni A, Vaughan C, Katsafanas E, Khushalani S et al (2014b) Antibodies to Toxoplasma gondii in individuals with mania. Bipolar Disord 16(2):129–136

Dickerson F, Severance E, Yolken R (2017) The microbiome, immunity, and schizophrenia and bipolar disorder. Brain Behav Immun 62:46–52

Dong XH, Zhen XC (2015) Glial pathology in bipolar disorder: potential therapeutic implications. CNS Neurosci Ther 21(5):393–397

Elisa B, Beny L (2010) Induction of manic switch by the tumour necrosis factor-alpha antagonist infliximab. Psychiatry Clin Neurosci 64(4):442–443

Estes ML, McAllister AK (2016) Maternal immune activation: implications for neuropsychiatric disorders. Science 353(6301):772–777

Evans SJ, Bassis CM, Hein R, Assari S, Flowers SA, Kelly MB et al (2017) The gut microbiome composition associates with bipolar disorder and illness severity. J Psychiatr Res 87:23–29

Eyre H, Baune BT (2012) Neuroplastic changes in depression: a role for the immune system. Psychoneuroendocrinology 37(9):1397–1416

Fillman SG, Sinclair D, Fung SJ, Webster MJ, Shannon WC (2014) Markers of inflammation and stress distinguish subsets of individuals with schizophrenia and bipolar disorder. Transl Psychiatry 4:e365

Forsythe P, Sudo N, Dinan T, Taylor VH, Bienenstock J (2010) Mood and gut feelings. Brain Behav Immun 24(1):9–16

Frick LR, Williams K, Pittenger C (2013) Microglial dysregulation in psychiatric disease. Clin Dev Immunol 2013:608654

Giakoumaki SG, Roussos P, Rogdaki M, Karli C, Bitsios P, Frangou S (2007) Evidence of disrupted prepulse inhibition in unaffected siblings of bipolar disorder patients. Biol Psychiatry 62(12):1418–1422

Gibbons AS, Scarr E, McLean C, Sundram S, Dean B (2009) Decreased muscarinic receptor binding in the frontal cortex of bipolar disorder and major depressive disorder subjects. J Affect Disord 116(3):184–191

Giridharan VV, Sayana P, Pinjari OF, Ahmad N, da Rosa MI, Quevedo J et al (2019) Postmortem evidence of brain inflammatory markers in bipolar disorder: a systematic review. Mol Psychiatry:1–20

Goldstein BI, Kemp DE, Soczynska JK, McIntyre RS (2009) Inflammation and the phenomenology, pathophysiology, comorbidity, and treatment of bipolar disorder: a systematic review of the literature. J Clin Psychiatry 70(8):1078–1090

Grande I, Berk M, Birmaher B, Vieta E (2016) Bipolar disorder. Lancet 387(10027):1561–1572

Grayson DR, Kundakovic M, Sharma RP (2010) Is there a future for histone deacetylase inhibitors in the pharmacotherapy of psychiatric disorders? Mol Pharmacol 77(2):126–135

Guo R, Gu J, Zong S, Wu M, Yang M (2018) Structure and mechanism of mitochondrial electron transport chain. Biom J 41(1):9–20

Haddad EB, Rousell J, Lindsay MA, Barnes PJ (1996) Synergy between tumor necrosis factor alpha and interleukin 1beta in inducing transcriptional down-regulation of muscarinic M2 receptor gene expression. Involvement of protein kinase A and ceramide pathways. J Biol Chem 271 (51):32586–32592

Hamdani N, Daban-Huard C, Lajnef M, Gadel R, Le Corvoisier P, Delavest M et al (2015) Cognitive deterioration among bipolar disorder patients infected by Toxoplasma gondii is correlated to interleukin 6 levels. J Affect Disord 179:161–166

Harrison PJ (2016) Molecular neurobiological clues to the pathogenesis of bipolar disorder. Curr Opin Neurobiol 36:1–6

Hillegers MH, Reichart CG, Wals M, Verhulst FC, Ormel J, Nolen WA et al (2007) Signs of a higher prevalence of autoimmune thyroiditis in female offspring of bipolar parents. Eur Neuropsychopharmacol 17(6–7):394–399

Hope S, Hoseth E, Dieset I, Morch RH, Aas M, Aukrust P et al (2015) Inflammatory markers are associated with general cognitive abilities in schizophrenia and bipolar disorder patients and healthy controls. Schizophr Res 165(2–3):188–194

Hoseth EZ, Westlye LT, Hope S, Dieset I, Aukrust P, Melle I et al (2016) Association between cytokine levels, verbal memory and hippocampus volume in psychotic disorders and healthy controls. Acta Psychiatr Scand 133(1):53–62

Houenou J, d'Albis MA, Daban C, Hamdani N, Delavest M, Lepine JP et al (2014) Cytomegalovirus seropositivity and serointensity are associated with hippocampal volume and verbal memory in schizophrenia and bipolar disorder. Prog Neuro-Psychopharmacol Biol Psychiatry 48:142–148

Hsu CC, Chen SC, Liu CJ, Lu T, Shen CC, Hu YW et al (2014) Rheumatoid arthritis and the risk of bipolar disorder: a nationwide population-based study. PLoS One 9(9):e107512

Huang TT, Lai JB, Du YL, Xu Y, Ruan LM, Hu SH (2019) Current understanding of gut microbiota in mood disorders: an update of human studies. Front Genet 10:98

Hwang N, Eom T, Gupta SK, Jeong SY, Jeong DY, Kim YS et al (2017) Genes and gut bacteria involved in luminal butyrate reduction caused by diet and loperamide. Genes (Basel) 8(12)

Isgren A, Jakobsson J, Palsson E, Ekman CJ, Johansson AG, Sellgren C et al (2015) Increased cerebrospinal fluid interleukin-8 in bipolar disorder patients associated with lithium and antipsychotic treatment. Brain Behav Immun 43:198–204

Isgren A, Sellgren C, Ekman CJ, Holmen-Larsson J, Blennow K, Zetterberg H et al (2017) Markers of neuroinflammation and neuronal injury in bipolar disorder: relation to prospective clinical outcomes. Brain Behav Immun 65:195–201

Iwasaki A, Medzhitov R (2010) Regulation of adaptive immunity by the innate immune system. Science 327(5963):291–295

Jakobsson J, Bjerke M, Ekman CJ, Sellgren C, Johansson AG, Zetterberg H et al (2014) Elevated concentrations of neurofilament light chain in the cerebrospinal fluid of bipolar disorder patients. Neuropsychopharmacology 39(10):2349–2356

Jakobsson J, Bjerke M, Sahebi S, Isgren A, Ekman CJ, Sellgren C et al (2015) Monocyte and microglial activation in patients with mood-stabilized bipolar disorder. J Psychiatry Neurosci 40 (4):250–258

Jones KA, Porjesz B, Almasy L, Bierut L, Goate A, Wang JC et al (2004) Linkage and linkage disequilibrium of evoked EEG oscillations with CHRM2 receptor gene polymorphisms: implications for human brain dynamics and cognition. Int J Psychophysiol 53(2):75–90

Kapczinski NS, Mwangi B, Cassidy RM, Librenza-Garcia D, Bermudez MB, Kauer-Sant'anna M et al (2017) Neuroprogression and illness trajectories in bipolar disorder. Expert Rev Neurother 17(3):277–285

Kapur N, Barker S, Burrows EH, Ellison D, Brice J, Illis LS et al (1994) Herpes simplex encephalitis: long term magnetic resonance imaging and neuropsychological profile. J Neurol Neurosurg Psychiatry 57(11):1334–1342

Kato T (2007) Mitochondrial dysfunction as the molecular basis of bipolar disorder: therapeutic implications. CNS Drugs 21(1):1–11

Kauer-Sant'Anna M, Kapczinski F, Andreazza AC, Bond DJ, Lam RW, Young LT et al (2009) Brain-derived neurotrophic factor and inflammatory markers in patients with early- vs. late-stage bipolar disorder. Int J Neuropsychopharmacol 12(4):447–458

Khairova RA, Machado-Vieira R, Du J, Manji HK (2009) A potential role for pro-inflammatory cytokines in regulating synaptic plasticity in major depressive disorder. Int J Neuropsychopharmacol 12(4):561–578

Khan D, Fernando P, Cicvaric A, Berger A, Pollak A, Monje FJ et al (2014) Long-term effects of maternal immune activation on depression-like behavior in the mouse. Transl Psychiatry 4:e363

Khandaker GM, Zimbron J, Dalman C, Lewis G, Jones PB (2012) Childhood infection and adult schizophrenia: a meta-analysis of population-based studies. Schizophr Res 139(1–3):161–168

Kilbourne ED (1987) Influenza. Plenum Medical Book, New York

Kim YK, Myint AM, Lee BH, Han CS, Lee SW, Leonard BE et al (2004) T-helper types 1, 2, and 3 cytokine interactions in symptomatic manic patients. Psychiatry Res 129(3):267–272

Kim YK, Jung HG, Myint AM, Kim H, Park SH (2007) Imbalance between pro-inflammatory and anti-inflammatory cytokines in bipolar disorder. J Affect Disord 104(1–3):91–95

Koehler K, Guth W (1979) The mimicking of mania in "benign" herpes simplex encephalitis. Biol Psychiatry 14(2):405–411

Konradi C, Eaton M, MacDonald ML, Walsh J, Benes FM, Heckers S (2004) Molecular evidence for mitochondrial dysfunction in bipolar disorder. Arch Gen Psychiatry 61(3):300–308

Kupferschmidt DA, Zakzanis KK (2011) Toward a functional neuroanatomical signature of bipolar disorder: quantitative evidence from the neuroimaging literature. Psychiatry Res 193(2):71–79

Lavagnino L, Cao B, Mwangi B, Wu MJ, Sanches M, Zunta-Soares GB et al (2015) Changes in the corpus callosum in women with late-stage bipolar disorder. Acta Psychiatr Scand 131 (6):458–464

Leboyer M, Oliveira J, Tamouza R, Groc L (2016) Is it time for immunopsychiatry in psychotic disorders? Psychopharmacology 233(9):1651–1660

Machado-Vieira R, Andreazza AC, Viale CI, Zanatto V, Cereser V Jr, da Silva VR et al (2007) Oxidative stress parameters in unmedicated and treated bipolar subjects during initial manic episode: a possible role for lithium antioxidant effects. Neurosci Lett 421(1):33–36

Marrie RA, Hitchon CA, Walld R, Patten SB, Bolton JM, Sareen J et al (2018a) Increased burden of psychiatric disorders in rheumatoid arthritis. Arthritis Care Res (Hoboken) 70(7):970–978

Marrie RA, Walld R, Bolton JM, Sareen J, Walker JR, Patten SB et al (2018b) Physical comorbidities increase the risk of psychiatric comorbidity in immune-mediated inflammatory disease. Gen Hosp Psychiatry 51:71–78

Maurer IC, Schippel P, Volz HP (2009) Lithium-induced enhancement of mitochondrial oxidative phosphorylation in human brain tissue. Bipolar Disord 11(5):515–522

Merikangas KR, Akiskal HS, Angst J, Greenberg PE, Hirschfeld RM, Petukhova M et al (2007) Lifetime and 12-month prevalence of bipolar spectrum disorder in the National Comorbidity Survey replication. Arch Gen Psychiatry 64(5):543–552

Meyer U (2014) Prenatal poly(i:C) exposure and other developmental immune activation models in rodent systems. Biol Psychiatry 75(4):307–315

Meyer U, Feldon J, Fatemi SH (2009) In-vivo rodent models for the experimental investigation of prenatal immune activation effects in neurodevelopmental brain disorders. Neurosci Biobehav Rev 33(7):1061–1079

Miller CL, Llenos IC, Dulay JR, Weis S (2006) Upregulation of the initiating step of the kynurenine pathway in postmortem anterior cingulate cortex from individuals with schizophrenia and bipolar disorder. Brain Res 1073-1074:25–37

Miller BJ, Buckley P, Seabolt W, Mellor A, Kirkpatrick B (2011) Meta-analysis of cytokine alterations in schizophrenia: clinical status and antipsychotic effects. Biol Psychiatry 70 (7):663–671

Modabbernia A, Taslimi S, Brietzke E, Ashrafi M (2013) Cytokine alterations in bipolar disorder: a meta-analysis of 30 studies. Biol Psychiatry 74(1):15–25

Mook-Kanamori BB, Geldhoff M, van der Poll T, van de Beek D (2011) Pathogenesis and pathophysiology of pneumococcal meningitis. Clin Microbiol Rev 24(3):557–591

Moore PB, El-Badri SM, Cousins D, Shepherd DJ, Young AH, McAllister VL et al (2001) White matter lesions and season of birth of patients with bipolar affective disorder. Am J Psychiatry 158(9):1521–1524

Munkholm K, Vinberg M, Vedel KL (2013) Cytokines in bipolar disorder: a systematic review and meta-analysis. J Affect Disord 144(1–2):16–27

Mwangi B, Wu MJ, Cao B, Passos IC, Lavagnino L, Keser Z et al (2016) Individualized prediction and clinical staging of bipolar disorders using neuroanatomical biomarkers. Biol Psychiatry Cogn Neurosci Neuroimag 1(2):186–194

Myint AM, Kim YK, Verkerk R, Park SH, Scharpe S, Steinbusch HW et al (2007) Tryptophan breakdown pathway in bipolar mania. J Affect Disord 102(1–3):65–72

Nery FG, Monkul ES, Hatch JP, Fonseca M, Zunta-Soares GB, Frey BN et al (2008) Celecoxib as an adjunct in the treatment of depressive or mixed episodes of bipolar disorder: a double-blind, randomized, placebo-controlled study. Hum Psychopharmacol 23(2):87–94

Nordentoft M, Mortensen PB, Pedersen CB (2011) Absolute risk of suicide after first hospital contact in mental disorder. Arch Gen Psychiatry 68(10):1058–1064

Orlovska-Waast S, Kohler-Forsberg O, Brix SW, Nordentoft M, Kondziella D, Krogh J et al (2019) Cerebrospinal fluid markers of inflammation and infections in schizophrenia and affective disorders: a systematic review and meta-analysis. Mol Psychiatry 24(6):869–887

Ortiz-Dominguez A, Hernandez ME, Berlanga C, Gutierrez-Mora D, Moreno J, Heinze G et al (2007) Immune variations in bipolar disorder: phasic differences. Bipolar Disord 9(6):596–602
Painold A, Morkl S, Kashofer K, Halwachs B, Dalkner N, Bengesser S et al (2019) A step ahead: exploring the gut microbiota in inpatients with bipolar disorder during a depressive episode. Bipolar Disord 21(1):40–49
Panizzutti B, Gubert C, Schuh AL, Ferrari P, Bristot G, Fries GR et al (2015) Increased serum levels of eotaxin/CCL11 in late-stage patients with bipolar disorder: an accelerated aging biomarker? J Affect Disord 182:64–69
Parboosing R, Bao Y, Shen L, Schaefer CA, Brown AS (2013) Gestational influenza and bipolar disorder in adult offspring. JAMA Psychiat 70(7):677–685
Perry W, Minassian A, Feifel D, Braff DL (2001) Sensorimotor gating deficits in bipolar disorder patients with acute psychotic mania. Biol Psychiatry 50(6):418–424
Post RM (2007) Kindling and sensitization as models for affective episode recurrence, cyclicity, and tolerance phenomena. Neurosci Biobehav Rev 31(6):858–873
Rao JS, Harry GJ, Rapoport SI, Kim HW (2010) Increased excitotoxicity and neuroinflammatory markers in postmortem frontal cortex from bipolar disorder patients. Mol Psychiatry 15 (4):384–392
Reinares M, Papachristou E, Harvey P, Mar Bonnin C, Sanchez-Moreno J, Torrent C et al (2013) Towards a clinical staging for bipolar disorder: defining patient subtypes based on functional outcome. J Affect Disord 144(1–2):65–71
Rolstad S, Jakobsson J, Sellgren C, Ekman CJ, Blennow K, Zetterberg H et al (2015a) Cognitive performance and cerebrospinal fluid biomarkers of neurodegeneration: a study of patients with bipolar disorder and healthy controls. PLoS One 10(5):e0127100
Rolstad S, Jakobsson J, Sellgren C, Isgren A, Ekman CJ, Bjerke M et al (2015b) CSF neuroinflammatory biomarkers in bipolar disorder are associated with cognitive impairment. Eur Neuropsychopharmacol 25(8):1091–1098
Ronovsky M, Berger S, Molz B, Berger A, Pollak DD (2016) Animal models of maternal immune activation in depression research. Curr Neuropharmacol 14(7):688–704
Rosa AR, Magalhaes PV, Czepielewski L, Sulzbach MV, Goi PD, Vieta E et al (2014) Clinical staging in bipolar disorder: focus on cognition and functioning. J Clin Psychiatry 75(5):e450–e456
Rosenblat JD, McIntyre RS (2015) Are medical comorbid conditions of bipolar disorder due to immune dysfunction? Acta Psychiatr Scand 132(3):180–191
Rosenblat JD, Cha DS, Mansur RB, McIntyre RS (2014) Inflamed moods: a review of the interactions between inflammation and mood disorders. Prog Neuro-Psychopharmacol Biol Psychiatry 53:23–34
Samalin L, de Chazeron I, Vieta E, Bellivier F, Llorca PM (2016) Residual symptoms and specific functional impairments in euthymic patients with bipolar disorder. Bipolar Disord 18 (2):164–173
Savitz J, Preskorn S, Teague TK, Drevets D, Yates W, Drevets W (2012) Minocycline and aspirin in the treatment of bipolar depression: a protocol for a proof-of-concept, randomised, double-blind, placebo-controlled, 2x2 clinical trial. BMJ Open 2(1):e000643
Savitz JB, Price JL, Drevets WC (2014) Neuropathological and neuromorphometric abnormalities in bipolar disorder: view from the medial prefrontal cortical network. Neurosci Biobehav Rev 42:132–147
Sayana P, Colpo GD, Simoes LR, Giridharan VV, Teixeira AL, Quevedo J et al (2017) A systematic review of evidence for the role of inflammatory biomarkers in bipolar patients. J Psychiatr Res 92:160–182
Schlitt M, Lakeman FD, Whitley RJ (1985) Psychosis and herpes simplex encephalitis. South Med J 78(11):1347–1350
Sellner J, Tauber MG, Leib SL (2010) Pathogenesis and pathophysiology of bacterial CNS infections. Handb Clin Neurol 96:1–16

Soderlund J, Olsson SK, Samuelsson M, Walther-Jallow L, Johansson C, Erhardt S et al (2011) Elevation of cerebrospinal fluid interleukin-1ss in bipolar disorder. J Psychiatry Neurosci 36 (2):114–118

Steiner J, Walter M, Gos T, Guillemin GJ, Bernstein HG, Sarnyai Z et al (2011) Severe depression is associated with increased microglial quinolinic acid in subregions of the anterior cingulate gyrus: evidence for an immune-modulated glutamatergic neurotransmission? J Neuroinflammation 8:94

Stephenson J, Nutma E, van der Valk P, Amor S (2018) Inflammation in CNS neurodegenerative diseases. Immunology 154(2):204–219

Stertz L, Magalhaes PV, Kapczinski F (2013) Is bipolar disorder an inflammatory condition? The relevance of microglial activation. Curr Opin Psychiatry 26(1):19–26

Stich O, Andres TA, Gross CM, Gerber SI, Rauer S, Langosch JM (2015) An observational study of inflammation in the central nervous system in patients with bipolar disorder. Bipolar Disord 17 (3):291–302

Strakowski SM, DelBello MP, Zimmerman ME, Getz GE, Mills NP, Ret J et al (2002) Ventricular and periventricular structural volumes in first- versus multiple-episode bipolar disorder. Am J Psychiatry 159(11):1841–1847

Strasser HC, Lilyestrom J, Ashby ER, Honeycutt NA, Schretlen DJ, Pulver AE et al (2005) Hippocampal and ventricular volumes in psychotic and nonpsychotic bipolar patients compared with schizophrenia patients and community control subjects: a pilot study. Biol Psychiatry 57 (6):633–639

Sun X, Wang JF, Tseng M, Young LT (2006) Downregulation in components of the mitochondrial electron transport chain in the postmortem frontal cortex of subjects with bipolar disorder. J Psychiatry Neurosci 31(3):189–196

Tedla Y, Shibre T, Ali O, Tadele G, Woldeamanuel Y, Asrat D et al (2011) Serum antibodies to Toxoplasma gondii and Herpesvidae family viruses in individuals with schizophrenia and bipolar disorder: a case-control study. Ethiop Med J 49(3):211–220

Tiosano S, Nir Z, Gendelman O, Comaneshter D, Amital H, Cohen AD et al (2017) The association between systemic lupus erythematosus and bipolar disorder - a big data analysis. Eur Psychiatry 43:116–119

Traynelis SF, Wollmuth LP, McBain CJ, Menniti FS, Vance KM, Ogden KK et al (2010) Glutamate receptor ion channels: structure, regulation, and function. Pharmacol Rev 62(3):405–496

Vitaliani R, Mason W, Ances B, Zwerdling T, Jiang Z, Dalmau J (2005) Paraneoplastic encephalitis, psychiatric symptoms, and hypoventilation in ovarian teratoma. Ann Neurol 58 (4):594–604

Wadee AA, Kuschke RH, Wood LA, Berk M, Ichim L, Maes M (2002) Serological observations in patients suffering from acute manic episodes. Hum Psychopharmacol 17(4):175–179

Yolken RHDF (2011) Neurotropic viruses and cognitive impairment. Schizophr Bull 37:235–236

Yolken RH, Torrey EF (2008) Are some cases of psychosis caused by microbial agents? A review of the evidence. Mol Psychiatry 13(5):470–479

Zou J, Wang YX, Dou FF, Lu HZ, Ma ZW, Lu PH et al (2010) Glutamine synthetase down-regulation reduces astrocyte protection against glutamate excitotoxicity to neurons. Neurochem Int 56(4):577–584

Neuropsychology of Bipolar Disorder

Peter Gallagher

Contents

Abstract Bipolar disorder is associated with significant dysfunction in a broad range of neuropsychological domains and processes. Deficits have been reported to occur in symptomatic states (depression, [hypo]mania) as well as in remission (euthymia), having consequences for psychological well-being and social and occupational functioning. The profile and magnitude of neuropsychological deficits in bipolar disorder have been explored in a number of systematic reviews and meta-analyses. After discussing these briefly, this chapter will focus on examining the clinical and demographic factors that influence and modify the pattern and magnitude of deficits, as well as reviewing methods of assessment and analysis approaches which may improve our understanding of these problems.

P. Gallagher (✉)
Faculty of Medical Sciences, Newcastle University – Translational and Clinical Research Institute, Newcastle upon Tyne, UK
e-mail: peter.gallagher@ncl.ac.uk

Curr Topics Behav Neurosci (2021) 48: 239–254
https://doi.org/10.1007/7854_2020_148
Published Online: 16 July 2020

Keywords Attention · Attention · Bipolar disorder · Depression · Mania · Memory · Neuropsychology · Processing speed

1 Pattern and Magnitude of Impairment

Early studies of neuropsychological impairment in bipolar disorder were typically small and clinically heterogeneous samples (Henry et al. 1973; Johnson and Magaro 1987). Assessing the verbal memory and fluency of 12 participants with bipolar depression, Wolfe et al. (1987) reported significantly poorer performance compared to those with major depression and with memory dysfunction which was qualitatively similar to patients with early Huntingdon's disease. While in 20 bipolar patients in the manic state, Morice (1990) reported significant deficits in cognitive flexibility (Wisconsin Card Sorting Test performance) which were similar to those in schizophrenia. An initial narrative review of the literature highlighted the difficulties inherent in comparing neuropsychological profiles across groups that differ, in terms of symptoms and concomitant treatments (Murphy and Sahakian 2001).

Around this time, an increasing interest emerged in understating the 'trait' aspects of neuropsychological dysfunction in bipolar disorder.[1] One of the earliest systematic assessments by Ferrier et al. (1999) suggested that executive function/working memory dysfunction was evident in euthymic patients compared to matched, healthy controls, including when factors such as age, IQ and residual depressive symptoms were accounted for in the analysis. Subsequent work sought to minimise the potential confounding effects of residual symptoms in the study design, for example, through prospective verification of euthymia. Thompson et al. (2005) assessed 63 patients with bipolar disorder (where euthymia was confirmed through clinical ratings over the month prior to testing) and a matched control group with a wide-ranging neuropsychological test battery. While the patient group was found to have performed statistically worse than controls across multiple cognitive domains (executive function and attention, working memory, verbal and visuospatial memory and psychomotor/processing speed), clinically significant deficits – defined as performance below the fifth percentile of controls on any outcome – were also noted in a high proportion of patients (e.g. 36% in processing/psychomotor speed and 19–34% on a number of the attentional, executive and memory measures).

[1]Unlike many other clinical conditions in which neuropsychological problems have been characterised, the focus of a great many studies in bipolar disorder has been when individuals are asymptomatic or euthymic. This is most likely a consequence of challenging the Kraepelinian dichotomy, in which cognitive decline was believed to occur in dementia praecox (schizophrenia), but not in manic-depressive psychosis (bipolar disorder) Kraepelin (1899) Psychiatrie. Ein Lehrbuch für Studierende und Ärzte, sixth edn. Barth, Leipzig, Germany.

Following the increase in studies focussed on the neuropsychological profile of euthymia, Robinson et al. (2006) published the first systematic meta-analysis synthesising the results of 26 studies, involving a total of 689 patients with bipolar disorder and 721 controls. Statistically significant differences were found in every measure assessed. The largest effect sizes were observed in measures of executive function (category fluency, mental manipulation) and in verbal learning, while medium effect sizes ($0.5 \leq d < 0.8$) were observed for short-term and delayed verbal memory, other executive measures (abstraction and set-shifting), sustained attention, response inhibition and psychomotor speed. Small effect sizes ($0.2 \leq d < 0.5$) were observed in verbal (letter) fluency and immediate memory. Subsequent meta-analyses supported this initial synthesis, reporting a similar pattern of results (Arts et al. 2008; Bora et al. 2009). In a recent systematic review of 250 studies of neuropsychological function across all illness phases (which included a summary of previous meta-analyses), Tsitsipa and Fountoulakis (2015) reported that across these studies, there is evidence that almost every cognitive domain that has been assessed has found poorer performance in BD compared to controls, in the 'medium' range (Cohen 1988) in euthymia, but of greater magnitude in acute episodes. However, it has also been consistently noted that this conclusion lies very much 'at the group level' and there is notable heterogeneity of neuropsychological performance in BD (Bourne et al. 2013; Cullen et al. 2016; Douglas et al. 2018; Iverson et al. 2011; Krabbendam et al. 2005; Lima et al. 2019; Russo et al. 2017). When more stringent criteria are used to define 'impairment' (i.e. the proportion of BD falling below a specific healthy control-derived cutoff, such as <fifth percentile or <1.5 standard deviations), the majority of many BD samples fall out of this range, with only a minority of individuals demonstrating global impairment (Douglas et al. 2018; Iverson et al. 2011, 2009).

A number of diagnostic/clinical features and illness-related physical symptoms have been found to affect the pattern and, particularly, severity of impairment.

2 What Are the Factors that Affect Cognition?

As previously outlined, there is a great deal of interest in neuropsychological heterogeneity and, further, in determining whether cognitive deficits are a consequence or simply covary with clinical or illness-related features of bipolar disorder.

2.1 *Diagnostic Features*

One area of focus has been in determining the profile of bipolar subtypes, i.e. BD-I and BD-II. In general, the neuropsychological performance of individuals with a history of full-manic episodes is worse than those with a history of hypomania (Bora 2018; Kessler et al. 2013; Schenkel et al. 2012; Torrent et al. 2006) although there are inconsistencies across specific domains (Harkavy-Friedman et al. 2006; Solé et al. 2012; Tsitsipa and Fountoulakis 2015). In a meta-analysis focussed on six executive function processes, Dickinson et al. (2017) found that while BD-II was associated with significant impairment in four of six measures compared to controls, BD-I was associated with impairment in six of six. However, direct comparison of subtypes revealed significant variability in effects across studies, with some processes (e.g. planning) being more impaired in BD-II. This heterogeneity has beset attempts to identify specific differences in the neuropsychological profile of BD-I and BD–II (Solé et al. 2011), although it has been suggested that there may be latent cognitive subgroups across the bipolar spectrum, especially in terms of impaired verbal memory (Aminoff et al. 2013).

More consistently, it has been demonstrated that psychotic symptoms are associated with worse neuropsychological function in bipolar disorder (Allen et al. 2010; Bora 2018; Glahn et al. 2006; Martinez-Aran et al. 2008; Tsitsipa and Fountoulakis 2015). This has been supported by meta-analysis, with performance in individuals with a history of psychosis being significantly worse than those without in four of six cognitive domains: planning and reasoning, working memory, verbal memory and processing speed (Bora et al. 2010). However, it should be noted that psychosis may not influence the neuropsychological profile of first-episode BD (Demmo et al. 2016).

2.2 *Sleep*

Sleep and circadian rhythm disturbance is a commonly reported clinical feature of bipolar disorder (Bradley et al. 2017; Eidelman et al. 2010; Geoffroy et al. 2015; Gruber et al. 2009; Harvey et al. 2005; Kelly et al. 2013; Millar et al. 2004). It is notable that the pattern and magnitude of neuropsychological dysfunction as a consequence of primary sleep disorder closely resembles that described in BD (Lim and Dinges 2010; Waters and Bucks 2011), with 'moderate' general deficits across a range of domains but with larger effects in processing speed/attention – this closely resembles that seen in BD depression (Boland and Alloy 2013; Gallagher et al. 2014, 2015b). Interestingly, recent work incorporating contiguous assessment of sleep and neuropsychological function in BD has found that, rather than exacerbating neuropsychological deficits, only individuals with sleep disturbance exhibited deficits (particularly processing speed and attentional deficits), while those with 'normal' sleep did not differ from controls (Bradley et al. 2020). This raises the possibility that sleep problems may be a primary driver of neuropsychological dysfunction and therefore opens up novel treatment possibilities targeting sleep and circadian rhythm disturbance (Harvey et al. 2015; Jansson-Fröjmark and Norell-Clarke 2016).

2.3 Physical Health

There are a number of concomitant physical health-related illness features which may contribute to or exacerbate neuropsychological dysfunction in bipolar disorder. Obesity, metabolic syndrome and cardiovascular risk have all been found to be increased in bipolar disorder (Czepielewski et al. 2013; Silarova et al. 2015), which in turn may negatively affect neuropsychological function (McIntyre et al. 2017; Mora et al. 2017). Although the mechanism is complex, hyperactivity of the hypothalamic-pituitary-adrenal axis may be implicated (Gallagher et al. 2009). There is a wealth of evidence that these problems are linked to neuropsychological impairment independently of mood disorder and possibly that they are worsened by some medications (Mackin et al. 2007); therefore more work is needed to understand the temporal relationship between these observations.

2.4 Medication

A long-standing question is the degree to which treatment with psychotropic medication contributes to neuropsychological dysfunction in bipolar disorder. Some systematic reviews have suggested that there may be some evidence of poorer performance in those treated with antipsychotics (Cullen et al. 2016). However, others have suggested that the effects are limited and may be confounded by the clinical symptoms leading to their use (Tsitsipa and Fountoulakis 2015). A recent analysis of data from UK Biobank analysed a sample of $n = 2{,}709$ individuals characterised as having bipolar disorder revealed small neuropsychological effects restricted to visuospatial memory with around a quarter of this effect attributable to psychotropic medication (Cullen et al. 2019); however the method of ascertaining diagnosis and the restricted cognitive testing protocol may have limited these findings. It should also be noted that several primary data studies have reported evidence of widespread neuropsychological impairment in medication-free samples (Goswami et al. 2009, 2006; Pavuluri et al. 2006). Moreover, in a pooled analysis of data from $n = 1{,}267$ BD patients, regression analysis of specific medication classes revealed few effects on performance other than subtle effects on verbal learning, with the majority of contrasts suggesting no relationship (Bourne et al. 2013). Thus, it appears that broadly, the neuropsychological deficits seen in bipolar disorder are not iatrogenic.

3 Summary

It is clear that individuals with a diagnosis of bipolar disorder, when symptomatic and when euthymic, exhibit neuropsychological dysfunction. However, it is also clear that this conclusion relates specifically to the 'cohort level' – there is considerable heterogeneity in the actual profile of deficits, with numerous diagnostic and clinical features that are deleterious to performance. In the next section, the focus will be on exploring a range of methods of assessment (both in terms of design and analysis) that may provide a better understanding of the neuropsychology of bipolar disorder.

4 Methods of Assessment

One of the most pressing questions for this field of research is whether there are specific methods or approaches at our disposal that may take us closer to establishing a cognitive profile of bipolar disorder – or even to understand to what extent this is possible?

4.1 Longitudinal Changes

The majority of studies conducted which examined neuropsychological function in bipolar disorder are cross-sectional and, therefore, cannot further our understanding of the stability and temporal trajectory of cognitive deficits (Ryan et al. 2016). Findings from a recent meta-analysis suggest that there is no relative cognitive decline between bipolar disorder and controls in either short-term (~1.5 years) or longer-term (~5.5 years) follow-up studies (Bora and Özerdem 2017). In terms of short-term changes in response to treatment, the extent to which neuropsychological processes improve is domain-specific. Xu et al. (2012) found that in the depressed phase, while the predicted processing speed, memory and executive deficits were observed, after treatment for 6 weeks, those who remitted continued to exhibit impairments in processing speed and memory. Diagnostic subtype has also been observed to affect changes during early remission, with psychotic symptoms leading to higher rates of residual symptoms, neuropsychological dysfunction and poorer functional recovery (Levy et al. 2013). Examination of longer timeframes has demonstrated an association between neuropsychological performance and 1-year functional outcome in bipolar disorder (Tabarés-Seisdedos et al. 2008). However, it is important to understand the temporal trajectories as recent work utilizing a cross-lagged panel model approach suggested that while the neuropsychological function was causally primary and moderately predictive of subsequent functional outcome (1 year later), the converse did not hold – psychosocial functioning did not predict subsequent neuropsychological performance (Ehrminger et al. 2019).

4.2 Identifying Neuropsychological Phenotypic Clusters

Several different approaches have been utilized better to understand the specific neuropsychological profile of bipolar disorder. Burdick et al. (2014) applied hierarchical cluster analysis to data from the MATRICS test battery in $n = 136$ participants proposing three specific clusters – globally impaired, globally intact and an intermediate group with selective deficits in processing speed, attention, verbal learning and social cognition. This pattern has also been observed in other studies (Russo et al. 2017) where it has been suggested that such clusters are actually representative of subsections of a continuum (Lima et al. 2019; Van Rheenen et al. 2017). Similar approaches have been applied to executive processes and imaging data, also producing three clusters (Kollmann et al. 2019), and to reward processing (Jimenez et al. 2018). In a larger dataset of general neuropsychological measures from $n = 258$ euthymic patients, Roux et al. (2017) proposed a four-cluster pattern, with a globally impaired cluster, a globally intact (above average performance) cluster and two further clusters that were normal with the exception of impaired or superior verbal performance. This pattern is very similar to that found in an earlier study in individuals with psychosis, including $n = 73$ with bipolar disorder (Lewandowski et al. 2014) which was later replicated (Lewandowski et al. 2018). Interestingly, in a cross-diagnostic cluster analysis, Lee et al. (2017) found only two clusters – impaired and intact/superior – but this did not map onto clinical diagnosis (although poorer social functioning appeared to differentiate those with a diagnosis of schizophrenia from bipolar disorder in the 'impaired' cluster). Collectively, this approach appears to confirm the heterogeneity described previously, when examining the proportions of bipolar samples falling below percentile cutoffs. However, it is also uncertain whether these clusters represent clear, clinically independent subgroups or are simply categories of severity along a continuum.

A related approach has focussed on attempts to elaborate on the factor structure of cognition within bipolar disorder and other related groups and to explore their differences from healthy controls. In a large dataset from up to $n = 5{,}414$ individuals with a diagnosis of bipolar disorder BPI and $n = 3{,}942$ schizophrenia, Harvey et al. (2016) used principal components (PCA) and factor analysis to determine that neuropsychological performance and functional capacity measures combined (as well as the neuropsychological measures or the diagnoses independently) could be reduced to a single principal component that explained most of the variation in the original variables. This is of note as the authors point to earlier studies that have identified as many as six components, consistent between bipolar and schizophrenia samples (Czobor et al. 2007; Schretlen et al. 2013).

Other studies have used PCA as a data reduction technique (acknowledging that resultant component solutions are frequently dataset specific) to explore the cognitive process loadings within each component between bipolar disorder and healthy controls (Gallagher et al. 2014). In controls, there was a clear delineation between components along theoretically derived lines (e.g. visuospatial, verbal memory). However, there were fewer extracted components in the bipolar sample suggesting greater functional homogeneity, particularly of visuospatial processes. It is also of note that the individual variables that loaded into these components were less specific in terms of modality, with every one containing combinations of both verbal and visuospatial measures. In bipolar disorder, some measures loaded heavily across all components, such as processing speed. This pattern was interpreted as being similar to that seen in cognitive ageing, where *dedifferentiation* also leads to a loss of process specificity; notably, previously functionally discrete processes become more amorphous and less differentiated through decline in neural connectivity (Dolcos et al. 2002). Another parallel was highlighted, that of *cognitive scaffolding*, whereby interindividual adaptive changes may occur in underlying neural circuitry engaged in the 'normal' performance of cognitive tasks, resulting in the recruitment of alternative circuits or supportive processes than those typically used (Park and Reuter-Lorenz 2009). There is some suggestion that this may occur in bipolar depression, where it has been shown that deficits in facets of visuospatial memory may be compensated through verbal memory scaffolding (Gallagher et al. 2015a).

Therefore, any attempt to capture the specific neuropsychological profile of bipolar disorder needs to consider this heterogeneity – that while subgrouping by cognitive phenotype (and with a better understanding of the clinical and illness correlates of these) we may come closer to a 'profile', further heterogeneity may be introduced from other adaptive cognitive changes that might occur, closer to the individual level.

4.3 Hierarchical Organization of Cognition

As already discussed, it is commonly reported that neurocognitive deficits in BD at the group level are relatively 'broad' and of moderate effect size. However, this does not account for both the hierarchical organisation of human cognitive functions and the complex interplay between different cognitive processes. The conceptualisation of any observed profile of deficits is changed fundamentally if we consider that neuropsychological functions do not operate independently and, further, that some may be subordinate to impairments in more circumscribed but functionally primary processes. This approach has been applied to the neuropsychology of major depression in older adults, where hierarchical regression modelling of cognitive processes has revealed that broader deficits in episodic memory and visuospatial processes may be mediated by decreased fundamental processing resources (Nebes et al. 2000). Similar approaches applied in younger depressed patients have found that primary attentional deficits may similarly account for deficits in some executive processes (Nilsson et al. 2016). Such methods have been used extensively to develop a better understanding of the neuropsychology of typical and pathological ageing (Clarys et al. 2009), especially on the role of information processing speed and efficiency (Joy et al. 2000, 2004; Salthouse 1996, 2000, 2017). Here, it is also important to note the potential of applying approaches used in experimental neuropsychology studies to understand better the role of specific (primary) cognitive processes in common task performance (e.g. Cepeda et al. 2013; Davis and Pierson 2012; Tam and Schmitter-Edgecombe 2013) and complimentary task design aimed at manipulating specific processes or cognitive load during active task performance. Collectively this may lead to greater insights into the organisation of cognition in bipolar disorder if applied to narrower well-defined clinical phenotypes.

One approach that may also facilitate a better understanding of individual profiles is the application of finite *partially ordered sets (posets)* as classification models (Tatsuoka 2002; Tatsuoka and Ferguson 2003). The approach involves the statistical modelling of cognitive processes in a manner which closely resembles that done during single-case clinical neuropsychological assessment. This has been applied to neuropsychological data for individuals with a diagnosis of schizophrenia (Jaeger et al. 2006a, b) and would be of great interest to apply to bipolar disorder. Given the group-level heterogeneity described in the neuropsychological literature of bipolar disorder, it would similarly be of interest to explore a variety of available methods to assess patterns of performance at an individual level (Crawford and Garthwaite 2002; Crawford et al. 2009a, b). Some of these methods could overcome methodological issues that have not received a great deal of attention, such as comparisons against small sample control norms (Crawford and Garthwaite 2002; Crawford et al. 2009b), quantifying deficits when the psychometric properties of tests differ (Chapman and Chapman 1973) and assessing whether deficits qualify as differential deficits (Crawford et al. 2000), the latter having revealed hierarchical organisation of the cognitive profile of euthymic bipolar disorder (Thompson et al. 2006, 2009).

4.4 Experimental Analysis Methods

There are a growing number of examples of the utility of applying novel analysis methods to refine the measurement of selective cognitive processes in bipolar disorder. One specific example is the assessment of attentional processes. By fitting reaction time (RT) data from sustained attention tests to non-Gaussian distributions (e.g. a mathematically convolved Gaussian and exponential; the ex-Gaussian), differential RT profiles were observed between major depression and bipolar disorder, resulting in larger, statistically significant effect size differences which typical measures of central tendency failed to detect (Gallagher et al. 2015b; Moss et al. 2016). Given the potential importance of attention and processing speed within the cognitive hierarchy, and the relationship with structural and functional connectivity (Pavuluri et al. 2009; Poletti et al. 2015), such methods may offer unique insights into the cognitive profile of bipolar disorder. Similar approaches, coupled with an assessment of intra-individual RT using fast Fourier transform, have been applied successfully to explore candidate cognitive endophenotypes in ADHD (Vaurio et al. 2009), while drift-diffusion modelling and Bayesian approaches have revealed differences in information processing efficiency and reward learning parameters in psychosis (Mathias et al. 2017; Moustafa et al. 2015).

5 Conclusions

Overall, we see clear evidence for neuropsychological impairment in bipolar disorder. Much of the evidence is at the group level, and from numerous sources, it is apparent that specific demographic-, diagnostic- and illness-related features can influence the profile and/or severity of the observed deficits. To further our understanding of neuropsychological processes within bipolar disorder, it is suggested that studies should be more cognizant of the hierarchical organisation of cognition and the existing methods of analysis and task design which might provide unique insights in future work. Such approaches hold promise in deriving a more refined conceptualisation of specific phenotypic presentations – beyond the 'group level' – which could ultimately aid illness stratification.

Acknowledgements I am grateful to Susanna Carella for assistance with the literature searches supporting this manuscript.

References

Allen DN, Randall C, Bello D, Armstrong C, Frantom L, Cross C, Kinney J (2010) Are working memory deficits in bipolar disorder markers for psychosis? Neuropsychology 24:244–254

Aminoff SR, Hellvin T, Lagerberg TV, Berg AO, Andreassen OA, Melle I (2013) Neurocognitive features in subgroups of bipolar disorder. Bipolar Disord 15:272–283

Arts B, Jabben N, Krabbendam L, van Os J (2008) Meta-analyses of cognitive functioning in euthymic bipolar patients and their first-degree relatives. Psychol Med 38:771–785

Boland EM, Alloy LB (2013) Sleep disturbance and cognitive deficits in bipolar disorder: toward an integrated examination of disorder maintenance and functional impairment. Clin Psychol Rev 33:33–44

Bora E (2018) Neurocognitive features in clinical subgroups of bipolar disorder: a meta-analysis. J Affect Disord 229:125–134

Bora E, Özerdem A (2017) Meta-analysis of longitudinal studies of cognition in bipolar disorder: comparison with healthy controls and schizophrenia. Psychol Med 47:2753–2766

Bora E, Yücel M, Pantelis C (2010) Neurocognitive markers of psychosis in bipolar disorder: a meta-analytic study. J Affect Disord 127:1–9

Bora E, Yucel M, Pantelis C (2009) Cognitive endophenotypes of bipolar disorder: a meta-analysis of neuropsychological deficits in euthymic patients and their first-degree relatives. J Affect Disord 113(1–2):1–20

Bourne C, Aydemir O, Balanzá-Martínez V, Bora E, Brissos S, Cavanagh JTO, Clark L, Cubukcuoglu Z, Dias VV, Dittmann S, Ferrier IN, Fleck DE, Frangou S, Gallagher P, Jones L, Kieseppä T, Martínez-Aran A, Melle I, Moore PB, Mur M, Pfennig A, Raust A, Senturk V, Simonsen C, Smith DJ, Soares D, Soeiro-de-Souza MG, Stoddart SDR, Sundet K, Szöke A, Thompson JM, Torrent C, Zalla T, Craddock N, Andreassen OA, Leboyer M, Vieta E, Bauer M, Worhunsky P, Tzagarakis C, Rogers RD, Geddes JR, Goodwin GM (2013) Neuropsychological testing of cognitive impairment in euthymic bipolar disorder: an individual patient data meta-analysis. Acta Psychiatr Scand 128:149–162

Bradley AJ, Webb-Mitchell R, Hazu A, Slater N, Middleton B, Gallagher P, McAllister-Williams H, Anderson KN (2017) Sleep and circadian rhythm disturbance in bipolar disorder. Psychol Med 47:1678–1689

Bradley AJ, Anderson KN, Gallagher P, McAllister-Williams RH (2020) The association between sleep and cognitive abnormalities in bipolar disorder. Psychol Med 50:125–132

Burdick KE, Russo M, Frangou S, Mahon K, Braga RJ, Shanahan M, Malhotra AK (2014) Empirical evidence for discrete neurocognitive subgroups in bipolar disorder: clinical implications. Psychol Med 44:3083–3096

Cepeda NJ, Blackwell KA, Munakata Y (2013) Speed isn't everything: complex processing speed measures mask individual differences and developmental changes in executive control. Dev Sci 16:269–286

Chapman LJ, Chapman JP (1973) Problems in the measurement of cognitive deficit. Psychol Bull 79:380–385

Clarys D, Bugaiska A, Tapia G, Alexia Baudouin A (2009) Ageing, remembering, and executive function. Memory 17:158–168

Cohen J (1988) Statistical power analysis for the behavioral sciences, 2nd edn. Lawrence Erlbaum Associates, Hillsdale

Crawford JR, Garthwaite PH (2002) Investigation of the single case in neuropsychology: confidence limits on the abnormality of test scores and test score differences. Neuropsychologia 40:1196–1208

Crawford JR, Blackmore LM, Lamb AE, Simpson SA (2000) Is there a differential deficit in fronto-executive functioning in Huntington's disease? Clin Neuropsychol Assess 1:4–20

Crawford JR, Garthwaite PH, Betkowska K (2009a) Bayes' theorem and diagnostic tests in neuropsychology: interval estimates for post-test probabilities. Clin Neuropsychol 23:624–644

Crawford JR, Garthwaite PH, Howell DC (2009b) On comparing a single case with a control sample: an alternative perspective. Neuropsychologia 47:2690–2695

Cullen B, Ward J, Graham NA, Deary IJ, Pell JP, Smith DJ, Evans JJ (2016) Prevalence and correlates of cognitive impairment in euthymic adults with bipolar disorder: a systematic review. J Affect Disord 205:165–181

Cullen B, Smith DJ, Deary IJ, Pell JP, Keyes KM, Evans JJ (2019) Understanding cognitive impairment in mood disorders: mediation analyses in the UK Biobank Cohort. Br J Psychiatry:1–8

Czepielewski L, Daruy Filho L, Brietzke E, Grassi-Oliveira R (2013) Bipolar disorder and metabolic syndrome: a systematic review. Revista Brasileira de Psiquiatria 35:88–93

Czobor P, Jaeger J, Berns SM, Gonzalez C, Loftus S (2007) Neuropsychological symptom dimensions in bipolar disorder and schizophrenia. Bipolar Disord 9:71–92

Davis AS, Pierson EE (2012) The relationship between the WAIS-III digit symbol coding and executive functioning. Appl Neuropsychol:1–6

Demmo C, Lagerberg TV, Aminoff SR, Hellvin T, Kvitland LR, Simonsen C, Andreassen OA, Melle I, Ueland T (2016) History of psychosis and previous episodes as potential explanatory factors for neurocognitive impairment in first-treatment bipolar I disorder. Bipolar Disord 18:136–147

Dickinson T, Becerra R, Coombes J (2017) Executive functioning deficits among adults with bipolar disorder (types I and II): a systematic review and meta-analysis. J Affect Disord 218:407–427

Dolcos F, Rice HJ, Cabeza R (2002) Hemispheric asymmetry and aging: right hemisphere decline or asymmetry reduction. Neurosci Biobehav Rev 26:819–825

Douglas KM, Gallagher P, Robinson LJ, Carter JD, McIntosh VV, Frampton CM, Watson S, Young AH, Ferrier IN, Porter RJ (2018) Prevalence of cognitive impairment in major depression and bipolar disorder. Bipolar Disord 20:260–274

Ehrminger M, Brunet-Gouet E, Cannavo A-S, Aouizerate B, Cussac I, Azorin J-M, Bellivier F, Bougerol T, Courtet P, Dubertret C, Etain B, Kahn J-P, Leboyer M, Olié E, Passerieux C, Roux P (2019) Longitudinal relationships between cognition and functioning over 2 years in euthymic patients with bipolar disorder: a cross-lagged panel model approach with the FACE-BD cohort. Br J Psychiatry:1–8

Eidelman P, Talbot LS, Gruber J, Harvey AG (2010) Sleep, illness course, and concurrent symptoms in inter-episode bipolar disorder. J Behav Ther Exp Psychiatry 41:145–149

Ferrier IN, Stanton BR, Kelly TP, Scott J (1999) Neuropsychological function in euthymic patients with bipolar disorder. Br J Psychiatry 175:246–251

Gallagher P, Reid KS, Ferrier IN (2009) Neuropsychological functioning in health and mood disorder: modulation by glucocorticoids and their receptors. Psychoneuroendocrinology 34:196–207

Gallagher P, Gray JM, Watson S, Young AH, Ferrier IN (2014) Neurocognitive functioning in bipolar depression: a component structure analysis. Psychol Med 44:961–974

Gallagher P, Gray JM, Kessels RPC (2015a) Fractionation of visuo-spatial memory processes in bipolar depression: a cognitive scaffolding account. Psychol Med 45:545–558

Gallagher P, Nilsson J, Finkelmeyer AE, Goshawk M, Macritchie KA, Lloyd AJ, Thompson JM, Porter RJ, Young AH, Ferrier IN, McAllister-Williams RH, Watson S (2015b) Neurocognitive intra-individual variability in mood disorders: effects on attentional response time distributions. Psychol Med 45:2985–2997

Geoffroy PA, Scott J, Boudebesse C, Lajnef M, Henry C, Leboyer M, Bellivier F, Etain B (2015) Sleep in patients with remitted bipolar disorders: a meta-analysis of actigraphy studies. Acta Psychiatr Scand 131:89–99

Glahn DC, Bearden CE, Cakir S, Barrett JA, Najt P, Monkul ES, Maples N, Velligan DI, Soares JC (2006) Differential working memory impairment in bipolar disorder and schizophrenia: effects of lifetime history of psychosis. Bipolar Disord 8:117–123

Goswami U, Sharma AN, Khastigir U, Ferrier IN, Young AH, Gallagher P, Thompson JM, Moore PB (2006) Neuropsychological dysfunction, soft neurological signs, and social disability in euthymic bipolar subjects. Br J Psychiatry 188:366–373

Goswami U, Sharma A, Varma A, Gulrajani C, Ferrier IN, Young AH, Gallagher P, Thompson JM, Moore PB (2009) The neurocognitive performance of drug-free and medicated euthymic bipolar patients does not differ. Acta Psychiatr Scand 120:456–463

Gruber J, Harvey AG, Wang PW, Brooks Iii JO, Thase ME, Sachs GS, Ketter TA (2009) Sleep functioning in relation to mood, function, and quality of life at entry to the systematic treatment enhancement program for bipolar disorder (STEP-BD). J Affect Disord 114:41–49

Harkavy-Friedman JM, Keilp JG, Grunebaum MF, Sher L, Printz D, Burke AK, Mann JJ, Oquendo M (2006) Are BPI and BPII suicide attempters distinct neuropsychologically? J Affect Disord 94:255–259

Harvey AG, Schmidt DA, Scarna A, Semler CN, Goodwin GM (2005) Sleep-related functioning in euthymic patients with bipolar disorder, patients with insomnia, and subjects without sleep problems. Am J Psychiatry 162:50–57

Harvey AG, Soehner AM, Kaplan KA, Hein K, Lee J, Kanady J, Rabe-Hesketh S, Neylan TC, Li D, Ketter TA, Buysse DJ (2015) Treating insomnia improves mood state, sleep, and functioning in bipolar disorder: a pilot randomized controlled trial. J Consult Clin Psychol 83:564–577

Harvey PD, Aslan M, Du M, Zhao H, Siever LJ, Pulver A, Gaziano JM, Concato J (2016) Factor structure of cognition and functional capacity in two studies of schizophrenia and bipolar disorder: implications for genomic studies. Neuropsychology 30:28–39

Henry GM, Weingartner H, Murphy DL (1973) Influence of affective states and psychoactive drugs on verbal learning and memory. Am J Psychiatry 130:966–971

Iverson GL, Brooks BL, Young AH (2009) Rapid computerized assessment of neurocognitive deficits in bipolar disorder. Appl Neuropsychol 16:207–213

Iverson GL, Brooks BL, Langenecker SA, Young AH (2011) Identifying a cognitive impairment subgroup in adults with mood disorders. J Affect Disord 132:360–367

Jaeger J, Tatsuoka C, Berns S, Varadi F, Czobor P, Uzelac S (2006a) Associating functional recovery with neurocognitive profiles identified using partially ordered classification models. Schizophr Res 85:40–48

Jaeger J, Tatsuoka C, Berns SM, Varadi F (2006b) Distinguishing neurocognitive functions in schizophrenia using partially ordered classification models. Schizophr Bull 32:679–691

Jansson-Fröjmark M, Norell-Clarke A (2016) Cognitive Behavioural therapy for insomnia in psychiatric disorders. Curr Sleep Med Rep 2:233–240
Jimenez E, Sole B, Arias B, Mitjans M, Varo C, Reinares M, Bonnin CM, Salagre E, Ruiz V, Torres I, Tomioka Y, Saiz PA, Garcia-Portilla MP, Buron P, Bobes J, Martinez-Aran A, Torrent C, Vieta E, Benabarre A (2018) Characterizing decision-making and reward processing in bipolar disorder: a cluster analysis. Eur Neuropsychopharmacol 28:863–874
Johnson MH, Magaro PA (1987) Effects of mood and severity on memory processes in depression and mania. Psychol Bull 101:28–40
Joy S, Fein D, Kaplan E, Freedman M (2000) Speed and memory in WAIS–R–NI digit symbol performance among healthy older adults. J Int Neuropsychol Soc 6:770–780
Joy S, Kaplan E, Fein D (2004) Speed and memory in the WAIS-III digit symbol-coding subtest across the adult lifespan. Arch Clin Neuropsychol 19:759–767
Kelly T, Douglas L, Denmark L, Brasuell G, Lieberman DZ (2013) The high prevalence of obstructive sleep apnea among patients with bipolar disorders. J Affect Disord 151:54–58
Kessler U, Schoeyen HK, Andreassen OA, Eide GE, Hammar Å, Malt UF, Oedegaard KJ, Morken G, Sundet K, Vaaler AE (2013) Neurocognitive profiles in treatment-resistant bipolar I and bipolar II disorder depression. BMC Psychiatry 13:105
Kollmann B, Yuen K, Scholz V, Wessa M (2019) Cognitive variability in bipolar I disorder: a cluster-analytic approach informed by resting-state data. Neuropharmacology 156:107585
Krabbendam L, Arts B, van Os J, Aleman A (2005) Cognitive functioning in patients with schizophrenia and bipolar disorder: a quantitative review. Schizophr Res 80:137–149
Kraepelin E (1899) Psychiatrie. Ein Lehrbuch für Studierende und Ärzte, 6th edn. Barth, Leipzig
Lee J, Rizzo S, Altshuler L, Glahn DC, Miklowitz DJ, Sugar CA, Wynn JK, Green MF (2017) Deconstructing bipolar disorder and schizophrenia: a cross-diagnostic cluster analysis of cognitive phenotypes. J Affect Disord 209:71–79
Levy B, Medina AM, Weiss RD (2013) Cognitive and psychosocial functioning in bipolar disorder with and without psychosis during early remission from an acute mood episode: a comparative longitudinal study. Compr Psychiatry 54:618–626
Lewandowski KE, Sperry SH, Cohen BM, Ongur D (2014) Cognitive variability in psychotic disorders: a cross-diagnostic cluster analysis. Psychol Med 44:3239–3248
Lewandowski KE, Baker JT, McCarthy JM, Norris LA, Ongur D (2018) Reproducibility of cognitive profiles in psychosis using cluster analysis. J Int Neuropsychol Soc 24:382–390
Lim J, Dinges DF (2010) A meta-analysis of the impact of short-term sleep deprivation on cognitive variables. Psychol Bull 136:375–389
Lima F, Rabelo-da-Ponte FD, Bücker J, Czepielewski L, Hasse-Sousa M, Telesca R, Solé B, Reinares M, Vieta E, Rosa AR (2019) Identifying cognitive subgroups in bipolar disorder: a cluster analysis. J Affect Disord 246:252–261
Mackin P, Bishop D, Watkinson H, Gallagher P, Ferrier IN (2007) Metabolic disease and cardiovascular risk in people treated with antipsychotics in the community. Br J Psychiatry 191:23–29
Martinez-Aran A, Torrent C, Tabares-Seisdedos R, Salamero M, Daban C, Balanza-Martinez V, Sanchez-Moreno J, Manuel Goikolea J, Benabarre A, Colom F, Vieta E (2008) Neurocognitive impairment in bipolar patients with and without history of psychosis. J Clin Psychiatry 69:233–239
Mathias SR, Knowles EEM, Barrett J, Leach O, Buccheri S, Beetham T, Blangero J, Poldrack RA, Glahn DC (2017) The processing-speed impairment in psychosis is more than just accelerated aging. Schizophr Bull 43:814–823
McIntyre RS, Mansur RB, Lee Y, Japiassu L, Chen K, Lu R, Lu W, Chen X, Li T, Xu G, Lin K (2017) Adverse effects of obesity on cognitive functions in individuals at ultra high risk for bipolar disorder: results from the global mood and brain science initiative. Bipolar Disord 19:128–134
Millar A, Espie CA, Scott J (2004) The sleep of remitted bipolar outpatients: a controlled naturalistic study using actigraphy. J Affect Disord 80:145–153

Mora E, Portella MJ, Martinez-Alonso M, Teres M, Forcada I, Vieta E, Mur M (2017) The impact of obesity on cognitive functioning in euthymic bipolar patients: a cross-sectional and longitudinal study. J Clin Psychiatry 78:e924–e932

Morice R (1990) Cognitive inflexibility and pre-frontal dysfunction in schizophrenia and mania. Br J Psychiatry 157:50–54

Moss RA, Finkelmeyer A, Robinson LJ, Thompson JM, Watson S, Ferrier IN, Gallagher P (2016) The impact of target frequency on intra-individual variability in euthymic bipolar disorder: a comparison of two sustained attention tasks. Front Psychiatry 7

Moustafa AA, Kéri S, Somlai Z, Balsdon T, Frydecka D, Misiak B, White C (2015) Drift diffusion model of reward and punishment learning in schizophrenia: modeling and experimental data. Behav Brain Res 291:147–154

Murphy FC, Sahakian BJ (2001) Neuropsychology of bipolar disorder. Br J Psychiatry 178:S120–S127

Nebes RD, Butters MA, Mulsant BH, Pollock BG, Zmuda MD, Houck PR, Reynolds CF (2000) Decreased working memory and processing speed mediate cognitive impairment in geriatric depression. Psychol Med 30:679–691

Nilsson J, Thomas AJ, Stevens LH, McAllister-Williams RH, Ferrier IN, Gallagher P (2016) The interrelationship between attentional and executive deficits in major depressive disorder. Acta Psychiatr Scand 134:73–82

Park DC, Reuter-Lorenz P (2009) The adaptive brain: aging and neurocognitive scaffolding. Annu Rev Psychol 60:173–196

Pavuluri MN, Schenkel LS, Aryal S, Harral EM, Hill SK, Herbener ES, Sweeney JA (2006) Neurocognitive function in unmedicated manic and medicated euthymic pediatric bipolar patients. Am J Psychiatry 163:286–293

Pavuluri MN, Yang S, Kamineni K, Passarotti AM, Srinivasan G, Harral EM, Sweeney JA, Zhou XJ (2009) Diffusion tensor imaging study of white matter fiber tracts in pediatric bipolar disorder and attention-deficit/hyperactivity disorder. Biol Psychiatry 65:586–593

Poletti S, Bollettini I, Mazza E, Locatelli C, Radaelli D, Vai B, Smeraldi E, Colombo C, Benedetti F (2015) Cognitive performances associate with measures of white matter integrity in bipolar disorder. J Affect Disord 174:342–352

Robinson LJ, Thompson JM, Gallagher P, Goswami U, Young AH, Ferrier IN, Moore PB (2006) A meta-analysis of cognitive deficits in euthymic bipolar subjects. J Affect Disord 93:105–115

Roux P, Raust A, Cannavo AS, Aubin V, Aouizerate B, Azorin JM, Bellivier F, Belzeaux R, Bougerol T, Cussac I, Courtet P, Etain B, Gard S, Job S, Kahn JP, Leboyer M, Olie E, Henry C, Passerieux C (2017) Cognitive profiles in euthymic patients with bipolar disorders: results from the FACE-BD cohort. Bipolar Disord 19:146–153

Russo M, Van Rheenen TE, Shanahan M, Mahon K, Perez-Rodriguez MM, Cuesta-Diaz A, Larsen E, Malhotra AK, Burdick KE (2017) Neurocognitive subtypes in patients with bipolar disorder and their unaffected siblings. Psychol Med 47:2892–2905

Ryan KA, Assari S, Pester BD, Hinrichs K, Angers K, Baker A, Marshall DF, Stringer D, Saunders EF, Kamali M, McInnis MG, Langenecker SA (2016) Similar trajectory of executive functioning performance over 5 years among individuals with bipolar disorder and unaffected controls using latent growth modeling. J Affect Disord 199:87–94

Salthouse TA (1996) The processing-speed theory of adult age differences in cognition. Psychol Rev 103:403–428

Salthouse TA (2000) Aging and measures of processing speed. Biol Psychol 54:35–54

Salthouse TA (2017) Shared and unique influences on age-related cognitive change. Neuropsychology 31:11–19

Schenkel LS, West AE, Jacobs R, Sweeney JA, Pavuluri MN (2012) Cognitive dysfunction is worse among pediatric patients with bipolar disorder type I than type II. J Child Psychol Psychiatry 53:775–781

Schretlen DJ, Pena J, Aretouli E, Orue I, Cascella NG, Pearlson GD, Ojeda N (2013) Confirmatory factor analysis reveals a latent cognitive structure common to bipolar disorder, schizophrenia, and normal controls. Bipolar Disord 15:422–433
Silarova B, Giltay EJ, Van Reedt DA, Van Rossum EF, Hoencamp E, Penninx BW, Spijker AT (2015) Metabolic syndrome in patients with bipolar disorder: comparison with major depressive disorder and non-psychiatric controls. J Psychosom Res 78:391–398
Solé B, Martínez-Arán A, Torrent C, Bonnin CM, Reinares M, Popovic D, Sánchez-Moreno J, Vieta E (2011) Are bipolar II patients cognitively impaired? A systematic review. Psychol Med 41:1791–1803
Solé B, Bonnin CM, Torrent C, Martinez-Aran A, Popovic D, Tabarés-Seisdedos R, Vieta E (2012) Neurocognitive impairment across the bipolar Spectrum. CNS Neurosci Ther 18:194–200
Tabarés-Seisdedos R, Balanzá-Martínez V, Sánchez-Moreno J, Martinez-Aran A, Salazar-Fraile J, Selva-Vera G, Rubio C, Mata I, Gómez-Beneyto M, Vieta E (2008) Neurocognitive and clinical predictors of functional outcome in patients with schizophrenia and bipolar I disorder at one-year follow-up. J Affect Disord 109:286–299
Tam JW, Schmitter-Edgecombe M (2013) The role of processing speed in the brief Visuospatial memory test–revised. Clin Neuropsychol 27:962–972
Tatsuoka C (2002) Data analytic methods for latent partially ordered classification models. J R Stat Soc Ser C 51:337–350
Tatsuoka C, Ferguson T (2003) Sequential classification on partially ordered sets. J R Stat Soc Ser B 65:143–157
Thompson JM, Gallagher P, Hughes JH, Watson S, Gray JM, Ferrier IN, Young AH (2005) Neurocognitive impairment in euthymic bipolar disorder. Br J Psychiatry 186:32–40
Thompson J, Hamilton CJ, Gray J, Quinn JG, Mackin P, Young A, Nicol Ferrier I (2006) Executive and visuospatial sketchpad resources in euthymic bipolar disorder: implications for visuospatial working memory architecture. Memory 14:437–451
Thompson JM, Gray JM, Crawford JR, Hughes JH, Young AH, Ferrier IN (2009) Differential deficit in executive control in euthymic bipolar disorder? J Abnorm Psychol 118:146–160
Torrent C, Martinez-Aran A, Daban C, Sanchez-Moreno J, Comes M, Goikolea JM, Salamero M, Vieta E (2006) Cognitive impairment in bipolar II disorder. Br J Psychiatry 189:254–259
Tsitsipa E, Fountoulakis KN (2015) The neurocognitive functioning in bipolar disorder: a systematic review of data. Ann General Psychiatry 14:42
Van Rheenen TE, Lewandowski KE, Tan EJ, Ospina LH, Ongur D, Neill E, Gurvich C, Pantelis C, Malhotra AK, Rossell SL, Burdick KE (2017) Characterizing cognitive heterogeneity on the schizophrenia-bipolar disorder spectrum. Psychol Med 47:1848–1864
Vaurio RG, Simmonds DJ, Mostofsky SH (2009) Increased intra-individual reaction time variability in attention-deficit/hyperactivity disorder across response inhibition tasks with different cognitive demands. Neuropsychologia 47:2389–2396
Waters F, Bucks RS (2011) Neuropsychological effects of sleep loss: implication for neuropsychologists. J Int Neuropsychol Soc 17:571–586
Wolfe J, Granholm E, Butters N, Saunders E, Janowsky D (1987) Verbal memory deficits associated with major affective disorders: a comparison of unipolar and bipolar patients. J Affect Disord 13:83–92
Xu G, Lin K, Rao D, Dang Y, Ouyang H, Guo Y, Ma J, Chen J (2012) Neuropsychological performance in bipolar I, bipolar II and unipolar depression patients: a longitudinal, naturalistic study. J Affect Disord 136:328–339

The Kindling/Sensitization Model and Early Life Stress

Robert M. Post

Contents

Abstract *Introduction*: Few animal models address the characteristics of the longitudinal course of bipolar disorder. However, behavioral sensitization (to recurrent stressors and psychomotor stimulants) and kindling of seizures both provide clues to mechanisms in the progressive course of bipolar disorder.

Methods: We describe aspects of bipolar illness that show sensitization and kindling-like increases reactivity to the recurrence of stressors, mood episodes, and bouts of substance abuse. Mechanisms of these events and clinical implications for treatment are discussed.

Results: Early life stress is a risk factor for the development of episodes of unipolar depression and bipolar disorder and the acquisition of substance abuse. Initial affective episodes are often triggered by the recurrence of psychosocial stressors in adulthood, but after many episodes have occurred, episodes may also begin to emerge spontaneously in a kindling-like progression. More prior episodes

R. M. Post (✉)
Clinical Professor of Psychiatry, George Washington University School of Medicine, Bipolar Collaborative Network, Bethesda, MD, USA

Curr Topics Behav Neurosci (2021) 48: 255–276
https://doi.org/10.1007/7854_2020_172
Published Online: 12 January 2021

are associated with faster recurrences, dysfunction, disability, frontal cortical abnormalities, cognitive impairment, shorter telomeres, treatment refractoriness, and an increased risk of a diagnosis of dementia in old age. Sensitization to stressors, episodes, and substances of abuse each appear driven by epigenetic mechanisms and their accumulation on DNA, histones, and microRNA. Patients with bipolar illness in the USA are more ill than those from Europe and experience more sensitization to stressors, episodes, substance abuse, as well as more genetic vulnerability across four generations.

Discussion: The sensitization and kindling models highlight the importance of early intervention and prevention in order to limit or halt the downhill progression of bipolar disorder and its multiple comorbidities toward treatment refractoriness. Clinical data support this conclusion as well but have not been sufficient to change practice in the direction of early intervention. It is hoped that a better understanding of sensitization and kindling-like mechanisms will add neurobiological rationales for the importance of prevention and sustained prophylactic intervention in rendering bipolar disorder a more benign illness.

Keywords Anticonvulsants · Atypical antipsychotics · Complex combination treatment · Depression · Epigenetics · Genetic vulnerability · Lithium · Stressors · Substance abuse

1 Introduction

Early life stress is a well-known risk factor for a variety of psychiatric illnesses, but especially the mood disorders (Post 1992; Danese et al. 2009). Kraepelin described how early life stressors could predispose to stress later precipitating affective episodes. He then captured the essence of the sensitization/kindling model with the observations that after a number of stress-induced episodes, a person might be so sensitized that the mere anticipation of a stressor or no obvious stressor at all might be sufficient to trigger affective episodes.

This sequence of occurrences also bears some resemblance to the longitudinal unfolding of seizures in the kindling model (Post 2007a). How about: Repeated, once-daily electrical stimulation of say the amygdala for 1 s produces little effect initially. Then after-discharges and seizures begin to emerge, first unilaterally then bilaterally, culminating in full-blown seizures with rearing and falling. Following enough of these precipitated seizures, spontaneous seizures (with no amygdala stimulation) develop. This seizure model is a non-homologous one for the affective disorders, but helps crystalize the longitudinal unfolding of how repeated stimuli can have progressively greater consequences resulting in triggered full-blown episodes that then can go on to occur more spontaneously in the absence of stimulation.

In this chapter we provide some details of: the data confirming sensitization and kindling-like occurrences in the longitudinal course of the affective disorders; predictions that might be derived from them; and possible underlying mechanisms. We focus on the therapeutic implication of the sensitization/kindling model with the over-riding emphasis of the importance of early, sustained intervention to prevent the progressive unfolding of symptoms and the engendering of a down-hill course of illness.

2 Stress Sensitization

In animals recurrence of stressors can lead to sensitization or increasing behavioral reactivity rather than tolerance (Antelman 1988). There is a very wide literature supporting this phenomenon, but perhaps the best example in humans is the seminal work of Caspi et al. (2003) showing that increasing numbers of adversities in childhood were associated with later stressors in adulthood precipitating an increased incidence of depressive episodes. This work also showed a gene–environmental interaction in which those individuals with the short form of the serotonin transporter were the most prone to the depressive recurrences. While a recent meta-analysis consisting of very large numbers of depressive and control subjects failed to find validation of any gene associated with depression or any gene–environmental interaction, they did find overwhelming support for the findings that early stressors sensitized to later stressors that would then precipitate depressions in adulthood (Border et al. 2019).

In bipolar disorder there is evidence for both stress sensitization and stress accumulation. Early life stressors are associated with an earlier onset of illness and a more pernicious course of illness. In addition those with early life adversity experience more stressors over their course of illness, both at illness onset and prior to the last episode that occurred prior to entering a longitudinal follow-up study at average age 40 (Post et al. 2014). Danese et al. (2009) have shown that early life adversity in childhood has multiple consequences in adulthood, including not only depression, but also inflammation and metabolic abnormalities with subsequent high risks for cardiovascular disease. We saw similar effects of childhood adversity on an adverse course of bipolar disorder (Post et al. 2015a) and the occurrence of medical comorbidities in adulthood (Post et al. 2013a) (Table 1).

One of the best animal models of depression, repeated exposure to social defeat stress by a larger animal shows sensitization in that depression does not develop after the first defeat stress, but requires several such exposures and 10 days of defeat stress is the typical paradigm used (Tidey and Miczek 1997; Hollis and Kabbaj 2014; Bernton et al. 2006; Tsankova et al. 2006). Moreover, Weber et al. (2017) describe how one series of defeat stress can sensitize to a second round of defeat stress experiences, in part by priming of monocytes stored in the spleen which then access the brain and increase inflammatory responses. Defeat stress can also sensitize to abnormal cocaine-induced behaviors and immune activation.

Table 1 Consequences in adulthood of childhood adversity

I. Psychiatric	II. Medical[a]	III. Observed changes
Retrospective findings:	Allergies	Altered central nervous system development, structure, function, function, and neurotrophic factors
A. Early onset BP	Arthritis	Inflammation
B. Anxiety disorder	Asthma	Oxidative stress
C. Substance abuse	Chronic fatigue	Telomere shortening
D. Dysphoric mania	Chronic menstrual	Cortisol dysregulation irregularities
E. ≥20 episodes	Fibromyalgia	↓ BDNF in prefrontal cortex and hippocampus
Prospective findings:	Head injury	↓ 5HT1A receptors
	Hypertension	
	Hypotension	
F. More time depressed	Irritable bowel	
	Migraine headache	Reduced GABA transition from excitatory to inhibitory transmission
	Obesity	
G. More ultradian cycling		Corticotropin-releasing hormone: transition from appetitive to aversive properties
		Low oxytocin
Migraine headache		Obesity and metabolic syndrome

[a]See Post et al. (2013a)

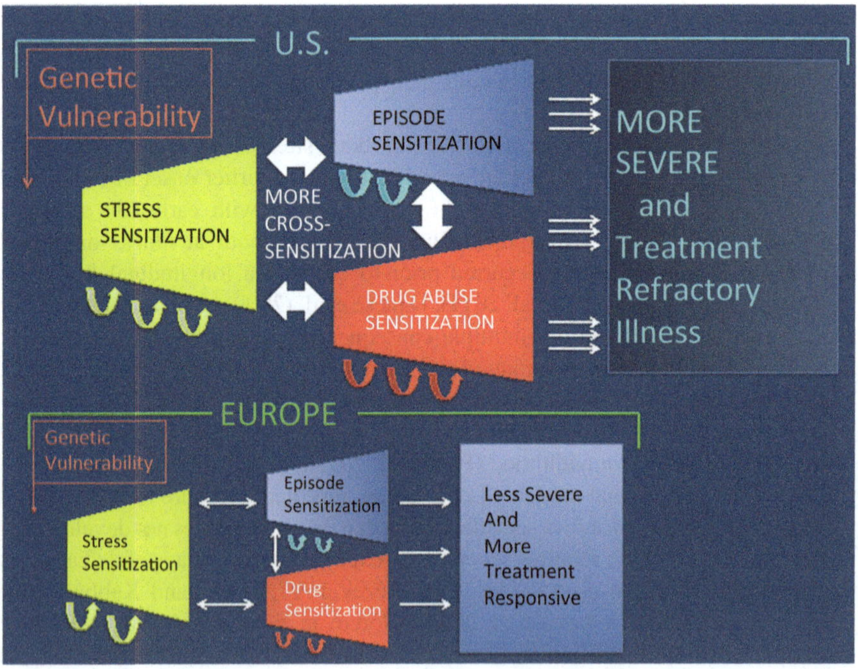

3 Stimulant-Induced Behavioral Sensitization

Repeated administration of the same dose of psychomotor stimulant, amphetamine or cocaine, results in increasing motor activity and stereotypy. There is also a conditioned component to this sensitization, as animals given the drug in the same environment show marked sensitization, while those treated in one cage and tested in another fail to show the sensitization (Post et al. 1987a, b). Interestingly, stress sensitization shows cross reactivity to stimulant-induced sensitization and vice versa.

4 Episode-Induced Sensitization

Multiple studies have revealed the pattern seen by Kraepelin for repeated episodes to occur faster (with a shorter well interval between episodes) (Post 1992; Kessing and Andersen 2017). Kessing et al. further demonstrated that this occurred for both unipolar and bipolar depression and there was a trend for subsequent episodes to be increased in severity. Kendler et al. (2000, 2001) validated the kindling model showing that the first five to seven episodes of unipolar depression were associated with psychosocial stressors but these became less necessary with subsequent recurrences. Thus, the fundamental predictions of the sensitization/kindling model have been repeated and validated with episodes coming more readily with shorter well intervals and also more autonomously from the requirement of precipitation by psychosocial stress.

Other derivative predictions have only been partially validated. The initial or (A) developmental stages of kindling development have a different pharmacological responsivity compared to (B) the mid phase full-blown seizures, and then again (C) the late spontaneous seizures are prevented by different agents than those of the full-blown mid phase seizures (Post 2007a, b). This leads to the suggestion that the underlying neurochemistry of kindling differs as a function of stage of kindling evolution. A modicum of clinical data support parallel observations in patients with bipolar disorder where lithium appears most effective in early full-blown affective episodes, but less effective with the occurrence of multiple episodes and rapid cycling. Although it is not well documented, the dihydropyridine calcium channel blocker nimodipine may be more effective against late stage, ultra-rapid, and ultradian cycling than earlier in the course.

In stimulant-induced behavioral sensitization the dopamine blocking antipsychotics are effective in blocking the initial development of sensitization, but do not block the expression of sensitization once has been induced. These observations parallel the observations that antipsychotic drugs in the treatment of psychotic episodes in schizophrenia become less effective after multiple episodes and relapses have occurred. Thus the kindling/sensitization models at least raise the question for further examination as to whether the neurochemistry and pharmacology of the recurrent affective disorders differs as a function of the stage of illness evolution, which if validated would have major implications for choice of treatment.

Moreover, it is easier to block full-blown kindled seizures in their earliest manifestation compared to intervening late in their evolution when they are harder to stop and tolerance to the anticonvulsant effects of drugs develops more readily. The derivative prediction that episodes late in the development of the longitudinal course of the affective disorders are harder to treat than initial episodes and are associated with increased proneness to tolerance development has a modicum of support in the literature (Post and Weiss 2011).

All of these observations in sensitized and kindled animals indicate the importance of intervening early in the evolution of these syndromes, and, accordingly, in the affective disorders. Earlier treatment is easier and less complicated and helps stop the evolution and progression of the next stages of more rapid recurrences, easier triggering, and spontaneity (Post 2015).

5 Cross-Sensitization: Neurochemical Commonalities and Inflammatory Mechanisms

Each type of sensitization to stressors, episodes, and substances of abuse contribute to illness progression and increased behavioral pathology (Post 2007a; Post and Kalivas 2013). In addition, each type shows cross-sensitization to the other two, implying some mechanisms in common and providing a basis for additional positive feedback cycles and spiraling toward a downhill course of deterioration. Stressors cross-sensitize to the occurrence of recurrent episodes and the initiation or relapse into bouts of substance use (Tidey and Miczek 1997). Lin et al. (2006) reported a genetic link between early onset bipolar disorder and substance use. The literature in both animals and humans is robust indicating that early adversity is a major risk factor not only for future episodes but also for adopting substance abuse (Post and Kalivas 2013). Psychological trauma can increase the rate of acquisition of cocaine self-administration (Brodnik et al. 2019). Conversely, episodes of illness also increase the risk of substance abuse and are generators of further stressors.

One example of the common mechanism involved in each type of sensitization is that each is associated with increases in brain derived neurotropic factor (BDNF) in the nucleus accumbens (Bernton et al. 2006). Depressed patients show evidence of BDNF increases in the nucleus accumbens (and decreases in BDNF in the hippocampus), changes which are mirrored by animals in the social defeat stress paradigm. Since BDNF is associated with long-term memory and the accumbens is associated with habits and reward, it would appear as if sensitized animals and humans acquire a habit for depressive behaviors as well as substances of abuse. This striatal substrate provides a mechanism for habit memory that is hard to overcome with voluntary representational memory strategies of the hippocampal and cortex (Post and Kalivas 2013; Fournier et al. 2017). Whittaker et al. (2018) reported increased functional

Table 2 Common effects of stress-, episode-, and cocaine-induced sensitization on BDNF in brain systems mediating habit and representational memory

Type of memory:	Habit memory	Representational memory
Substrate:	Striatum	Medial temporal lobe
Components:	1. Dorsal striatum	1. Hippocampus
	2. Ventral striatum (i.e., nucleus accumbens)	2. Amydala
Process:	Repetition and procedural learning	Declarative memory and single-trial learning
Awareness:	Unconscious	Conscious
BDNF:	BDNF is *increased* in: nucleus accumbens with defeat stress, human depression (suicide), and cocaine sensitization	BDNF is *decreased* in: hippocampus with defeat stress, human depression (suicide), and cocaine sensitization
Consequences:	Repetition of stress, depression, and cocaine may result in overlearned habits based in the striatum "…chronic stress leads to a bias in behavioral strategies toward habit." Dias-Ferreira et al. (2009)	Hippocampal dysfunction may induce biased retrieval of long-term memories and disinhibit the ventral tegmental area (VTA) pathway to nucleus accumbens, in turn increasing burst firing of the VTA, increasing BDNF, depression, and cocaine sensitization

connectivity of the n. accumbens and the mPFC as an endophenotype of bipolar disorder (Table 2).

Blocking the BDNF increases in the nucleus accumbens or the decreases in BDNF in the hippocampus prevent the development of social defeat stress behaviors and the development of stimulant-induced behavioral sensitization (Bernton et al. 2006; Tsankova et al. 2006, 2007). Cross-sensitization of stress to cocaine effects has elegantly been shown by Lo Iacono et al. (2018) who demonstrated a key role for immune activation in the sensitization. They demonstrated that early life social stress leads to persistent increases in peripheral inflammation (splenocytes) and brain microglia morphology typical of a "primed" state of hyper-reactivity, and then increased immune responses to cocaine, including increases in TNF alpha, Il-1b, and Il-6. Interestingly, immune system inhibition with minocycline during the early life stress in this mouse model blocked the preference for and hyper-reactivity to cocaine in adulthood.

Deighton et al. (2018) reviewed 14 studies in humans which all showed a link between adverse childhood experiences (ACEs) and increased inflammatory markers, including Il-6, TNF alpha, and CRP. Five studies found a relationship of ACEs to increased body mass index and others to shorter telomeres and increased DNA methylation of the 5HT-T promoter region.

6 An Epigenetic Basis for Sensitization

Epigenetic refers to mechanisms not involved in the nucleotide sequence conveying classical genetic inheritance. Events in the environment affect how easily gene are turned on and off based on modifications added to DNA such as methylation, acetylation or methylation of histones, or alterations in microRNA (Post 2016a, 2018a). If one blocks DNA methylation with the inhibitor zebularine, cocaine-induced sensitization does not occur (Anier et al. 2010). Animals stressed early in life are vulnerable not only to stresses in adulthood but also to lifelong decreases in BDNF in frontal cortex. If zebularine is given, the BDNF and behavioral changes are prevented (Roth et al. 2009). Similarly, epigenetic mechanisms are evident in the defeat stress paradigm, and if such changes are prevented with molecular modifications, defeat stress behaviors do not occur (Tsankova et al. 2006).

Evidences that epigenetic mechanisms are at play in humans with depression and childhood adversity are the increases in epigenetic marks found in brain and in white cells in these individuals (Labonte et al. 2013; McGowan et al. 2009; Meaney 2005). Thus, sensitization and cross-sensitization to stressors, episodes, and abuse substances would appear to involve the accumulation of pathological epigenetic marks on DNA, histones, or microRNA.

This provides an additional molecular rationale for early prevention of episodes and substance abuse and minimization of stressors in order to not only inhibit these mechanisms of illness progression, but also to inhibit the accumulation of the associated adverse epigenetic marks. This neurobiological rationale for clinical treatment and prevention of the three types of environmentally mediated sensitization converges with the clinical imperative of early sustained prophylaxis. We know that early, expert intervention decreases the risk for further recurrences of bipolar disorder and modifies the long-term course of illness in a more positive direction (Kessing et al. 2013). Kessing et al. (2013) randomized youngsters with an early manic hospitalization to 2 years of expert clinic treatment versus 2 years of treatment as usual (TAU). The clinic patients had fewer relapses not only over these 2 years, but also over the next 4 years, even though all patients had returned to TAU after 2 years. One would predict that these individuals so treated and having fewer relapses and better compliance would also have longer telomeres and fewer epigenetic marks than those initially randomized to TAU. In any case, excellent comprehensive treatment that prevents episodes changes the long-term course of bipolar disorder in a more positive direction.

7 One Genetic and Two Epigenetic Bases for Illness Vulnerability

One cannot do much currently about modifying one's genetic inheritance and vulnerability, but modifying the environmentally driven epigenetic mechanisms becomes a major target of clinical therapeutics. Adverse parental experiences can have a transgenerational impact based on offspring exposure to the associated adversity. Mice reared by high licking dams become high lickers themselves, as well as having less anxiety in open field testing. However, if these animals are cross-fostered by low licking dams, they become low lickers with higher anxiety in the open field (Meaney 2005). This first kind of epigenetic transgenerational transmission of traits is based on exposure of the offspring to the parental behaviors.

In addition, there is a second kind of transmission that is not based on exposure to the parental behaviors. It appears that some of the epigenetic marks of the parents are not erased at fertilization and are transferred to the next generation (as reviewed in Post (2016a, 2018a)). The data supporting these observations are striking and non-intuitive. One of the best examples is if a rodent father is conditioned to a smell and a shock in adulthood, his offspring (with which he has no contact) are more responsive to that same odor and not others. Increased number of neurons and receptors for that specific smell are also seen in the rat pup anatomically.

Similarly, parental exposure to drugs of abuse can alter vulnerability in the next generation. Cocaine exposure in the parent does not result in increased cocaine liking in the next generation but does result in cognitive deficits. In the case of parental exposure to THC, there is an interesting twist. The offspring subsequently prefer opiates. One wonders how much of the current opiate epidemic is contributed to by parental use of marijuana. These and multiple other examples in the literature indicate that some parental experiences can have an impact on the next generation through epigenetic marks transmitted in the gametes (Chan et al. 2017). These changes have been shown at the level of epigenetic marks in eggs and sperm on DNA, histones, and in composition of microRNA.

Thus parents convey vulnerability to psychiatric illness via classical genetic inheritance as well as the well-known epigenetic mechanisms based on exposure of the offspring to behaviors and now also the novel mechanism of transgenerational transmission in the absence of exposure to the parental behavior. There appears to be new meaning to the old saw of "choose your parents well" not only for their genes, but also for the way they behaved during their life that affects their epigenome. This has implications for children who are adopted. Not only do they inherit their biological parents' genetic vulnerabilities, but can be influenced by the lifetime parental behaviors, stresses, and substances of abuse even if the adoptee never has much or any contact with the biological parents. Good parenting by the adoptive parents may be able to overcome or reverse some of the biological parental vulnerability, but how much may persist based on genetic and non-exposure epigenetic is uncertain. One promising note is that if adult male mice which are conditioned to an odor and a shock are given extinction training to extinguish the fear of the odor, transmission of the fear of the odor to the offspring via the parental germline no longer occurs (Aoued et al. 2019). One might quip that "sperm can both learn and unlearn the intergenerational transmission of stress."

8 More Stress, Episodes, and Substance Abuse in the USA Than Europe Driving Sensitization

Considerable evidence indicates that there is an excess of childhood onset bipolar illness in the USA compared to European countries (Etain et al. 2012; Belleivier et al. 2014; Post et al. 2017a) The findings in our series that one-fourth of adults with bipolar disorder in the USA had their first episode prior to age 13 are also mirrored by the findings by Perlis et al. (2004) in the NIMH STEP-BD cohort emanating from all different academic centers and cities in the USA compared to our Bipolar Collaborative Network (BCN) which included patients from Los Angeles, Dallas, Cincinnati, and Bethesda. Our European sites were those from Utrecht, The Netherlands and Freiburg and Munich, Germany.

Etain et al. (2012) used a French comparison and Belleivier et al. (2014) included ten different European countries, such that the US data that two-thirds of adults in the USA had their first episode of depression or mania as children or adolescents (prior to age 19) (Post et al. 2017a; Perlis et al. 2004) compared to only about one-third to 40% in Europe are likely to be representative of large numbers of patients studied in multiple sites. James et al. (2014) reported a 70-fold increase in children in the USA with a discharge diagnosis of bipolar disorder compared to children from Britain. While some diagnostic ambiguity may be involved in these data, the phenomenally increased need for hospitalization of children in the USA for something resembling bipolar disorder is striking and disturbing.

The implications of these data are profound. Compared to those with adult onsets, early onset illness is associated with a much more adverse outcome in adulthood (Post et al. 2017a; Perlis et al. 2004). In addition, early onset illness is associated with longer delay to first treatment, and longer delays to first treatment are independently associated with more time and severity of depression and cycling in adulthood (Post et al. 2010). Also, early onset illness is associated with deficiencies in acquisition of educational, social, emotional/communication, and employment skills and with an increased incidence of substance abuse and suicide. Early onset cannot be readily modified, but treatment delay is a remedial risk factor, such that there needs to be a new focus on early illness recognition and treatment. This is woefully lacking in the USA, where some 80% of adolescents aged 13–18 who screen positive for a bipolar spectrum disorder are not in any kind of treatment (Merikangas et al. 2010). This is a recipe for disaster and needs to be addressed with new and widespread public health initiatives.

The origins of the excess of early onset bipolar disorder in the USA appear to be multifactorial, but are especially related to increases in both genetic and environmental risk factors (Post et al. 2016a). In the USA there is more familial illness in four generations of patients from the USA compared to Europe (Post et al. 2015b, 2017a). This includes patients' grandparent, parents, the patients and their spouses and siblings, and the patients' offspring where each have a higher incidence of almost all psychiatric problems, including depression, bipolar disorder, suicide attempts, alcohol abuse, substance abuse, and "other" illnesses. The degree of family loading in parents and grandparents is directed related to earlier onset of bipolar disorder in the patients (Post et al. 2015b, 2016a, b).

Another major risk factor for early onset bipolar illness is the occurrence of psychosocial adversity in childhood which is also almost twofold greater for physical, sexual, or verbal abuse in the USA compared to Europe (Table 3). We found that the greater was the history of adversity, the earlier the onset of illness. If one does a heat map and combines both the degree of familial burden and psychosocial adversity in childhood, there is an additive effect on age of onset of bipolar disorder where those who have high levels of both have average ages of onset largely prior to age 13. In the much larger group of patients who have no familial or psychosocial

Table 3 Incidence type of abuse in childhood by continent

Incidence of childhood abuse							
	Total	Overall		USA		Europe	
	N	Yes	Yes %	Yes	Yes%	Yes (*N*)	Yes %
No abuse	948	412	43.5	233	35.5	176	61.5
Verbal only	948	201	21.2	156	23.7	45	15.7
Any verbal	942	474	50.3	386	58.8	88	46.2
Any physical	942	231	24.5	191	29	40	14
Any sexual	936	205	21.9	168	36	37	12.9
All three	943	100	10.6	88	13.4	12	4.2

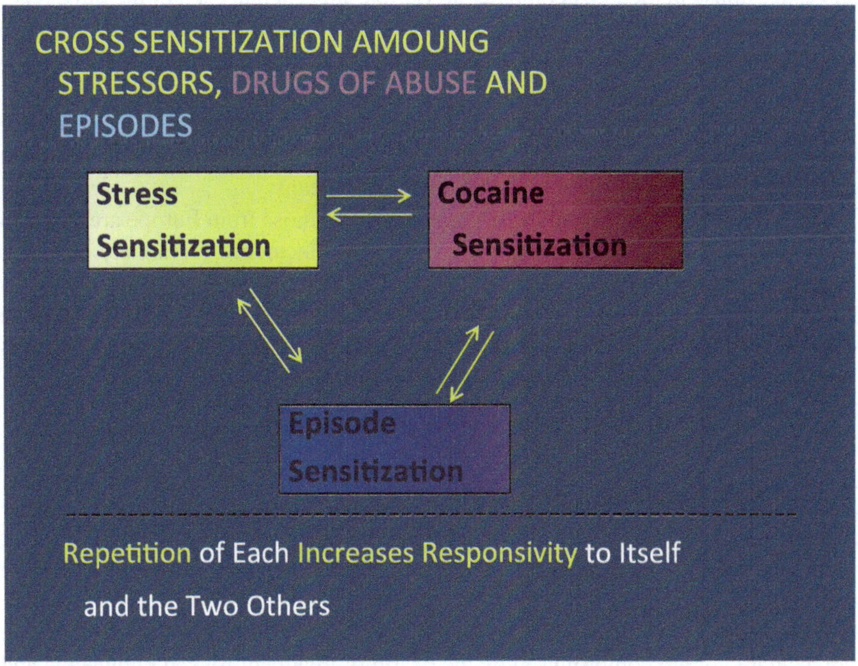

adversity, the average age of onset was 26 years of age (Post et al. 2016a). Other potential risk factors that are higher in the USA than in Europe include: a greater incidence of overweight/obesity; poorer diet; poor access to psychiatric care; more medical comorbidities, and longer delays to first treatment.

Among the three types of abuse in childhood that are associated with an earlier age of onset and a more adverse course of illness is what some might consider the more benign form of verbal abuse compared to physical or sexual abuse. However, when we examined the occurrence of verbal abuse in the absence of the other two forms, it was apparent that verbal abuse alone had a pernicious effect on early onset and subsequent adverse course of bipolar illness (Post et al. 2015a). This too has major implication for treatment and public health initiatives. Dealing with online and in school bullying and infra-familial belittling, abuse, and anger become major targets for therapeutic intervention.

Adversity in childhood was also related to more psychosocial stress and a greater number of negative life events in the year prior to the onset of bipolar disorder as well as in the year prior to a patient's last episode before entry into our network at average 40 (Post et al. 2013a). "In all three general categories of stressors evaluated (i.e. those related to loss of social support; financial/occupational difficulties; and medical/healthcare access), patients from the U.S. compared to Europe not only experienced a higher incidence of these in the year prior to illness onset, but also a significantly greater increase in these later in the illness course, i.e. prior to their latest episode."

Thus the occurrence of more psychosocial adversity, episodes of illness, and substance abuse in the USA compared to Europe suggest that each of the three types of sensitization (to stressors, episodes, and substance abuse) occur more readily in the USA (Fig. 1). Their sensitization and cross-sensitization to the others drive a more adverse progression and evolution of illness, culminating in a higher incidence of treatment resistance. When patients were treated naturalistically in a prospective fashion in the BCN, those from the USA had a higher incidence of treatment resistance (51.7%) compared to 31.1% in the Europeans.

Thus, bipolar patients from the USA compared to those from Europe are sicker in all respects (Post et al. 2017a). They have a higher incidence of: anxiety disorders (46.6% vs 28.1%); alcohol abuse (33.1% vs 14.7%); substance abuse (38.3% vs 17.8%); rapid cycling of 4 or more episodes/year (74.1% vs 41.5%); 20 or more episodes (59.0% vs 23.3%); yet fewer hospitalizations. As noted above their offspring were also sicker in almost all respects (Post et al. 2016c).

9 Implications for Treatment

Given the severity and magnitude of the problem seen in four generations of those from the USA, it would appear that major changes in treatment and prevention are indicated. Most of the needed remedies such as earlier recognition and treatment are not available or are compromised by the dearth of treatment-related knowledge and

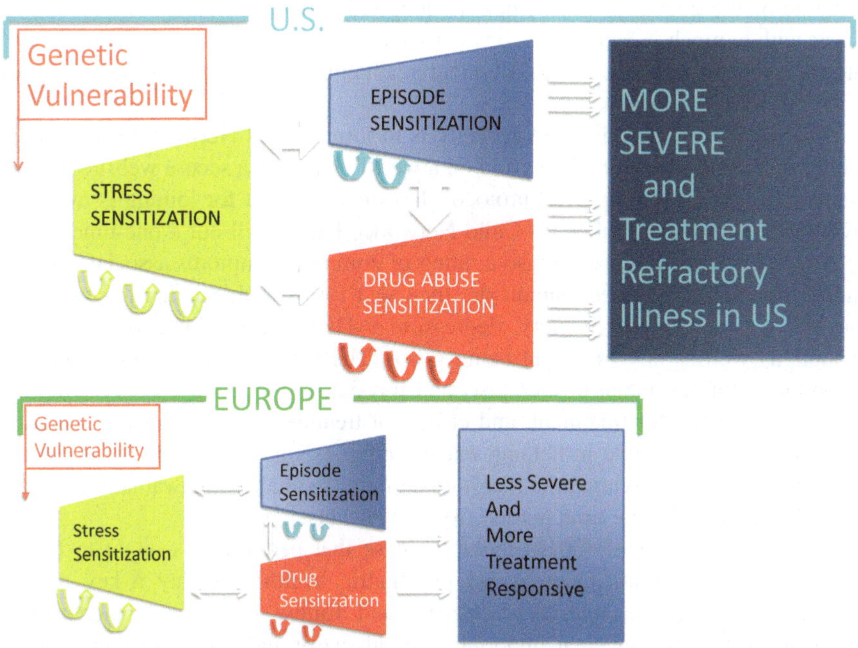

Fig. 1 Cross sensitization among stressors, mood episodes, and bouts of substance abuse: increased behavioral reactivity to the recurrence of each and to the others. Compared to bipolar patients from the Netherlands and Germany, those from the US experience more stressors, mood episodes, and substance abuse. The recurrence of each shows sensitization or increased reactivity which drives illness progression. Each type of sensitization appears to be based on epigenetic mechanisms. Prevention of stressors, episodes, and substance abuse would thus have the dual benefits of preventing sensitization and illness progression and limiting the accumulation of adverse epigenetic changes in the brain and body

early intervention studies in the USA. A clinical trials network for the assessment and children with early onset bipolar and related disorders would be ideal but has evoked no interest or support for initiating and funding in the NIMH and other funding bodies.

Since major treatment and public health initiatives do not appear in the offing, one must consider other ways of proceeding. We suggest that this can be done by all the principals involved including: children and their pediatricians; child psychiatrists and primary care physicians who see the majority of children with psychiatric illness in the USA (Anderson et al. 2015); parents and their therapists and physicians; adult psychiatrists; and psychologists, nurses, and social workers involved in treating children with psychiatric illness.

Since there is a shortage of child psychiatrists and most children are seen by primary care physicians in medical practices, there is a crucial role for parents to provide consistent longitudinal information about their child's symptoms. Creating a graphic of the course of key symptoms longitudinally rated on a daily or weekly basis will do much to help inform physicians about the problem areas, their severity, need for psychosocial or pharmacological treatment, and response to any treatments given (Post et al. 2017b).

One way of doing this systematically is to join a Child Network for parents of children aged 2–12 to rate their child on a weekly basis on a secure website under a John's Hopkins IRB approved protocol. Informed consent for joining is available www.bipolarnews.org (click on Child Network). Parents fill out a one-time demographics form and a more extensive rating of common symptoms associated childhood psychiatric illnesses. Parents are then sent forms each Sunday for rating the severity of their child's anxiety, depression, ADHD, oppositional behavior, and mania, as well as any treatment being given. These longitudinal ratings can then be printed out and taken to therapists and physicians for readily seeing symptom fluctuations, need for treatment, and effects of treatments utilized. Such symptom monitoring is invaluable in helping make a diagnosis and assessing the effects of treatment. Parents providing the longitudinal ratings enhance the evaluation process and also remove a large time burden from physicians.

Shonkoff and Garner (2012) make the case that pediatricians ought to be the guardians of children's medical and psychiatric health and play a key role in assessment, treatment, referral if needed, and follow up. This is increasingly recognized as being of great importance as adversity and toxic stress in children are key mediators of not only multiple psychiatric illnesses, but also a vast array of medical disorders that emerge in adulthood (Danese et al. 2009; Shonkoff and Garner 2012). Convergent with these findings in general medical practices, we found that if patients with bipolar illness had histories of childhood adversity, they had more than 13 different medical conditions than those without adversity. Moreover, patients from the USA had a higher incidence of nine different medical conditions than the Europeans.

These included significantly more cases of allergy, arthritis, asthma, fibromyalgia, head injury (without loss of consciousness), hyperglycemia, irritable bowel syndrome, migraine headaches, and seizures (Post et al. 2014). They also had an increased incidence of obesity.

Adult psychiatrists should ask about the children of their adult patients, and if needed evaluate and treat the children themselves or refer them for more expert assessment. Again encouraging the parents of these children to provide systematic numerical and graphic data should facilitate this process. If a child has a parent with bipolar disorder and is followed up longitudinally for 8 years, 74% will receive a major childhood psychiatric diagnosis (Axelson et al. 2015). The risks are thus high, and vigilance is indicated. About 20% of the offspring had a bipolar spectrum diagnosis, but even a higher percentage had an anxiety disorder, depression, ADHD, or a disruptive behavioral disorder, i.e., syndromes that also impair functioning and deserve treatment. The high risk for children in the USA is further emphasized by the findings of this same group that in the general population, children of the control parents (who did not have a bipolar diagnosis) also had a psychiatric diagnosis upon the 8 years of follow up in just under 50% of the time, suggesting that psychiatric illness in children in the general US population may be much more common than expected or appreciated.

10 High Risk Children Deserve Special Attention and Treatment

If a child has: (1) a loaded family history of psychiatric illness, (2) a history of adversity in childhood, and (3) prodromal symptoms, they are at a very high risk of developing a full-blown psychiatric disorder and efforts at primary or secondary prevention would appear indicated (Post et al. 2013b). Some interventions have been well studied such as the effectiveness of family focused therapy (FFT) or a related interventions focused on illness education, enhanced communication in the family, problem solving, stress coping mechanisms, and lowering the level of expressed emotion (Miklowitz et al. 2013; Fristad et al. 2009). The data are strong for this FFT intervention in children and adults with bipolar disorder and those at high risk in a total of 10 controlled clinical trials (Miklowitz and Chung 2016). Thus, some form of psychotherapy preferably with a family focus should be recommended for all prodromal children.

Those presenting with BP-NOS which is most common in the youngest children is associated with almost as much dysfunction as those with BP I or II and it takes longer, more than a year, to achieve mood stabilization with active psychopharmacology (Birmaher et al. 2009). Moreover, 30–50% of those with BP-NOS will convert to a full bipolar syndrome up 3–4 years of follow up.

11 Measures Considered for Primary and Secondary Prevention

In children at high risk by virtue of a loaded family history, even before they become subsyndromal, one might consider active recommendations including encouraging a good diet and regular sleep habits and exercise. Hudziak (Hudziak et al. 2014) has recommended a universal school program of head start-like activities for brain development including all children participating in team sports, the practice of mindfulness meditation, and encouragement of playing a musical instrument, which has been shown to increase brain growth and increase cognitive abilities. Eckenrode et al. (2001) and Olds et al. (1999) have shown the lasting value of home visits prenatally and in infancy of families at high risk.

For those at high risk who do have prodromal symptoms, the field needs studies of the effectiveness of interventions with supplements such as folate, vitamin D3, omega-3-fatty acids, N-acetylcysteine (NAC), and acetyl-L-carnitine which have some positive data in adults and may be worthy of individual clinical trials in those with depressive and anxious prodromes (Post et al. 2013). Acetyl-L-carnitine (LAC) deserves special attention in those with a history of childhood adversity as new data in humans indicated that LAC is low in the blood of depressed adults, particularly in those with a history of early onset and childhood adversity (Nasca et al. 2018). LAC has preliminary evidence of antidepressant effects in humans, and in animal models of depression (inescapable shock stress) LAC works more rapidly than regular antidepressants. It appears to do so by an epigenetic mechanism by which the acetyl group binds to the DNA promoter for the inhibitory metabotropic glutamate receptor mGluR-2, which inhibits glutamate release.

For children with evidence of inflammation (increases in Il-1, Il-6, TNF alpha, or CRP) trials with anti-inflammatories such as minocycline or celecoxib also deserve consideration. The social defeat stress model is ideal for studying the role of inflammation in the induction of depressive behaviors (Hodes et al. 2016; Niraula et al. 2018). If Il-6 secreted from lymphocytes is blocked depressive behaviors do not occur. Similarly, in Il-6 secreted from monocytes in the bone marrow, defeat stress behaviors are blocked. In a second defeat stress experience, primed monocytes are stored in the spleen, and if Il-6 is blocked from these splenic cells, depression-like symptoms do not occur. The storage of primed monocytes suggests the possibility that this type of second wave inflammatory mechanism could play a role in some types of sensitization.

It appears that Il-6 secreted in the periphery accesses brain by traversing endothelial cells, activating microglia, and generating NLRP3 inflammasomes. Assembly of the NLRP3 inflammasome leads to caspase 1-dependent release of the pro-inflammatory cytokines IL-1β and IL-18, as well as to pyroptotic cell death with subsequent negative effects on brain functioning, especially in the hippocampus. Both children with bipolar disorder and those with PTSD have evidence of inflammation and the therapeutic effects of anti-inflammatory manipulations deserve concerted study. Hinwood et al. (2013) reported that the effects of chronic stress could be inhibited by co-treatment with minocycline, which inhibits microglia activation. Others (Lo Iacono et al. 2018) have reported that minocycline inhibits some of the effects of early stress that then leads to increased sensitivity to stressors and cocaine later in life. Whether minocycline could similarly inhibit some of the sensitizing effects of early stress in children would deserve further study. For adults with multi-episode bipolar disorder driven by sensitization, inflammation, and other mechanisms, it may take highly complex combination treatment to bring the illness back under control (Post and Leverich 2008; Post 2016b, 2017a, 2019).

12 Conclusions

Early life stress primes the brain and body for increased reactivity to later stresses resulting in multiple adverse psychiatric and medical consequences. Episodes are triggered and substances are abused. Stressors, episodes, and bouts of substance abuse each show sensitization to themselves and cross-sensitization to the others deriving a down hill spiral of illness progression. This is seen in common effects on BDNF in nucleus accumbens and hippocampus, as well as on inflammatory and epigenetic mechanisms which convey some of the long-lasting sensitization. Childhood stressors are not only the precursors to multiple medical and psychiatric illness, but are also associated with lesser degrees of self-control which in turn augur poorer health, socioeconomic status, and even public safety into adulthood (Moffitt et al. 2011).

New data indicate that early effective long-term treatment must start after a first manic episode. The consequences of not doing this are highlighted by the data of Kozicky et al. (2014) and Yatham et al. (2017) that cognition and brain abnormalities after a first manic episode return to normal after 1 year only on the condition that there are no further episodes. More episodes are associated with dysfunction, disability, cognitive dysfunction, treatment resistance, prefrontal abnormalities, and an increased incidence of a diagnosis of dementia in old age (Post et al. 2012, 2017b).

The sensitization/kindling models make clear some of the mechanisms involved, including the accrual of epigenetic marks on DNA and histones fundamentally altering gene transcription. The models highlight the imperative of intervening early to stop the cascades of illness progression (Post 2018b). The clinical and neurobiological data indicating the progressive course of untreated bipolar illness are clear, as are the consequences for human dysfunction and suffering. The clinical data, now bolstered by neurobiological conceptualizations, should help propel a new round of treatment-related studies in order to see how one can intervene earlier and more optimally in order to prevent or delay illness progression.

References

Anderson LE, Chen ML, Perrin JM, Van Cleave J (2015) Outpatient visits and medication prescribing for US children with mental health conditions. Pediatrics 136:e1178–e1185

Anier K, Malinovskaja K, Aunurm-Helm A, Zharkovsky A, Kalda A (2010) DNA methylation regulates cocaine-induced behavioral sensitization in mice. Neuropsychopharmacology 35 (12):2450–2461

Antelman SM (1988) Stressor-induced sensitization to subsequent stress: implications for the development and treatment of clinical disorders. Telford Press, Caldwell, pp 227–256

Aoued HSSS, Doshi N, Morrison FG, Linsenbaum H, Hunter SC, Walum H, Baman J, Yao B, Jin P, Ressler KJ, Dias BG (2019) Reversing behavioral, neuroanatomical, and germline influences of intergenerational stress. Biol Psychiatry 85:248–256

Axelson D, Goldstein B, Goldstein T et al (2015) Diagnostic precursors to bipolar disorder in offspring of parents with bipolar disorder: a longitudinal study. Am J Psychiatry 172 (7):638–646

Belleivier F, Etain B, Malafosse A, Henry C, Kahn JP, Elgrabli-Wajsbrot O, Jamain S, Azorin JM, Frank E, Scott J, Grochocinski V, Kupfer DJ, Golmard JL, Leboyer M (2014) Age at onset in bipolar I affective disorder in the USA and Europe. World J Biol Psychiatry 15:369–376

Bernton O, McClung CA, Dileone RJ, Krishman V, Renthal W, Russon SJ, Graham D, Tsankova NM, Bolanos CA, Rios M, Monteggia LM, Self DW, Nestler EJ (2006) Essential role of BDNF in the mesolimbic dopamine pathway in social defeat stress. Science 311:864–868

Birmaher B, Axelson D, Goldstein B et al (2009) Four-year longitudinal course of children and adolescents with bipolar spectrum disorders: the course and outcome of bipolar youth (COBY) study. Am J Psychiatry 166(7):795–804

Border EJ, Evans LM, Smolen A, Berley N, Sullivan PF, Keller MC (2019) No support for historical candidate gene or candidate gene-by-interaction hypotheses for major depression across multiple large samples. Am J Psychiatry 176(5):376–387

Brodnik Z, Black E, España R (2019) Accelerated development of cocaine-associated dopamine transients and cocaine use vulnerability following traumatic stress. Neuropsychopharmacology:1–10

Caspi A, Sugden K, Moffitt TE et al (2003) Influence of life stress on depression: moderation by a polymorphism in the 5-HTT gene. Science 301(5631):386–389

Chan JC, Nugent BM, Bale TL (2017) Parental advisory: maternal and paternal stress can impact offspring neurodevelopment. Biol Psychiatry 83:886–894

Danese A, Moffitt TE, Harrington H, Milne BJ, Polanczyk G, Pariante C, Poulton R, Caspi A (2009) Adverse childhood experiences and adult risk factors for age-related disease: depression, inflammation, and clustering of metabolic risk markers. Pediatr Adolesc Med 163:1135–1143

Deighton A, Neville A, Pusch D, Dobson K (2018) Biomarkers of adverse childhood experiences: a scoping review. Psychiatry Res 269:719–732

Dias-Ferreira E, Sousa JC, Melo I, Morgado P, Mesquita AR, Cerqueira JJ, Costa RM, Sousa N (2009) Chronic stress causes frontostriatal reorganization and affects decision-making. Science 325(5940):621–625. https://doi.org/10.1126/science.1171203

Eckenrode ZD, Smith E, Marcynyszyn LA, Henderson CR Jr, Kitzman H, Cole R, Powers J, Olds DL (2001) Child maltreatment and the early onset of problem behaviors: can a program of nurse home visitation break the link? Dev Psychopathol 13(4):873–890

Etain B, Lajnef M, Bellivier F et al (2012) Clinical expression of bipolar disorder type I as a function of age and polarity at onset: convergent findings in samples from France and the United States. J Clin Psychiatry 73(4):e561–e566

Fournier M, d'Arripe-Longueville F, Radel R (2017) Effects of psychosocial stress on the goal-directed and habit memory systems during learning and later execution. Psychoneuroendocrinology 77:275–283

Fristad MA, Verducci JS, Walters K, Young ME (2009) Impact of multifamily psychoeducational psychotherapy in treating children aged 8 to 12 years with mood disorders. Arch Gen Psychiatry 66(9):1013–1021

Hinwood MTR, Charnley JL, Beynon SB, Day TA, Walker FR (2013) Chronic stress induced remodeling of the prefrontal cortex: structural re-organization of microglia and the inhibitory effect of minocycline. Cereb Cortex 23:1784–1797

Hodes GE, Menard C, Russo SJ (2016) Integrating interleukin-6 into depression diagnosis and treatment. https://www.sciencedirect.com/science/article/pii/S2352289515300461?via%3Dihub. Cited 8 Oct 2019

Hollis F, Kabbaj M (2014) Social defeat stress as a model of depression. ILAR 55(2):221–232

Hudziak JJ, Albaugh MD, Ducharme S et al (2014) Cortical thickness maturation and duration of music training: health-promoting activities shape brain development. J Am Acad Child Adolesc Psychiatry 53(11):1153–1161

James A, Hoang U, Seagroatt V, Clacey J, Goldacre M, Leibenluft E (2014) A comparison of American and English hospital discharge rates for pediatric bipolar disorder, 2000 to 2010. J Am Acad Child Adolesc Psychiatry 53(6):614–624

Kendler KS, Thornton LM, Gardner CO (2000) Stressful life events and previous episodes in the etiology of major depression in women: an evaluation of the "kindling" hypothesis. Am J Psychiatr 157(8):1243–1251

Kendler KS, Thornton LM, Gardner CO (2001) Genetic risk, number of previous depressive episodes, and stressful life events in predicting onset of major depression. Am J Psychiatr 158 (4):582–586

Kessing LV, Andersen PK (2017) Evidence for clinical progression of unipolar and bipolar disorders. Acta Psychiatr Scand 135:51–64

Kessing LV, Hansen HV, Hvenegaard A et al (2013) Treatment in a specialised out-patient mood disorder clinic v. standard out-patient treatment in the early course of bipolar disorder: randomised clinical trial. Br J Psychiatry 202(3):212–219

Kozicky J, Torres IJ, Silviera LE, Bond DJ, Lam RW, Yatham LN (2014) Cognitive change in the year after a first manic episode: association between clinical outcome and cognitive performance early in the course of bipolar I disorder. J Clin Psychiatry 75(6):587–593

Labonte B, Suderman M, Maussion G, Lopez JP, Navarro-Sanchez L, Yerko V, Mechawar N, Szyf M, Meaney MJ, Turecki G (2013) Genome-wide methylation changes in the brains of suicide completers. Am J Psychiatry 170:511–520

Lin PI, McInnis MG, Potash JB et al (2006) Clinical correlates and familial aggregation of age at onset in bipolar disorder. Am J Psychiatry 163(2):240–246

Lo Iacono LCC, Martini A, Valzania A, Viscomi MT, Chiurchiù V, Guatteo E, Bussone S, Perrone F, Di Sabato P, Aricò E, D'Argenio A, Troisi A, Mercuri NB, Maccarrone M, Puglisi-Allegra S, Casella P, Carola V (2018) From traumatic childhood to cocaine abuse: the critical function of the immune system. Biol Psychiatry 84:905–916

McGowan PO, Sasaki A, D'Alessio AC, Dymov S, Labonte B, Szyf M, Turecki G, Meaney MJ (2009) Epigenetic regulation of the glucocorticoid receptor in human brain associates with childhood abuse. Nat Neurosci 12:342–348

Meaney MJSM (2005) Environmental programming of stress responses through DNA methylation: life at the interface between a dynamic environment and a fixed genome. Dialogues Clin Neurosci 7(2):103–123

Merikangas KR, He JP, Burstein M, Swanson SA, Avenevoli S, Cui L, Benjet C, Georgiades K, Swendsen J (2010) Lifetime prevalence of mental disorders in U.S. adolescents: results from the National Comorbidity Survey Replication--Adolescent Supplement (NCS-A). J Am Acad Child Adolesc Psychiatry 49:990–989

Miklowitz DJ, Chung B (2016) Family-focused therapy for bipolar disorder: reflections on 30 years of research. Fam Process 55(3):483–499

Miklowitz DJ, Schneck CD, Singh MK et al (2013) Early intervention for symptomatic youth at risk for bipolar disorder: a randomized trial of family-focused therapy. J Am Acad Child Adolesc Psychiatry 52(2):121–131

Moffitt T, Arseneault L, Belsky D, Dickson N, Hancox RJ, Harrington H, Houts R, Poulton R, Roberts BW, Ross S, Sears MR, Thomson WM, Caspi A (2011) A gradient of childhood selfcontrol predicts health, wealth, and public safety. Proc Natl Acad Sci U S A 108 (7):2693–2698

Nasca C, Bigio B, Lee F et al (2018) Acetyl-L-carnitine deficiency in patients with major depressive disorder. Proc Natl Acad Sci 115:8627–8632

Niraula A, Witcher KG, Sheridan JF, Godbout JP (2018) Interleukin-6 induced by social stress promotes a unique transcriptional signature in the monocytes that facilitate anxiety. http://biologicalpsychiatryjournal.com/retrieve/pii/S0006322318319292. Cited 8 Oct 2019

Olds DL, Henderson CR Jr, Kitzman HJ, Eckenrode JJ, Cole RE, Tatelbaum RC (1999) Prenatal and infancy home visitation by nurses: recent findings. Future Child 9(1):44–65

Perlis RH, Miyahara S, Marangell LB et al (2004) Long-term implications of early onset in bipolar disorder: data from the first 1000 participants in the systematic treatment enhancement program for bipolar disorder (STEP-BD). Biol Psychiatry 55(9):875–881

Post RM (1992) Transduction of psychosocial stress into the neurobiology of recurrent affective disorder. Am J Psychiatry 149(8):999–1010

Post RM (2007a) Kindling and sensitization as models for affective episode recurrence, cyclicity, and tolerance phenomena. Neurosci Biobehav Rev 31(6):858–873

Post RM (2007b) Role of BDNF in bipolar and unipolar disorder: clinical and theoretical implications. J Psychiatr Res 41(12):979–990

Post RM (2015) In: Kapczinski F (ed) Clinical staging in bipolar disorder: a historical perspective. Oxford University Press, New York

Post RM (2016a) Epigenetic basis of sensitization to stress, affective episodes, and stimulants: implications for illness progression and prevention. Bipolar Disord 18(4):315–324

Post RM (2016b) Treatment of bipolar depression: evolving recommendations. Psychiatr Clin North Am 39(1):11–33

Post RM (2017a) New perspectives on the course and treatment of bipolar disorder. Minerva Med 58(1):40–53

Post RM (2017b) The new news about lithium: an underutilized treatment in the United States. Neuropsychopharmacology 43:1174–1179

Post RM (2018a) The neurochemistry and epigenetics of PTSD: implications for therapeutics in trauma- and stressor-related disorders. In: Stoddard FJB, David M, Milad M, Ursano RJ (eds) Trauma- and stressor-related disorders. Oxford University Press, New York, pp 161–176

Post R (2018b) Disturbing lack of early intervention studies in bipolar disorder. JAMA Psychiat 75 (11):1201–1202

Post RM, Kalivas P (2013) Bipolar disorder and substance misuse: pathological and therapeutic implications of their comorbidity and cross-sensitisation. Br J Psychiatry J Ment Sci 202 (3):172–176

Post RM, Leverich GS (2008) Treatment of bipolar illness: a case book for clinicians and patients. WW Norton and Company, New York

Post RM, Weiss SR (2011) Tolerance to the prophylactic effects of carbamazepine and related mood stabilizers in the treatment of bipolar disorders. CNS Neurosci Ther 17(6):649–660

Post RM, Weiss SR, Pert A (1987a) The role of context and conditioning in behavioral sensitization to cocaine. Psychopharmacol Bull 23(3):425–429

Post R, Weiss SR, Pert A, Uhde T (1987b) Chronic cocaine administration: sensitization and kindling effects. In: Raskin A, Fisher S (eds) Cocaine: clinical and biobehavioral aspects. Oxford University Press, New York, pp 109–173

Post RM, Leverich GS, Kupka RW et al (2010) Early-onset bipolar disorder and treatment delay are risk factors for poor outcome in adulthood. J Clin Psychiatry 71(7):864–872

Post RM, Fleming J, Kapczinski F (2012) Neurobiological correlates of illness progression in the recurrent affective disorders. J Psychiatr Res 46(5):561–573

Post RM, Altshuler L, Leverich G et al (2013a) More stressors prior to and during the course of bipolar illness in patients from the United States compared with the Netherlands and Germany. Psychiatry Res 210(3):880–886

Post R, Chang K, Frye M (2013b) Preliminary categorization of ultra-high risk for childhood bipolar disorder on clinical grounds so that preventive strategies can be studied. J Clin Psychiatry 74(2):167–169

Post RM, Altshuler LL, Leverich GS et al (2014) More medical comorbidities in patients with bipolar disorder from the United States than from the Netherlands and Germany. J Nerv Ment Dis 202(4):265–270

Post RM, Altshuler LL, Kupka R et al (2015a) Verbal abuse, like physical and sexual abuse, in childhood is associated with an earlier onset and more difficult course of bipolar disorder. Bipolar Disord 17(3):323–330

Post RM, Leverich GS, Kupka R et al (2015b) Increases in multiple psychiatric disorders in parents and grandparents of patients with bipolar disorder from the USA compared with the Netherlands and Germany. Psychiatr Genet 25(5):194–200

Post RM, Altshuler LL, Kupka R et al (2016a) Age of onset of bipolar disorder: combined effect of childhood adversity and familial loading of psychiatric disorders. J Psychiatr Res 81:63–70

Post R, Leverich G, Kupka R et al (2016b) Age of onset of bipolar disorder related to parental and grandparental illness burden. J Clin Psychiatry 77:e1309–e1315

Post RM, Altshuler LL, Kupka R, McElroy SL, Frye MA, Rowe M, Grunze H, Suppes T, Keck PE Jr, Leverich GS, Nolen WA (2016c) More illness in offspring of patients with bipolar illness from the us compared with The Netherlands and Germany. J Affect Disord 191:180–186

Post RM, Altshuler L, Kupka R et al (2017a) More childhood onset bipolar disorder in the United States than Canada or Europe: implications for treatment and prevention. Neurosci Biobehav Rev 74:204–213

Post RM, Rowe M, Kaplan D, Findling RL (2017b) A multisymptomatic child with bipolar disorder: how to track and sequence treatment. Prim Care Companion CNS Dis 19(4)

Post RM, Yatham LN, Vieta E, Berk M (2019) Beyond gold standard evidence-based treatment of bipolar disorder: rational pragmatic approaches to management. Bipolar Disord 21(7):650–659

Roth TL, Lubin FF, Funk AJ, Sweatt JD (2009) Lasting epigenetic influence of early-life adversity on the BDNF gene. Biol Psychiatry 65:790–769

Shonkoff JP, Garner AS (2012) The lifelong effects of early childhood adversity and toxic stress. Pediatrics 129(1):e232–e246

Tidey JW, Miczek KA (1997) Acquisition of cocaine self-administration after social stress: role of accumbens dopamine. Psychopharmacology 130(3):203–212

Tsankova NM, Berton O, Renthal W, Kumar A, Neve RL, Nestler EJ (2006) Sustained hippocampal chromatin regulation in a mouse model of depression and antidepressant action. Nat Neurosci 9(4):519–525

Tsankova N, Renthal W, Kumar A, Nestler EJ (2007) Epigenetic regulation in psychiatric disorders. Nat Rev Neurosci 8(5):355–367

Weber MD, Godbout JP, Sheridan JF (2017) Repeated social defeat, neuroinflammation, and behavior: monocytes carry the signal. Neuropsychopharmacology 42(1):46–61

Whittaker JR, Foley SF, Ackling E, Murphy K, Caseras X (2018) The functional connectivity between the nucleus accumbens and the ventromedial prefrontal cortex as an endophenotype for bipolar disorder. Biol Psychiatry 84:803–809

Yatham L, Mackala S, Basivireddy J, Ahn S, Walji N, Hu C (2017) Lurasidone versus treatment as usual for cognitive impairment in euthymic patients with bipolar I disorder: a randomised, open-label, pilot study. Lancet Psychiatry 4:208–217

Childhood Maltreatment in Bipolar Disorders

Bruno Etain and Monica Aas

Contents

B. Etain (✉)
Université de Paris, Paris, France

INSERM U1144, Faculté de Pharmacie de Paris, Université de Paris, Paris, France

Assistance Publique des Hôpitaux de Paris P-HP, GHU Saint-Louis – Lariboisière – F. Widal, Département de Psychiatrie et de Médecine Addictologique, Paris, France
e-mail: bruno.etain@inserm.fr

M. Aas
NORMENT, Division of Mental Health and Addiction, Oslo University Hospital, Oslo, Norway

Curr Topics Behav Neurosci (2021) 48: 277–302
https://doi.org/10.1007/7854_2020_149
Published Online: 12 July 2020

Abstract In this chapter, we will focus on childhood maltreatment and its role in the vulnerability to BD.

We will review how childhood maltreatment and trauma not only predispose to the development of BD but also to a more unstable, pernicious, and severe clinical expression of the disorder. This environmental risk factor is suggested to be part of a multiple hit model of vulnerability, involving not only early stressors (prenatal and postnatal ones) but also interactions with the genetic background of individuals and with other stressors occurring later in life. We will also review how childhood maltreatment and trauma may modify the brain functioning and circuits and alter some biological pathways in BD, hence leading to psychopathology. Finally, we will briefly discuss the implications for clinical practice and treatment.

Keywords Bipolar disorder · Childhood trauma · Mechanisms

1 Introduction

Early life stress is an umbrella term that refers to a wide range of early stressors such as childhood traumatic events, childhood maltreatment, childhood adversities, parental loss, or exposure to bullying at school. As defined by the World Health Organization (WHO) (WHO 2017), "child maltreatment is the abuse and neglect of people under 18 years of age. It includes all forms of physical and/or emotional ill-treatment, sexual abuse, neglect or negligent treatment or commercial or other exploitation, resulting in actual or potential harm to the child's health, survival, development or dignity in the context of a relationship of responsibility, trust or power. Four types of child maltreatment are generally recognized: physical abuse, sexual abuse, psychological (or emotional or mental) abuse, and neglect."

According to the US Department Human Services (Children's Bureau/ACYF/ACF/HHS 2017), approximately 683.000 children each year are victims of maltreatment. Most of these children are victims of neglect (75%), with physical abuse (17%), and sexual abuse (8%) being the next most frequent types of child maltreatment. According to the WHO (WHO 2017), a high number of children worldwide report that they have suffered some traumatic experiences or violence in the past year: physical abuse (23%), sexual abuse (8–18%), physical neglect (16%), or psychological/emotional abuse (36%).

Childhood maltreatment is an important risk factor for developing a wide range of severe mental disorders, including bipolar disorders (BD) (Lippard and Nemeroff 2020; Nemeroff 2016). Green et al., for example, support this statement (Green et al. 2010) in their study from the National Comorbidity Survey Replication showing that severe childhood adversity accounts for more than a quarter of adulthood psychiatric disorders and even more (around 45%) when considering specifically early-onset disorders (Green et al. 2010).

2 Childhood Maltreatment and (More Severe) Bipolar Disorders

A history of childhood adversity has been consistently associated with BD (Aas et al. 2016b), with a more severe clinical expression of the disorder being a function of number and severity of the exposure to traumatic events (Etain et al. 2013a).

2.1 *Childhood Maltreatment as a Risk Factor for Developing Bipolar Disorders*

About a decade ago, Fisher and Hosang identified only six studies suggesting childhood maltreatment as a risk factor for developing BD (Fisher and Hosang 2010), thus illustrating the relative paucity of good-quality studies in this field. Since then, numerous of studies have supported childhood maltreatment as part of a constellation of risk factors for BD, with a recent study suggesting childhood maltreatment as a causal factor for BD, independent of the genetic risk measured by the polygenic risk score for BD (Aas et al. 2020). Several studies have also shown that multiple childhood traumas are more frequent in patients with BD than in controls (63% versus 33%) (Etain et al. 2010), with a possible cumulative effect.

More recently, a systematic review of the literature and a meta-analysis of 19 studies showed that individuals with BD were 2.63 times (95% CI 2.00–3.47) more likely to report childhood adversity as compared to a non-clinical control group. This rate was similar across BD subtypes (type I and type II). When examining childhood maltreatment subtypes, emotional abuse showed the strongest effect (OR = 4.04, 95% CI 3.12–5.22). However, except for parental loss, all other subtypes of childhood maltreatment (emotional neglect, physical neglect, physical abuse, and sexual abuse) were also more frequently reported by individuals with BD as compared to the non-clinical group (Palmier-Claus et al. 2016).

Of note, most of the available studies had a retrospective design, thus preventing to infer any definite causality links. However, a few longitudinal studies have also been published to demonstrate that childhood adversities increase the risk of developing BD (Marangoni et al. 2016). As an example, in the National Epidemiologic Survey on Alcohol and Related Conditions, the analysis of the risk for initial-onset manic episodes during the study's 3-year follow-up period demonstrated that childhood physical maltreatment and sexual abuse were associated with a significantly higher risk of first-onset mania (Gilman et al. 2015).

2.2 Childhood Maltreatment and the Severity of the Clinical Expression of Bipolar Disorders

Similar to what is observed in patients with major depressive disorder (MDD), patients with BD and childhood maltreatment have a more severe illness over time, including earlier age at onset, an increased risk for suicide attempts, more frequent mood episodes, and poor response to treatment (Aas et al. 2016a; Etain et al. 2017b; Nemeroff 2016; Williams et al. 2016). Moreover, in BD, a history of childhood maltreatment has also been associated with an increased risk for rapid cycling course (Garno et al. 2005).

However, the quality of these studies, especially of the earliest ones, has been reported as poor to moderate. A review paper has highlighted this by Daruy-Filho et al. (2011), including 18 studies and underlining several study limitations, such as the lack of use of a structured clinical interview for diagnoses, the lack of use of a standardized childhood maltreatment assessment, small sample sizes (less than 100 patients), and the use of unsatisfactory measures of current mood states that represent a potential confounder in childhood maltreatment assessment (Daruy-Filho et al. 2011).

In 2013, in a large sample of patients with BD ($n = 587$), using a structured clinical interview for diagnoses and a validated standardized childhood maltreatment assessment, we confirmed the association between childhood maltreatment (and more specifically emotional abuse and sexual abuse) and several indicators of the complexity/severity of the clinical expression of BD, including earlier age of onset, an increased risk of at least one-lifetime suicide attempt, an increased risk of rapid cycling, and an increased number of mood episodes (Etain et al. 2013a).

In 2016, a meta-analysis of 30 studies confirmed that patients with BD and history of childhood maltreatment had greater severity for mania, depression, and psychosis, higher risk of comorbidity with several other psychiatric conditions (post-traumatic stress disorder, anxiety disorders, substance misuse disorders, alcohol misuse disorder), earlier age at onset of BD, higher risks of rapid cycling and suicide attempts, and a more unstable form of BD with greater numbers of manic and depressive episodes when compared to those patients without such childhood maltreatment. It should be mentioned here that, for this meta-analysis, the different subtypes of childhood maltreatment were not disentangled and several definitions were used for "maltreatment" (any maltreatment; total maltreatment score; sexual, physical, or emotional abuse; physical or emotional neglect) (Agnew-Blais and Danese 2016).

2.3 Childhood Maltreatment, Psychiatric, and Medical Comorbidities in Bipolar Disorders

Patients with a mood disorder, such as BD or MDD, and with a history of childhood maltreatment are more likely to have dual diagnoses as compared to patients without maltreatment. For example, substance use disorder is more commonly reported in BD with childhood maltreatment as compared to BD without childhood maltreatment (Etain et al. 2013a). These findings are consistent with the findings of Putnam and colleagues (Putnam et al. 2013), who found that multiple childhood traumatic events resulted in more complex adult psychopathology, as defined by higher rates of comorbidity and a greater number of symptoms. This is also supported by some of our studies showing that patients with BD and childhood maltreatment are more likely to also suffer from cannabis and alcohol misuse as compared to patients without childhood traumatic experiences (Aas et al. 2013a).

Patients with BD have around 10 years of shorter life expectancy than the general population (Hayes et al. 2015; Kessing et al. 2015). This is not only related to deaths due to suicide but also the heavy medical burden in BD. Recently, data have emerged about the associations between childhood maltreatment and medical comorbidities in BD. This suggests that childhood maltreatment not only associated with the complexity of BD in terms of psychiatric clinical expression but also with poor physical outcomes. In a cross-sectional study of 900 outpatients with BD, the number of medical conditions was significantly associated with childhood adversity (defined as a proxy, i.e., positive parental history of psychiatric illness) (Post et al. 2013). Another cross-sectional study compared 248 participants with recurrent major depressive disorders (MDD) to 72 participants with BD and showed that any type of childhood maltreatment, child abuse, and child neglect was significantly associated with an increased medical burden in individuals with BD, but not in individuals with MDD nor controls (Hosang et al. 2017, 2018). In a sample of patients with mood disorders, it has been found associations between childhood adversity and higher systolic and diastolic blood pressure, between sexual abuse and obesity (McIntyre et al. 2012). Finally, an association has been reported between childhood sexual abuse and higher body mass index (BMI) in individuals with BD, however only in a subgroup defined by late age at onset of BD (Leclerc et al. 2018).

2.4 *Issues About Childhood Maltreatment Subtypes, the Timing of Exposure, and Gender*

Until recently, the literature has been biased towards the study of childhood physical and sexual abuses in relation to BD. However, over the last decade, much more studies have suggested that emotional abuse might be of central importance in BD, with both higher prevalence as compared to controls, as well as the strongest links with the severe clinical expression of the disorder (Etain et al. 2008, 2010; Aas et al. 2016a) over other maltreatment subtypes. Indeed, the challenge is to separate the specific effects of each subtype of childhood maltreatment, given the fact that these different subtypes often covary together. For example, the use of measures of childhood maltreatment that incorporate the five different subtypes may be more relevant to be used as compared to scales that may focus more on physical or sexual

Table 1 Maltreatment subtypes assessed by the childhood maltreatment questionnaire (Bernstein et al. 2003)

Maltreatment subtype	Definition
Emotional neglect	Failure of caretakers to meet children's basic emotional and psychological needs, including love, belonging, nurturance, and support
Emotional abuse	Verbal assaults on a child's sense of worth or well-being or any humiliating or demeaning behavior directed towards a child by an adult or older person
Physical neglect	Failure of caretakers to provide for a child's basic physical needs, including food, shelter, clothing, safety, and health care
Physical abuse	Bodily assaults on a child by an adult or older person that posed a risk of or resulted in injury
Sexual abuse	Sexual contact or conduct between a child younger than 18 years of age and an adult or older person

abuse only. Such a comprehensive assessment can be performed using, for example, the Childhood Trauma Questionnaire by Bernstein et al. (2003) (see Table 1). Interestingly, a review of the literature across main psychiatric disorders suggested that physical abuse, sexual abuse, and unspecified neglect were associated with mood disorders and anxiety disorders; emotional abuse with personality disorders and schizophrenia; and physical neglect with personality disorders (Carr et al. 2013).

Another insufficiently addressed issue is the importance of the timing of exposure. It is suggested that the traumatic events may have different outcomes depending on whether they occurred early during childhood (before the age of 5), later during childhood (between the age of 5 and 12), or during the adolescence (between the age of 12 and 18). Indeed, the effects are expected to be different because, as an example, they do not occur in the same neurodevelopmental periods, thus not affecting the same brain circuits. Hence there are probably some sensitive periods when childhood maltreatment is more harmful. While there are no data specifically on BD, some articles have explored this issue in mood and psychotic disorders. As examples, early childhood sexual abuse (before the age of 5) and late childhood physical abuse (after the age of 13) were predictors of adult depression (Jaye Capretto 2017). Individuals who have been first exposed to childhood maltreatment during middle childhood had higher depressive symptoms and higher emotion dysregulation as compared to those who have been first exposed during later developmental stages (Dunn et al. 2017, 2018). The exposure to physical abuse during preschool was associated with a 77% increased risk of depression, and the exposure to sexual abuse during early childhood was associated with a 146% increased risk of suicidal ideation, when compared to the exposure in adolescence (Dunn et al. 2013). In individuals with early psychosis, the exposure to one or more subtypes of childhood trauma before the age of 16 was associated with higher levels of positive, depressive, manic, and negative symptoms (Alameda et al. 2016). Individuals with early sexual or physical abuse (before age 11 years) and psychotic symptoms showed poor functional outcome as compared to those with later sexual or physical abuse (between ages 12 and 15 years) (Alameda et al. 2015).

Finally, the effects of childhood maltreatment may differ according to gender, meaning that that the sensitivity to (specific) forms of childhood maltreatment would differ whether the individual is a female or a male. Females are more likely to suffer from mood disorders (including recurrent MDD or BD, especially type II) as compared to males, and females are also more likely to report having experienced childhood maltreatment than males, this being observed in the general population but also BD (Etain et al. 2013a; Fisher and Hosang 2010). We have suggested that the effect of specific subtypes of childhood maltreatment on the clinical expression of BD would differ according to gender. Indeed, the associations between sexual abuse and age at onset of BD, between sexual abuse and suicide attempts, between emotional abuse and rapid cycling, and between emotional abuse and the number of depressive episodes were all significant in females, but not significant or observed only at a trend level in males with BD (Etain et al. 2013a). A network analysis of data about childhood traumatic events performed in from a representative sample of 5,037 members of the general population living in a large metropolitan area suggested that "neglect" and "parental death" were more important for females than males. In contrast, "parental mental disorders" were more important for males (Coelho et al. 2018). However, the data are still inconsistent about gender sensitivity since a meta-analysis of gender differences of childhood maltreatment on adult depression and anxiety concluded that, even though associations were larger for females than for males, these gender differences were not statistically significant (Gallo et al. 2018).

3 Moving to Dimensions of Psychopathology in Association with Childhood Maltreatment

The recent literature has proposed that the "bipolar phenotype" can be disentangled in several dimensions of psychopathology that can be specific or not to BD and may be studied in association with both genetic and environmental risk factors. Several dimensions of psychopathology (or so-called "trait" dimensions) or cognitive traits have been studied in association with childhood maltreatment in patients with BD (Henry and Etain 2010). Such dimensions include affective regulation, impulsivity/hostility, and proneness to psychosis.

3.1 Childhood Maltreatment and Dimensions of Psychopathology in Bipolar Disorders

Patients with BD have a more unstable mood as compared to the general population and score higher on affective lability measures (Aminoff et al. 2012; Henry et al. 2008). Affective lability is characterized by intense and rapid fluctuations in affects in response to both pleasant and unpleasant events. In BD, studies have found heightened affective lability in both manic and mixed episodes (Henry et al. 2003) that also persist during euthymic periods (Henry et al. 2008). Affective lability is associated with greater severity of the clinical expression of BD, this including suicide attempts, earlier age at onset, and more frequent lifetime comorbidities (with anxiety disorders and substance use disorders) (Henry et al. 2008). Research in twins has demonstrated that the heritability of affective lability is relatively low (approximately 25% of the variance), hence suggesting that this dimension might be more driven by environmental risk factors than by genetic susceptibility (Coccaro et al. 2012). Other authors and we have previously published several papers showing that, in patients with BD, those individuals who report childhood maltreatment, especially emotional abuse, had a less stable mood that is characterized by higher scores on affect lability measures (Aas et al. 2014a; Etain et al. 2008; Marwaha et al. 2016).

Higher levels of impulsivity or hostility also characterize patients with BD as compared to non-clinical samples (Etain et al. 2013b; Saddichha and Schuetz 2014); this impulsivity may in turn associate with a worse clinical expression of the disorder in terms of suicide attempts or substance use disorders. Independently of the psychiatric diagnoses, a recent meta-analysis of 55 studies demonstrated a positive association between childhood maltreatment and trait impulsivity (Liu 2019).

More recent studies have attempted to model the complex links between childhood maltreatment and the clinical expression of BD, using dimensions of psychopathology as potential mediators. We demonstrated that affective lability mediates the association between childhood maltreatment and increased risk of suicide attempts, mixed episodes, and anxiety disorders supporting a pathway from childhood maltreatment, through dimensions of mood instability, to clinical indicators of the severity of BD (Aas et al. 2016b). The model has been further extended in a larger sample of euthymic patients with BD, showing two main paths: the first one is mainly going from emotional abuse to suicidal behavior through dimensions of affective dysregulation (affect intensity, affective lability, and emotional hostility), while the second one is mainly going from emotional abuse to substance misuse through dimensions of impulsivity and motor hostility (Etain et al. 2017c). Marwaha et al. have subsequently replicated this model in 2019 who demonstrated that affective instability significantly mediated the associations between childhood abuse and earlier age of onset, the number of depressive and manic episodes/illness year, anxiety disorders, and rapid cycling. Furthermore, impulsivity significantly mediated the associations between childhood abuse and manic episodes/illness year, anxiety disorders, rapid cycling, suicidal behavior, and substance misuse (Marwaha et al. 2019).

Finally, patients with BD frequently experience psychotic features (delusions or hallucinations) during mood episodes, this being potentially associated with the exposure to childhood maltreatment (Agnew-Blais and Danese 2016; Upthegrove et al. 2015). Incorporating a measure of delusional beliefs as a trait for proneness to psychosis, a path analysis demonstrated that emotional and physical abuse and cannabis misuse were each directly and independently associated with delusional beliefs that in turn were strongly associated with psychotic features during mood episodes (Etain et al. 2017a).

Of note, these associations between childhood maltreatment and dimensions of psychopathology are likely to be rather unspecific to BD, because it is also observed in other psychiatric conditions. As an example, patients with borderline personality or with attention deficit with hyperactivity disorder also presented with high levels of impulsivity and emotional deregulation but also with high rates of reported childhood maltreatment (Porter et al. 2020; Sugaya et al. 2012). Hence, a same environmental risk factor can lead to similar increases of dimensional traits in various psychiatric conditions that share these dimensions of psychopathology (Richard-Lepouriel et al. 2016, 2019).

3.2 Childhood Maltreatment and Cognition in Bipolar Disorders

A small number of independent studies has linked childhood maltreatment to worse cognitive functioning measured in adulthood in BD. Childhood maltreatment has been associated with decreased general cognitive abilities as measured by the Wechsler Abbreviated Scale of Intelligence (Martins et al. 2019), poor cognitive performance (Jimenez et al. 2017), poorer cognitive performance in patients on cognitive measures of IQ, auditory attention, and verbal and working memory (Bucker et al. 2013).

There are also two reviews on this topic in BD and schizophrenia. The first one suggested that patients who experienced childhood maltreatment displayed deficits in general cognitive ability as compared to those without such exposure (Dauvermann and Donohoe 2019). The second one focused on social cognition in association with early life stress in major psychiatric disorders (Rokita et al. 2018). It concluded in favor of an association between early childhood social experience, including both insecure attachment and adversity relating to neglect or abuse and poorer social cognitive performance.

4 Childhood Maltreatment as Part of a Multiple Hit Model of Susceptibility

When exposed to childhood maltreatment, a lot of individuals will fortunately not develop any severe psychiatric disorder, thus being qualified as "resilient." Those individuals who have been exposed to childhood maltreatment and are not resilient may develop a wide range of possible psychiatric disorders (Sugaya et al. 2012). Hence, it is not known (1) why only some exposed individuals go on to develop BD and (2) why they will develop BD and not another psychiatric condition, such as schizophrenia, for example. This heterogeneity in trajectories of susceptibility can be possibly explained by the combined occurrences: a constellation of different genetic and environmental factors and the complex interactions between all these risk factors. Different trajectories might thus be explained by the exposure to different environmental risk factors, a different genetic background, and the interactions between these. Multiple hit models of vulnerability have been thus proposed to take into account resilient trajectories, multiple exposures to various types of environmental factors at different periods of the neurodevelopment, and their interactions with the genetic background (Cannon et al. 2014).

4.1 Interactions Between Childhood Maltreatment and the Genetic Susceptibility to Bipolar Disorders

Genetic variants in multiple genes are supposed to interact with childhood maltreatment to increase the risk of developing BD and/or a more severe clinical expression of BD (Aas et al. 2016a; Etain et al. 2008). In this context, a moderation effect is hypothesized, according to which the genetic background of the individual conditions the response to the environmental factor (i.e., childhood trauma or maltreatment) on the phenotype.

A review by Rudolf Uher (2014) and a more recent one by Misiak et al. (2018) have suggested several interactions between variants of candidate genes and childhood maltreatment or early life stress on BD and/or on the clinical expression of BD (mainly age at onset of BD, suicidal behaviors, or cognitive functioning). Such proposed candidate genes were mainly the serotonin transporter gene (SLC6A4: solute carrier family 6 (neurotransmitter transporter, serotonin), member 4), the BDNF (brain-derived neurotrophic factor) gene and a few candidate genes involved in the HPA axis or immune-inflammatory processes. Several examples can highlight the fact that genetic variants in some candidate genes may moderate the effect of childhood maltreatment on the clinical expression of BD.

Earlier studies found complex interactions between sexual abuse, cannabis use, and SLC6A4 short/long variant on psychotic features in patients with BD (De Pradier et al. 2010). Etain et al. found that SLC6A4 interacted with childhood maltreatment to decrease the age at onset of BD in individuals who were homozygous for the short variant of this gene (5HTT-LPR) (Etain et al. 2015). Oliveira et al. found that some genetic variants of the TLR2 (toll-like receptor 2) gene interacted with sexual abuse to decrease the age at onset of BD (Oliveira et al. 2015). It has also been suggested that childhood sexual abuse associated with earlier BD onset age in BDNF Met (methionine) allele carriers (Miller et al. 2013). Finally, Anand et al. (2015) demonstrated a relationship between childhood maltreatment and SNPs in or near genes coding for calcium channel activity-related proteins and an earlier age at onset of BD (Anand et al. 2015).

Regarding suicidal behaviors in BD, it has been shown significant effects of 5HTT-LPR on the links between early life stress and suicidal behaviors among short allele carriers, but not among homozygotes for the long variant in patients with BD (Benedetti et al. 2014). A lifetime history of attempted suicide was associated with exposure to early life stress, with CLOCK (Circadian Locomotor Output Cycles Kaput) rs1801260*C carriers having the highest effects of early stress exposure on the probability of attempting suicide (Benedetti et al. 2015). Finally, no interaction was observed on suicidal behaviors in BD between childhood maltreatment and genetic variability in CRH-BP (corticotropin-releasing hormone-binding protein) and FKBP5 (FK506 binding protein 5) genes (two genes involved in the cortisol/stress pathway) (Segura et al. 2019).

Regarding cognitive functioning, the low-activity Met allele of the BDNF gene and the epsilon4 allele of the apolipoprotein E gene interacted with sexual abuse to decrease the performance to memory tests (Savitz et al. 2007). In a mixed sample of patients with schizophrenia spectrum disorders or BD, carriers of the low-activity Met allele of the BDNF gene and exposed to higher levels of childhood abuse (mainly physical abuse and emotional abuse) showed poorer cognitive functioning compared to the Val/Val (Valine) homozygotes (Aas et al. 2013b).

Beyond the candidate genes approach, genome-wide association studies (GWAS) in BD have identified many loci that are likely to contain variants in genes implicated in the vulnerability to BD. All these loci may provide potential new candidates to be studied in interaction with early life stress or childhood maltreatment in BD. However, the literature in this domain remains scarce. More recently, due to the limitations inherent to the selection of candidate genes, approaches that focus on polygenic risk scores (PRS) for various phenotypes (mainly BD and schizophrenia) have been proposed. The first study of this kind has included a sample of 402 patients with BD from Norway and France. All participants were assessed using the CTQ (Childhood Trauma Questionnaire) and characterized by a PRS for BD being calculated from previous GWAS data as part of the PGC (Psychiatric Genomics Consortium) (Aas et al. 2020). This study has suggested an interaction between childhood maltreatment and the PRS for BD on rapid cycling. No further interactions between PRS for BD and childhood maltreatment were observed for other clinical characteristics (age at onset, suicide attempts, number of mood episodes, mixed features, substance use disorders, and psychotic symptoms). To date, these results remain preliminary until replicated in independent samples but highlight the fact that PRS may offer a more comprehensive way to study the interactions between childhood maltreatment and the genetic background of individuals on the clinical expression of BD.

4.2 Interactions Between Childhood Maltreatment and Later Stressors in Bipolar Disorders

A two-hit model of vulnerability for developing a severe mental disorder has been proposed in psychosis (Pruessner et al. 2017), as well as in BD (Aas et al. 2016b). This model includes the exposure to prenatal or postnatal stressors interacting with genetic factors to "prime" the brain for psychopathology. Further stressors during adolescence or young adulthood (such as substance misuse or latter stressful events) may serve to convert the vulnerability into a disorder. In BD, the interaction between childhood maltreatment and susceptibility genes may predispose the individual to subtle changes in certain biological and physiopathological processes linked to the disorder (see the previous paragraph about maltreatment/gene interactions in BD). In this context, cannabis misuse during adolescence and later life stressful events may act together and reveal this susceptibility and/or increasing the severity of the disorder.

We have previously published a study showing that both childhood maltreatment and cannabis misuse increase the rates for rapid cycling, earlier age at onset, and suicide attempts beyond that of each risk factors taken alone (Aas et al. 2013a), thus supporting this two-hit model in BD. We have previously postulated that a putative mechanism of the co-occurrence of childhood maltreatment and cannabis abuse could be related to their opposite effects on the hypothalamic-pituitary-adrenal (HPA) axis (Heim et al. 2008; van Leeuwen et al. 2011), with a reduction of HPA axis functioning in substance abusers, while a history of childhood maltreatment has been linked to an increased HPA axis activity (Aas et al. 2019c). Hence, it could be that cannabis or substance use in individuals with a history of childhood maltreatment can be viewed as self-medication to "regulate" the HPA axis, in addition to the emotional turmoil (e.g., as shown by higher affect lability) caused by the childhood maltreatment events.

Numerous studies have also investigated whether stressful adulthood life events precede the onset of mood episodes in BD (Cohen et al. 2004; Johnson and Roberts 1995; Swendsen et al. 1995). For example, the study by Cohen et al. (2004) concluded that high levels of stressful life events in adulthood predict depressive recurrence over 12 months (Cohen et al. 2004). Also, Swendsen et al. (1995) found that stressful life events predicted more severe clinical features for over 12 months (Swendsen et al. 1995). We have recently shown that patients with BD and/or psychosis with childhood maltreatment experiences have higher cortisol in hair compared to patients without childhood maltreatment experiences, suggesting higher stress levels in adulthood in those with childhood maltreatment. Interestingly both a history of childhood maltreatment and higher hair cortisol were associated with more severe current symptoms and poorer functioning. It could be proposed that childhood maltreatment sensitizes the HPA axis in the way that later life experiences are perceived as more stressful. We hypothesize that this is mediated by long-term changes in affect regulation and poorer coping mechanisms following childhood maltreatment experiences in BD. As discussed in previous paragraphs, childhood maltreatment indeed is associated with long-term changes in affect regulation, impulse control, as well as cognition that might reduce in turn the ability to cope with later stressors.

5 Biological Correlates of Childhood Maltreatment in Bipolar Disorders

5.1 Childhood Maltreatment and Neuroimaging in Bipolar Disorders

Childhood maltreatment is, by definition, occurring during a neurodevelopmental period that is critical for brain maturation and the future effective regulation of emotional and cognitive processes. Several meta-analyses have been published to characterize the neural correlates of the exposure to childhood maltreatment in clinical and non-clinical samples.

A meta-analysis of published whole-brain voxel-based morphometry studies in childhood maltreatment showed that individuals who reported to have been exposed to childhood maltreatment had significantly smaller gray matter volumes in some brain areas (right orbitofrontal/superior temporal gyrus, amygdala, insula, and parahippocampal, middle temporal gyri, and left inferior frontal and postcentral gyri) but also larger gray matter volumes in other brain areas (right superior frontal and left middle occipital gyri) (Lim et al. 2014). A meta-analysis of structural magnetic resonance imaging studies of adults with a history of childhood maltreatment showed that individuals with childhood maltreatment had smaller hippocampus and amygdala volumes bilaterally and abnormal grey matter volumes in prefrontal-limbic brain regions (Paquola et al. 2016). Finally, the exposure to childhood maltreatment significantly correlated with an increased activation (left superior frontal gyrus, left middle temporal gyrus) and a decreased activation (left superior parietal lobule, left hippocampus) of some brain structures as measured by fMRI in adults (Heany et al. 2018).

The previously mentioned studies have been performed independently of the clinical status for any kind of psychiatric disorders. To date, only a few studies are available specifically in BD. A study using voxel-based morphometry compared grey matter volumes in association with the Childhood Trauma Questionnaire in euthymic BD type I patients and healthy controls. Grey matter volumes negatively correlated with several subtypes of childhood maltreatment (physical abuse, physical neglect, emotional neglect), mainly in the right dorsolateral prefrontal cortex and right thalamus (Duarte et al. 2016). A large sample of 105 outpatients with BD and 113 healthy controls have been assessed using the Childhood Trauma Questionnaire and high-resolution magnetic resonance imaging and demonstrated that childhood maltreatment was associated with increased bilateral volumes in patients with BD (amygdala, hippocampus) (Janiri et al. 2017).

A multimodal MRI (T1, diffusion weighted, and resting state fMRI) study was performed in patients with BD and healthy controls. The Childhood Trauma Questionnaires total score was negatively correlated with amygdala volumes, the prefronto-limbic functional connectivity, and the uncinate fractional anisotropy. Only physical and emotional neglects affected the neural parameters (Souza-Queiroz et al. 2016). Another study examined the association between childhood abuse and neglect on white matter integrity using diffusion tensor imaging (DTI), quantified as fractional anisotropy (FA), in 251 patients with BD type I and 163 healthy controls. Patients with BD and childhood abuse had lower FA in widespread regions of the brain as compared to patients without childhood abuse. Moreover, differences in mean FA significantly mediated the association between childhood abuse and BD (Stevelink et al. 2018). These studies might thus suggest lower integrity of white matter microstructure across the brain in patients with BD when exposed to childhood maltreatment.

In a mixed sample of patients with schizophrenia or bipolar spectrum disorders who have completed the Childhood Trauma Questionnaire, fMRI was used to measure brain activation during a presentation task of faces with negative or positive emotional expressions. After the scanner session, patients also performed emotional ratings of the same faces. Higher levels of childhood maltreatment were associated with stronger differentiation in brain responses to negative compared with positive faces (right angular gyrus, supramarginal gyrus, middle temporal gyrus, lateral occipital cortex), thus suggesting a negativity bias in the assessment of emotional valence of faces when exposed to childhood maltreatment (Aas et al. 2017b).

Interestingly, a few studies have used brain imaging (MRI) in association with genetic markers and measures of childhood maltreatment. In a mixed sample of patients with schizophrenia spectrum disorders or BD, BDNF gene Met carriers – when exposed to sexual abuse – had reduced right hippocampal volume and larger right and left lateral ventricles (Aas et al. 2013b). Furthermore, in the same sample, BDNF gene Met carriers reporting high levels of childhood maltreatment (specifically sexual or physical abuse) had significantly reduced hippocampal subfield volumes CA2/3 and CA4 dentate gyrus (Aas et al. 2014b).

Altogether, these studies highlight the fact that childhood maltreatment might deeply modify brain structures and circuits, possibly leading to an overactivation of the limbic system and a decreased ability of the prefrontal cortex to regulate emotions due to lower connectivity between limbic and prefrontal brain areas.

5.2 *Childhood Maltreatment and Peripheral Blood Biomarkers in Bipolar Disorders*

Several studies have attempted at identifying a peripheral signature of the exposure to childhood maltreatment in patients with BD. This has mainly involved the study of BDNF plasma levels, levels of several cytokines or interleukins, and telomere length.

Brain-derived neurotrophic factor (BDNF) promotes neuronal growth and differentiation during brain development and increase synaptic plasticity and maintenance of neurons in adult life (Nuernberg et al. 2016). It crosses the blood-brain barrier, and peripheral levels are highly correlated with levels in the cerebrospinal fluid (Fernandes et al. 2015). For example, early maltreatment was negatively associated with BDNF serum levels (Benedetti et al. 2017), and the 5HTT-LPR significantly influenced the relationship between early maltreatment and adult BDNF levels, due to a significant relationship between maltreatment and BDNF in 5HTT-LPR*short allele carriers, but not among long/long homozygotes (Benedetti et al. 2017). Patients with schizophrenia spectrum disorders or BD who have been exposed to childhood sexual abuse showed decreased plasma BDNF levels as compared to those without abuse (Aas et al. 2019a). Furthermore, a history of childhood maltreatment was associated with significantly reduced BDNF mRNA level in the blood (Aas et al. 2014b).

In a mixed sample of patients with schizophrenia/schizoaffective disorder, psychotic BD, and healthy controls, serum levels of interleukin 6 (IL-6), tumor necrosis factor-α (TNF-α), and C-reactive protein (CRP) were quantified and correlated with scores obtained with the Childhood Trauma Questionnaire. The exposure to sexual abuse was positively associated with levels of CRP in the schizophrenia group, but there were no significant associations between any form of maltreatment exposure and cytokine levels in the BD groups or healthy controls (Quide et al. 2019). This contrasts with the results obtained in a similar sample of patients with schizophrenia or BD spectrum disorders and healthy controls in which plasma levels of inflammatory markers (high-sensitivity C-reactive protein (hs-CRP), soluble tumor necrosis factor receptor type 1 (TNFR-R1), glycoprotein 130 (gp130)) were studied in association with the Childhood Trauma Questionnaire. The severity of childhood abuse (up to three types of abuse: sexual abuse, physical abuse, and emotional abuse) was associated with hs-CRP (Aas et al. 2017a). The levels of high-sensitivity C-reactive protein (hs-CRP) were also studied in association with the Childhood Trauma Questionnaire in a sample of 92 patients with BD type I and II and 142 healthy controls. Around 55% of the variance in hs-CRP was explained by cumulative and independent effects of age, BMI, and childhood maltreatment, especially sexual abuse (Moraes et al. 2017). Of note, independently of psychiatric diagnoses, childhood maltreatment has been shown to induce long-term modifications in inflammation processes in several independent meta-analyses (Baumeister et al. 2016; Coelho et al. 2014; Tursich et al. 2014).

Human telomeres are composed of tandem repeats of the TTAGGG sequence and average between 6 and 12 kilobases (kb) in length (Aubert and Lansdorp 2008; Yamaguchi et al. 2005). When telomeres become critically short, the cellular proliferation is stopped, and the risk of cellular apoptosis is increased, which eventually compromises tissue renewal capacity and function (Blasco 2007). Telomere length (TL) may, therefore, represent a biomarker of the "molecular clock" that contributes to cellular ageing and altered physical health. A study measured TL using quantitative polymerase chain reaction in a sample of 1,024 individuals (373 with schizophrenia, 249 with BD, and 402 healthy controls). Patients who reported a history of childhood sexual, physical, or emotional abuse as measured by the Childhood Trauma Questionnaire had shorter TL as compared to patients without such early life stress and to healthy controls (Aas et al. 2019b). These results are consistent with those reported by three previous meta-analyses performed independently of the psychiatric clinical status of the participants and demonstrating a significant association between childhood maltreatment or early life stress and accelerated telomere erosion in adulthood (Hanssen et al. 2017; Li et al. 2017; Ridout et al. 2018).

All these results may be interpreted as an imbalance between higher inflammatory processes and lower neuroplasticity in the peripheral blood, alongside with accelerated ageing in the case of exposure to childhood maltreatment. However, more studies are explicitly required in BD to better characterize the molecular signature of early life stress at the peripheral blood level, given the fact that studies remain scare. Indeed, the results might be not specific to BD since also reported in other psychiatric conditions but also the general population, and further studies should explore whether these phenomena are simply similar as those in the general population or amplified (i.e., of greater magnitude) in BD.

5.3 *Epigenetic Alterations Linked to Childhood Maltreatment in Bipolar Disorders*

Epigenetic mechanisms refer to a variety of molecular mechanisms that can modulate gene expression without any modifications of the DNA sequence. They represent adaptive mechanisms that are involved in response to stressors and lead to subtle modifications of gene expression, primarily through DNA methylation and histone modifications. Even though rare are the data coming specifically from samples of individuals with BD, a large body of evidence suggested that childhood maltreatment dramatically alters epigenetic processes and in turn can lead to modifications of gene expression.

A recent review of the literature has suggested that, in mood disorders, the methylation level of the glucocorticoid receptor NR3C1 gene and the correlation with the level of childhood maltreatment is robust. Alterations of the level of methylation in other genes were also suggested in association with childhood maltreatment, even though with less robust findings such as for the SLC6A4, BDNF, and FKBP5 genes (Nothling et al. 2019). Several studies have used large clinical or non-clinical samples and whole genome DNA methylation approaches to characterize the regions of the genome that were the most differentially methylated in the presence of an history of childhood maltreatment. In two large cohorts of women, it has been shown that several regions of the genome (obtained from peripheral blood or buccal cells) were differentially methylated according to several measures of childhood early stress (life adverse events or parental death as examples). However, the level of replication between cohorts was very low (Houtepen et al. 2018). A similar study quantified DNA methylation across the genome in buccal epithelial cell samples from a high-risk sample of inner-city youth and identified some highly differentially methylated regions of the genome as being associated with childhood maltreatment (as examples, physical abuse and PSEN2 gene or sexual abuse and GRIN2D gene). The results also demonstrated significant enrichment in differentially methylated genes encoding for proteins involved in neuronal systems (Cecil et al. 2016).

Of note, epigenetic signatures can also be used to calculate the epigenetic age that is a marker of accelerated ageing. Accelerated aging can indeed be defined by DNA methylation-based estimates of cellular age that exceed chronological age. Some data are suggesting that traumatic stress is associated with advanced epigenetic age (Wolf et al. 2018).

To date, these epigenetic mechanisms remain insufficiently characterized specifically in patients with BD. Indeed, while several reviews have highlighted that the levels of DNA methylation (whole genome or within candidate genes) may be abnormal in patients with BD as compared to healthy controls, they did not report the associations between these abnormalities and the exposure to early life stress (Teroganova et al. 2016). There are also some intriguing results suggesting that childhood adversity was associated with increased KITLG (KIT ligand) gene methylation in healthy individuals, but not patients with BD (He et al. 2018). Nevertheless, childhood maltreatment is suggested to induce long-lasting effects on several biological pathways that are important for the physiopathology of BD through deep modifications of epigenetic processes.

6 Implications for Clinical Practice and Treatment of Bipolar Disorders

Up to 60% of patients with BD have experienced a history of childhood maltreatment, which is associated with higher relapse rates, greater episode severity, and a higher risk of psychiatric and physical comorbidity. As such, these individuals would require close monitoring and more intensive care management because highly vulnerable. Although the long-term negative effects of childhood maltreatment in BD have received more attention over the last years, physical and sexual abuses (in childhood or adulthood) are often inadequately assessed in psychiatric clinics (Read and Fraser 1998). The main recommendation would be to propose a systematic assessment of childhood maltreatment for any patients with BD or at least for those with the most unstable and comorbid form of BD.

Regarding medications, there are almost no data about the response to psychotropic according to the exposure to childhood maltreatment in patients with BD. One study has found no association between childhood maltreatment and response to lithium in BD (Cakir et al. 2016), while a second one found that physical abuse correlated with a lower response to lithium in BD (Etain et al. 2017b). There are no data regarding the response to atypical antipsychotics and anticonvulsants according to a previous history of childhood maltreatment, hence preventing any recommendations that could be proposed for personalizing treatments for these patients who have suffered from childhood maltreatment.

Trauma-focused psychosocial and psychotherapeutically intervention strategies for BD are, therefore, pivotal, but studies of efficacy are currently almost absent in BD. Thompson et al. (2014) suggested including psychotherapy for trauma victims focusing on targeting the dissociative experiences that can follow childhood maltreatment experiences, specifically coping strategies, body awareness/mindfulness techniques, and stress management, was suggested. Interestingly a recent study suggested that eye movement desensitization and reprocessing (EMDR) can reduce affective episodes, affective symptoms, and functional, cognitive, and trauma symptoms (Moreno-Alcazar et al. 2017). Some study protocols have been published for the use of EMDR in individuals with BD, with planned secondary outcomes concerning functioning (Moreno-Alcazar et al. 2017). As discussed in Aas et al. (2016a), other types of psychotherapy for trauma victims would be trauma-focused cognitive behavioral therapy [CBT] for sexually abused children (Ehring et al. 2014). Based on the research showing that emotional dysregulation is one of the key long-term effects of childhood maltreatment in BD, we also suggest that interventions target affective lability in these patients specifically. The challenge for the next years will be to fill the gap between research on this field and routine practice since recommendations for managing this specific population are lacking. In particular, the knowledge is scare on which psychotherapies to provide or which targets therapists should focus on, as well as how childhood maltreatment could explain the resistance to mood stabilizers.

7 Conclusions

Patients with BD report more often childhood maltreatment experiences than the general population. They are also presenting, when exposed, a much more severe disorder over time, including an earlier onset, more mood episodes, more frequent suicide attempts, rapid cycling and higher risks of having a comorbid diagnosis with another psychiatric disorder, and a higher medical burden. The association between childhood maltreatment and poor outcomes in BD is suggested to be mediated through dimensions of psychopathology such as affective lability or impulsivity/hostility. Childhood maltreatment interacts with other environmental stressors (cannabis use, later stressors in life) but also with the genetic background of individuals to draw a trajectory of vulnerability that goes from early life stress to later severe BD. Childhood maltreatment could lead to psychopathology by altering brain structures and circuits, epigenetic mechanisms, and peripheral biomarkers. However, these long-lasting biological correlates of childhood maltreatment are probably not specific to BD but ubiquitously observed in many other psychiatric disorders. The specific recommendations for the care management of these vulnerable patients with BD and childhood maltreatment remain to be written. Given the heightened proneness to relapses and unstable BD, these patients should be the subjects of careful clinical scrutiny. However, to date, no specific recommendation can be made, except the systematic screening of childhood maltreatment, at least for those with a more severe, unstable, comorbid, or relapsing course. Given the devastating clinical consequences of childhood maltreatment in the prognosis and course of BD, we recommend need for further research about evidence-based interventions that target the consequences of childhood maltreatment in BD.

References

Aas M, Etain B, Bellivier F, Henry C, Lagerberg T, Ringen A, Agartz I, Gard S, Kahn JP, Leboyer M, Andreassen OA, Melle I (2013a) Additive effects of childhood abuse and cannabis abuse on clinical expressions of bipolar disorders. Psychol Med:1–10

Aas M, Haukvik UK, Djurovic S, Bergmann O, Athanasiu L, Tesli MS, Hellvin T, Steen NE, Agartz I, Lorentzen S, Sundet K, Andreassen OA, Melle I (2013b) BDNF val66met modulates the association between childhood trauma, cognitive and brain abnormalities in psychoses. Prog Neuro-Psychopharmacol Biol Psychiatry 46:181–188

Aas M, Aminoff SR, Vik Lagerberg T, Etain B, Agartz I, Andreassen OA, Melle I (2014a) Affective lability in patients with bipolar disorders is associated with high levels of childhood trauma. Psychiatry Res 218:252–255

Aas M, Haukvik UK, Djurovic S, Tesli M, Athanasiu L, Bjella T, Hansson L, Cattaneo A, Agartz I, Andreassen OA, Melle I (2014b) Interplay between childhood trauma and BDNF val66met variants on blood BDNF mRNA levels and on hippocampus subfields volumes in schizophrenia spectrum and bipolar disorders. J Psychiatr Res 59:14–21

Aas M, Henry C, Andreassen OA, Bellivier F, Melle I, Etain B (2016a) The role of childhood trauma in bipolar disorders. Int J Bipolar Disord 4:2

Aas M, Henry C, Bellivier F, Lajnef M, Gard S, Kahn JP, Lagerberg TV, Aminoff SR, Bjella T, Leboyer M, Andreassen OA, Melle I, Etain B (2016b) Affective lability mediates the association between childhood trauma and suicide attempts, mixed episodes and co-morbid anxiety disorders in bipolar disorders. Psychol Med:1–11

Aas M, Dieset I, Hope S, Hoseth E, Morch R, Reponen E, Steen NE, Laskemoen JF, Ueland T, Aukrust P, Agartz I, Andreassen OA, Melle I (2017a) Childhood maltreatment severity is associated with elevated C-reactive protein and body mass index in adults with schizophrenia and bipolar diagnoses. Brain Behav Immun 65:342–349

Aas M, Kauppi K, Brandt CL, Tesli M, Kaufmann T, Steen NE, Agartz I, Westlye LT, Andreassen OA, Melle I (2017b) Childhood trauma is associated with increased brain responses to emotionally negative as compared with positive faces in patients with psychotic disorders. Psychol Med 47:669–679

Aas M, Dieset I, Morch R, Steen NE, Hope S, Reponen EJ, Laskemoen JF, Ueland T, Aukrust P, Melle I, Agartz I, Andreassen OA (2019a) Reduced brain-derived neurotrophic factor is associated with childhood trauma experiences and number of depressive episodes in severe mental disorders. Schizophr Res 205:45–50

Aas M, Elvsashagen T, Westlye LT, Kaufmann T, Athanasiu L, Djurovic S, Melle I, van der Meer D, Martin-Ruiz C, Steen NE, Agartz I, Andreassen OA (2019b) Telomere length is associated with childhood trauma in patients with severe mental disorders. Transl Psychiatry 9:97

Aas M, Pizzagalli DA, Laskemoen JF, Reponen EJ, Ueland T, Melle I, Agartz I, Steen NE, Andreassen OA (2019c) Elevated hair cortisol is associated with childhood maltreatment and cognitive impairment in schizophrenia and in bipolar disorders. Schizophr Res 213:65

Aas M, Bellivier F, Bettella F, Henry C, Gard S, Kahn JP, Lagerberg TV, Aminoff SR, Melle I, Leboyer M, Jamain S, Andreassen OA, Etain B (2020) Childhood maltreatment and polygenic risk in bipolar disorders. Bipolar Disord 22(2):174–181

Agnew-Blais J, Danese A (2016) Childhood maltreatment and unfavourable clinical outcomes in bipolar disorder: a systematic review and meta-analysis. Lancet Psychiatry 3:342–349

Alameda L, Ferrari C, Baumann PS, Gholam-Rezaee M, Do KQ, Conus P (2015) Childhood sexual and physical abuse: age at exposure modulates impact on functional outcome in early psychosis patients. Psychol Med 45:2727–2736

Alameda L, Golay P, Baumann PS, Ferrari C, Do KQ, Conus P (2016) Age at the time of exposure to trauma modulates the psychopathological profile in patients with early psychosis. J Clin Psychiatry 77:e612–e618

Aminoff SR, Jensen J, Lagerberg TV, Hellvin T, Sundet K, Andreassen OA, Melle I (2012) An association between affective lability and executive functioning in bipolar disorder. Psychiatry Res 198:58–61

Anand A, Koller DL, Lawson WB, Gershon ES, Nurnberger JI, Bi GSC (2015) Genetic and childhood trauma interaction effect on age of onset in bipolar disorder: an exploratory analysis. J Affect Disord 179:1–5

Aubert G, Lansdorp PM (2008) Telomeres and aging. Physiol Rev 88:557–579

Baumeister D, Akhtar R, Ciufolini S, Pariante CM, Mondelli V (2016) Childhood trauma and adulthood inflammation: a meta-analysis of peripheral C-reactive protein, interleukin-6 and tumour necrosis factor-alpha. Mol Psychiatry 21:642–649

Benedetti F, Riccaboni R, Poletti S, Radaelli D, Locatelli C, Lorenzi C, Pirovano A, Smeraldi E, Colombo C (2014) The serotonin transporter genotype modulates the relationship between early stress and adult suicidality in bipolar disorder. Bipolar Disord 16:857–866

Benedetti F, Riccaboni R, Dallaspezia S, Locatelli C, Smeraldi E, Colombo C (2015) Effects of CLOCK gene variants and early stress on hopelessness and suicide in bipolar depression. Chronobiol Int 32:1156–1161

Benedetti F, Ambree O, Locatelli C, Lorenzi C, Poletti S, Colombo C, Arolt V (2017) The effect of childhood trauma on serum BDNF in bipolar depression is modulated by the serotonin promoter genotype. Neurosci Lett 656:177–181

Bernstein DP, Stein JA, Newcomb MD, Walker E, Pogge D, Ahluvalia T, Stokes J, Handelsman L, Medrano M, Desmond D, Zule W (2003) Development and validation of a brief screening version of the childhood trauma questionnaire. Child Abuse Negl 27:169–190

Blasco MA (2007) Telomere length, stem cells and aging. Nat Chem Biol 3:640–649

Bucker J, Kozicky J, Torres IJ, Kauer-Sant'anna M, Silveira LE, Bond DJ, Lam RW, Yatham LN (2013) The impact of childhood trauma on cognitive functioning in patients recently recovered from a first manic episode: data from the systematic treatment optimization program for early mania (STOP-EM). J Affect Disord 148:424–430

Cakir S, Tasdelen Durak R, Ozyildirim I, Ince E, Sar V (2016) Childhood trauma and treatment outcome in bipolar disorder. J Trauma Dissociation 17:397–409

Cannon M, Clarke MC, Cotter DR (2014) Priming the brain for psychosis: maternal inflammation during fetal development and the risk of later psychiatric disorder. Am J Psychiatry 171:901–905

Carr CP, Martins CM, Stingel AM, Lemgruber VB, Juruena MF (2013) The role of early life stress in adult psychiatric disorders: a systematic review according to childhood trauma subtypes. J Nerv Ment Dis 201:1007–1020

Cecil CA, Smith RG, Walton E, Mill J, McCrory EJ, Viding E (2016) Epigenetic signatures of childhood abuse and neglect: implications for psychiatric vulnerability. J Psychiatr Res 83:184–194

Children's Bureau/ACYF/ACF/HHS (2017) Department of Health & Human Services

Coccaro EF, Ong AD, Seroczynski AD, Bergeman CS (2012) Affective intensity and lability: heritability in adult male twins. J Affect Disord 136:1011–1016

Coelho R, Viola TW, Walss-Bass C, Brietzke E, Grassi-Oliveira R (2014) Childhood maltreatment and inflammatory markers: a systematic review. Acta Psychiatr Scand 129:180–192

Coelho BM, Santana GL, Duarte-Guerra LS, Viana MC, Neto FL, Andrade LH, Wang YP (2018) The role of gender in the structure of networks of childhood adversity. Psychiatry Res 270:348–356

Cohen AN, Hammen C, Henry RM, Daley SE (2004) Effects of stress and social support on recurrence in bipolar disorder. J Affect Disord 82:143–147

Daruy-Filho L, Brietzke E, Lafer B, Grassi-Oliveira R (2011) Childhood maltreatment and clinical outcomes of bipolar disorder. Acta Psychiatr Scand 124:427–434

Dauvermann MR, Donohoe G (2019) The role of childhood trauma in cognitive performance in schizophrenia and bipolar disorder – a systematic review. Schizophr Res Cogn 16:1–11

De Pradier M, Gorwood P, Beaufils B, Ades J, Dubertret C (2010) Influence of the serotonin transporter gene polymorphism, cannabis and childhood sexual abuse on phenotype of bipolar disorder: a preliminary study. Eur Psychiatry 25:323–327

Duarte DG, Neves Mde C, Albuquerque MR, de Souza-Duran FL, Busatto G, Correa H (2016) Gray matter brain volumes in childhood-maltreated patients with bipolar disorder type I: a voxel-based morphometric study. J Affect Disord 197:74–80

Dunn EC, McLaughlin KA, Slopen N, Rosand J, Smoller JW (2013) Developmental timing of child maltreatment and symptoms of depression and suicidal ideation in young adulthood: results from the National Longitudinal Study of adolescent health. Depress Anxiety 30:955–964

Dunn EC, Nishimi K, Powers A, Bradley B (2017) Is developmental timing of trauma exposure associated with depressive and post-traumatic stress disorder symptoms in adulthood? J Psychiatr Res 84:119–127

Dunn EC, Nishimi K, Gomez SH, Powers A, Bradley B (2018) Developmental timing of trauma exposure and emotion dysregulation in adulthood: are there sensitive periods when trauma is most harmful? J Affect Disord 227:869–877

Ehring T, Welboren R, Morina N, Wicherts JM, Freitag J, Emmelkamp PM (2014) Meta-analysis of psychological treatments for posttraumatic stress disorder in adult survivors of childhood abuse. Clin Psychol Rev 34:645–657

Etain B, Henry C, Bellivier F, Mathieu F, Leboyer M (2008) Beyond genetics: childhood affective trauma in bipolar disorder. Bipolar Disord 10:867–876

Etain B, Mathieu F, Henry C, Raust A, Roy I, Germain A, Leboyer M, Bellivier F (2010) Preferential association between childhood emotional abuse and bipolar disorder. J Trauma Stress 23:376–383

Etain B, Aas M, Andreassen OA, Lorentzen S, Dieset I, Gard S, Kahn JP, Bellivier F, Leboyer M, Melle I, Henry C (2013a) Childhood trauma is associated with severe clinical characteristics of bipolar disorders. J Clin Psychiatry 74:991–998

Etain B, Mathieu F, Liquet S, Raust A, Cochet B, Richard JR, Gard S, Zanouy L, Kahn JP, Cohen RF, Bougerol T, Henry C, Leboyer M, Bellivier F (2013b) Clinical features associated with trait-impulsiveness in euthymic bipolar disorder patients. J Affect Disord 144:240–247

Etain B, Lajnef M, Henrion A, Dargel AA, Stertz L, Kapczinski F, Mathieu F, Henry C, Gard S, Kahn JP, Leboyer M, Jamain S, Bellivier F (2015) Interaction between SLC6A4 promoter variants and childhood trauma on the age at onset of bipolar disorders. Sci Rep 5:16301

Etain B, Lajnef M, Bellivier F, Henry C, M'Bailara K, Kahn JP, Leboyer M, Fisher HL (2017a) Revisiting the association between childhood trauma and psychosis in bipolar disorder: a quasi-dimensional path-analysis. J Psychiatr Res 84:73–79

Etain B, Lajnef M, Brichant-Petitjean C, Geoffroy PA, Henry C, Gard S, Kahn JP, Leboyer M, Young AH, Bellivier F (2017b) Childhood trauma and mixed episodes are associated with poor response to lithium in bipolar disorders. Acta Psychiatr Scand 135:319–327

Etain B, Lajnef M, Henry C, Aubin V, Azorin JM, Bellivier F, Bougerol T, Courtet P, Gard S, Kahn JP, Passerieux C, Leboyer M, Center F.-B.C.C., Center F.-B.D.C., Sites F.-B.C., Principal

Collaborators in F (2017c) Childhood trauma, dimensions of psychopathology and the clinical expression of bipolar disorders: a pathway analysis. J Psychiatr Res 95:37–45

Fernandes BS, Molendijk ML, Kohler CA, Soares JC, Leite CM, Machado-Vieira R, Ribeiro TL, Silva JC, Sales PM, Quevedo J, Oertel-Knochel V, Vieta E, Gonzalez-Pinto A, Berk M, Carvalho AF (2015) Peripheral brain-derived neurotrophic factor (BDNF) as a biomarker in bipolar disorder: a meta-analysis of 52 studies. BMC Med 13:289

Fisher HL, Hosang GM (2010) Childhood maltreatment and bipolar disorder: a critical review of the evidence. Mind Brain J Psychiatry 1:75–85

Gallo EAG, Munhoz TN, Loret de Mola C, Murray J (2018) Gender differences in the effects of childhood maltreatment on adult depression and anxiety: a systematic review and meta-analysis. Child Abuse Negl 79:107–114

Garno JL, Goldberg JF, Ramirez PM, Ritzler BA (2005) Impact of childhood abuse on the clinical course of bipolar disorder. Br J Psychiatry 186:121–125

Gilman SE, Ni MY, Dunn EC, Breslau J, McLaughlin KA, Smoller JW, Perlis RH (2015) Contributions of the social environment to first-onset and recurrent mania. Mol Psychiatry 20:329–336

Green JG, McLaughlin KA, Berglund PA, Gruber MJ, Sampson NA, Zaslavsky AM, Kessler RC (2010) Childhood adversities and adult psychiatric disorders in the national comorbidity survey replication I: associations with first onset of DSM-IV disorders. Arch Gen Psychiatry 67:113–123

Hanssen LM, Schutte NS, Malouff JM, Epel ES (2017) The relationship between childhood psychosocial stressor level and telomere length: a meta-analysis. Health Psychol Res 5:6378

Hayes JF, Miles J, Walters K, King M, Osborn DP (2015) A systematic review and meta-analysis of premature mortality in bipolar affective disorder. Acta Psychiatr Scand 131:417–425

He Y, Vinkers CH, Houtepen LC, de Witte LD, Boks MP (2018) Childhood adversity is associated with increased KITLG methylation in healthy individuals but not in bipolar disorder patients. Front Psych 9:743

Heany SJ, Groenewold NA, Uhlmann A, Dalvie S, Stein DJ, Brooks SJ (2018) The neural correlates of childhood trauma questionnaire scores in adults: a meta-analysis and review of functional magnetic resonance imaging studies. Dev Psychopathol 30:1475–1485

Heim C, Mletzko T, Purselle D, Musselman DL, Nemeroff CB (2008) The dexamethasone/corticotropin-releasing factor test in men with major depression: role of childhood trauma. Biol Psychiatry 63:398–405

Henry C, Etain B (2010) New ways to classify bipolar disorders: going from categorical groups to symptom clusters or dimensions. Curr Psychiatry Rep 12:505–511

Henry C, Swendsen J, Van den Bulke D, Sorbara F, Demotes-Mainard J, Leboyer M (2003) Emotional hyper-reactivity as a fundamental mood characteristic of manic and mixed states. Eur Psychiatry 18:124–128

Henry C, Van den Bulke D, Bellivier F, Roy I, Swendsen J, M'Bailara K, Siever LJ, Leboyer M (2008) Affective lability and affect intensity as core dimensions of bipolar disorders during euthymic period. Psychiatry Res 159:1–6

Hosang GM, Fisher HL, Uher R, Cohen-Woods S, Maughan B, McGuffin P, Farmer AE (2017) Childhood maltreatment and the medical morbidity in bipolar disorder: a case- control study. Int J Bipolar Disord 5:30

Hosang GM, Fisher HL, Hodgson K, Maughan B, Farmer AE (2018) Childhood maltreatment and adult medical morbidity in mood disorders: comparison of unipolar depression with bipolar disorder. Br J Psychiatry 213:645–653

Houtepen LC, Hardy R, Maddock J, Kuh D, Anderson EL, Relton CL, Suderman MJ, Howe LD (2018) Childhood adversity and DNA methylation in two population-based cohorts. Transl Psychiatry 8:266

Janiri D, Sani G, Rossi P, Piras F, Iorio M, Banaj N, Giuseppin G, Spinazzola E, Maggiora M, Ambrosi E, Simonetti A, Spalletta G (2017) Amygdala and hippocampus volumes are differently affected by childhood trauma in patients with bipolar disorders and healthy controls. Bipolar Disord 19:353–362

Jaye Capretto J (2017) Developmental timing of childhood physical and sexual maltreatment predicts adult depression and Post-traumatic stress symptoms. J Interpers Violence. https://doi.org/10.1177/0886260517704963

Jimenez E, Sole B, Arias B, Mitjans M, Varo C, Reinares M, Bonnin CDM, Ruiz V, Saiz PA, Garcia-Portilla MP, Buron P, Bobes J, Amann BL, Martinez-Aran A, Torrent C, Vieta E, Benabarre A (2017) Impact of childhood trauma on cognitive profile in bipolar disorder. Bipolar Disord 19:363–374

Johnson SL, Roberts JE (1995) Life events and bipolar disorder: implications from biological theories. Psychol Bull 117:434–449

Kessing LV, Vradi E, Andersen PK (2015) Life expectancy in bipolar disorder. Bipolar Disord 17:543–548

Leclerc E, Mansur RB, Grassi-Oliveira R, Cordeiro Q, Kapczinski F, McIntyre RS, Brietzke E (2018) The differential association between history of childhood sexual abuse and body mass index in early and late stages of bipolar disorder. J Affect Disord 227:214–218

Li Z, He Y, Wang D, Tang J, Chen X (2017) Association between childhood trauma and accelerated telomere erosion in adulthood: a meta-analytic study. J Psychiatr Res 93:64–71

Lim L, Radua J, Rubia K (2014) Gray matter abnormalities in childhood maltreatment: a voxel-wise meta-analysis. Am J Psychiatry 171:854–863

Lippard ETC, Nemeroff CB (2020) The devastating clinical consequences of child abuse and neglect: increased disease vulnerability and poor treatment response in mood disorders. Am J Psychiatry 177:20–36

Liu RT (2019) Childhood maltreatment and impulsivity: a meta-analysis and recommendations for future study. J Abnorm Child Psychol 47:221–243

Marangoni C, Hernandez M, Faedda GL (2016) The role of environmental exposures as risk factors for bipolar disorder: a systematic review of longitudinal studies. J Affect Disord 193:165–174

Martins DS, Hasse-Sousa M, Petry-Perin C, Arrial-Cordeiro RT, Rabelo-da-Ponte FD, Lima FM, Rosa AR, Bucker J, Gama CS, Czepielewski LS (2019) Perceived childhood adversities: impact of childhood trauma to estimated intellectual functioning of individuals with bipolar disorder. Psychiatry Res 274:345–351

Marwaha S, Gordon-Smith K, Broome M, Briley PM, Perry A, Forty L, Craddock N, Jones I, Jones L (2016) Affective instability, childhood trauma and major affective disorders. J Affect Disord 190:764–771

Marwaha S, Briley PM, Perry A, Rankin P, DiFlorio A, Craddock N, Jones I, Broome M, Gordon-Smith K, Jones L (2019) Explaining why childhood abuse is a risk factor for poorer clinical course in bipolar disorder: a path analysis of 923 people with bipolar I disorder. Psychol Med:1–9

McIntyre RS, Soczynska JK, Liauw SS, Woldeyohannes HO, Brietzke E, Nathanson J, Alsuwaidan M, Muzina DJ, Taylor VH, Cha DS, Kennedy SH (2012) The association between childhood adversity and components of metabolic syndrome in adults with mood disorders: results from the international mood disorders collaborative project. Int J Psychiatry Med 43:165–177

Miller S, Hallmayer J, Wang PW, Hill SJ, Johnson SL, Ketter TA (2013) Brain-derived neurotrophic factor val66met genotype and early life stress effects upon bipolar course. J Psychiatr Res 47:252–258

Misiak B, Stramecki F, Gaweda L, Prochwicz K, Sasiadek MM, Moustafa AA, Frydecka D (2018) Interactions between variation in candidate genes and environmental factors in the etiology of schizophrenia and bipolar disorder: a systematic review. Mol Neurobiol 55:5075–5100

Moraes JB, Maes M, Barbosa DS, Ferrari TZ, Uehara MKS, Carvalho AF, Nunes SOV (2017) Elevated C-reactive protein levels in women with bipolar disorder may be explained by a history of childhood trauma, especially sexual abuse, body mass index and age. CNS Neurol Disord Drug Targets 16:514–521

Moreno-Alcazar A, Radua J, Landin-Romero R, Blanco L, Madre M, Reinares M, Comes M, Jimenez E, Crespo JM, Vieta E, Perez V, Novo P, Donate M, Cortizo R, Valiente-Gomez A, Lupo W, McKenna PJ, Pomarol-Clotet E, Amann BL (2017) Eye movement desensitization and reprocessing therapy versus supportive therapy in affective relapse prevention in bipolar patients with a history of trauma: study protocol for a randomized controlled trial. Trials 18:160

Nemeroff CB (2016) Paradise lost: the neurobiological and clinical consequences of child abuse and neglect. Neuron 89:892–909

Nothling J, Malan-Muller S, Abrahams N, Hemmings SMJ, Seedat S (2019) Epigenetic alterations associated with childhood trauma and adult mental health outcomes: a systematic review. World J Biol Psychiatry:1–20

Nuernberg GL, Aguiar B, Bristot G, Fleck MP, Rocha NS (2016) Brain-derived neurotrophic factor increase during treatment in severe mental illness inpatients. Transl Psychiatry 6:e985

Oliveira J, Etain B, Lajnef M, Hamdani N, Bennabi M, Bengoufa D, Sundaresh A, Chaabane AB, Bellivier F, Henry C, Kahn JP, Charron D, Krishnamoorthy R, Leboyer M, Tamouza R (2015) Combined effect of TLR2 gene polymorphism and early life stress on the age at onset of bipolar disorders. PLoS One 10:e0119702

Palmier-Claus JE, Berry K, Bucci S, Mansell W, Varese F (2016) Relationship between childhood adversity and bipolar affective disorder: systematic review and meta-analysis. Br J Psychiatry 209:454–459

Paquola C, Bennett MR, Lagopoulos J (2016) Understanding heterogeneity in grey matter research of adults with childhood maltreatment-A meta-analysis and review. Neurosci Biobehav Rev 69:299–312

Porter C, Palmier-Claus J, Branitsky A, Mansell W, Warwick H, Varese F (2020) Childhood adversity and borderline personality disorder: a meta-analysis. Acta Psychiatr Scand 141:6–20

Post RM, Altshuler LL, Leverich GS, Frye MA, Suppes T, McElroy SL, Keck PE Jr, Nolen WA, Kupka RW, Grunze H, Rowe M (2013) Role of childhood adversity in the development of medical co-morbidities associated with bipolar disorder. J Affect Disord 147:288–294

Pruessner M, Cullen AE, Aas M, Walker EF (2017) The neural diathesis-stress model of schizophrenia revisited: an update on recent findings considering illness stage and neurobiological and methodological complexities. Neurosci Biobehav Rev 73:191–218

Putnam KT, Harris WW, Putnam FW (2013) Synergistic childhood adversities and complex adult psychopathology. J Trauma Stress 26:435–442

Quide Y, Bortolasci CC, Spolding B, Kidnapillai S, Watkeys OJ, Cohen-Woods S, Berk M, Carr VJ, Walder K, Green MJ (2019) Association between childhood trauma exposure and pro-inflammatory cytokines in schizophrenia and bipolar-I disorder. Psychol Med 49:2736–2744

Read J, Fraser A (1998) Staff response to abuse histories of psychiatric inpatients. Aust N Z J Psychiatry 32:206–213

Richard-Lepouriel H, Etain B, Hasler R, Bellivier F, Gard S, Kahn JP, Prada P, Nicastro R, Ardu S, Dayer A, Leboyer M, Aubry JM, Perroud N, Henry C (2016) Similarities between emotional dysregulation in adults suffering from ADHD and bipolar patients. J Affect Disord 198:230–236

Richard-Lepouriel H, Kung AL, Hasler R, Bellivier F, Prada P, Gard S, Ardu S, Kahn JP, Dayer A, Henry C, Aubry JM, Leboyer M, Perroud N, Etain B (2019) Impulsivity and its association with childhood trauma experiences across bipolar disorder, attention deficit hyperactivity disorder and borderline personality disorder. J Affect Disord 244:33–41

Ridout KK, Levandowski M, Ridout SJ, Gantz L, Goonan K, Palermo D, Price LH, Tyrka AR (2018) Early life adversity and telomere length: a meta-analysis. Mol Psychiatry 23:858–871

Rokita KI, Dauvermann MR, Donohoe G (2018) Early life experiences and social cognition in major psychiatric disorders: a systematic review. Eur Psychiatry 53:123–133

Saddichha S, Schuetz C (2014) Is impulsivity in remitted bipolar disorder a stable trait? A meta-analytic review. Compr Psychiatry 55:1479–1484

Savitz J, van der Merwe L, Stein DJ, Solms M, Ramesar R (2007) Genotype and childhood sexual trauma moderate neurocognitive performance: a possible role for brain-derived neurotrophic factor and apolipoprotein E variants. Biol Psychiatry 62:391–399

Segura AG, Mitjans M, Jimenez E, Fatjo-Vilas M, Ruiz V, Saiz PA, Garcia-Portilla MP, Gonzalez-Blanco L, Bobes J, Vieta E, Benabarre A, Arias B (2019) Association of childhood trauma and genetic variability of CRH-BP and FKBP5 genes with suicidal behavior in bipolar patients. J Affect Disord 255:15–22

Souza-Queiroz J, Boisgontier J, Etain B, Poupon C, Duclap D, d'Albis MA, Daban C, Hamdani N, Le Corvoisier P, Delavest M, Bellivier F, Guevara P, Leboyer M, Henry C, Houenou J (2016) Childhood trauma and the limbic network: a multimodal MRI study in patients with bipolar disorder and controls. J Affect Disord 200:159–164

Stevelink R, Abramovic L, Verkooijen S, Begemann MJH, Sommer IEC, Boks MP, Mandl RCW, van Haren NEM, Vinkers CH (2018) Childhood abuse and white matter integrity in bipolar disorder patients and healthy controls. Eur Neuropsychopharmacol 28:807–817

Sugaya L, Hasin DS, Olfson M, Lin KH, Grant BF, Blanco C (2012) Child physical abuse and adult mental health: a national study. J Trauma Stress 25:384–392

Swendsen J, Hammen C, Heller T, Gitlin M (1995) Correlates of stress reactivity in patients with bipolar disorder. Am J Psychiatry 152:795–797

Teroganova N, Girshkin L, Suter CM, Green MJ (2016) DNA methylation in peripheral tissue of schizophrenia and bipolar disorder: a systematic review. BMC Genet 17:27

Tursich M, Neufeld RW, Frewen PA, Harricharan S, Kibler JL, Rhind SG, Lanius RA (2014) Association of trauma exposure with proinflammatory activity: a transdiagnostic meta-analysis. Transl Psychiatry 4:e413

Uher R (2014) Gene-environment interactions in severe mental illness. Front Psych 5:48

Upthegrove R, Chard C, Jones L, Gordon-Smith K, Forty L, Jones I, Craddock N (2015) Adverse childhood events and psychosis in bipolar affective disorder. Br J Psychiatry 206:191–197

van Leeuwen AP, Creemers HE, Greaves-Lord K, Verhulst FC, Ormel J, Huizink AC (2011) Hypothalamic-pituitary-adrenal axis reactivity to social stress and adolescent cannabis use: the TRAILS study. Addiction 106:1484–1492

WHO (2017) Violence info – childhood maltreatment

Williams LM, Debattista C, Duchemin AM, Schatzberg AF, Nemeroff CB (2016) Childhood trauma predicts antidepressant response in adults with major depression: data from the randomized international study to predict optimized treatment for depression. Transl Psychiatry 6:e799

Wolf EJ, Maniates H, Nugent N, Maihofer AX, Armstrong D, Ratanatharathorn A, Ashley-Koch AE, Garrett M, Kimbrel NA, Lori A, Va Mid-Atlantic Mirecc W, Aiello AE, Baker DG, Beckham JC, Boks MP, Galea S, Geuze E, Hauser MA, Kessler RC, Koenen KC, Miller MW, Ressler KJ, Risbrough V, Rutten BPF, Stein MB, Ursano RJ, Vermetten E, Vinkers CH, Uddin M, Smith AK, Nievergelt CM, Logue MW (2018) Traumatic stress and accelerated DNA methylation age: a meta-analysis. Psychoneuroendocrinology 9:123–134

Yamaguchi H, Calado RT, Ly H, Kajigaya S, Baerlocher GM, Chanock SJ, Lansdorp PM, Young NS (2005) Mutations in TERT, the gene for telomerase reverse transcriptase, in aplastic anemia. N Engl J Med 352:1413–1424

New Pharmacological Interventions in Bipolar Disorder

Mario F. Juruena, Luke A. Jelen, Allan H. Young, and Anthony J. Cleare

Contents

Abstract The biological bases of bipolar disorder include aspects related, among others, to neurohormonal pathways, neurotransmission, signal transduction, regulation of gene expression, oxidative stress, neuroplasticity, and changes in the immune system. There is still a gap in understanding its complex neurobiology and, consequently, developing new treatments. Multiple factors probably interact in this complex equation of pathophysiology of bipolar disorder, such as genetic, biochemical, psychosocial, and environmental stress events, correlating with the development and

M. F. Juruena (✉), L. A. Jelen, A. H. Young, and A. J. Cleare
Centre for Affective Disorders, Department of Psychological Medicine, Institute of Psychiatry, Psychology and Neuroscience, King's College London, London, UK
e-mail: Mario.Juruena@kcl.ac.uk

Curr Topics Behav Neurosci (2021) 48: 303–324
https://doi.org/10.1007/7854_2020_181
Published Online: 6 February 2021

severity of the bipolar disorder. These mechanisms can interact to exacerbate inflammation, impair neurogenesis, and increase oxidative stress damage, cellular mitochondrial dysfunction, changes in neurotrophins and in epigenetic mechanisms, neuroendocrine dysfunction, activation of neuronal death pathways, and dysfunction in neurotransmission systems. In this review, we explore the up-to-date knowledge of the neurobiological underpinnings of bipolar disorders. The difficulty in developing new drugs for bipolar disorder is very much associated with the lack of knowledge about the precise pathophysiology of this disorder. Pharmacological treatment for bipolar patients is vital; to progress to effective medications, it is essential to understand the neurobiology in bipolar patients better and identify novel therapeutic targets.

Keywords Bipolar disorder · Depression · Mania · Targets · Treatment

1 Introduction

Although current therapeutic agents for bipolar disorder are valuable, long-term response for most of the bipolar patients remains low (Gitlin et al. 1995). In general, treatment for bipolar disorder, can be challenging to manage, as treatments that act on manic/hypomanic episodes can lead to a depressive episode and the use of antidepressants can cause manic or hypomanic mood episodes (Geddes and Miklowitz 2013). Lithium is considered the gold standard drug for all phases of bipolar disorder (López-Muñoz et al. 2018).

Lithium, so far, is the only one with benefit in both poles and efficacy also for suicidal behavior (Geddes and Miklowitz 2013). Therefore, the investigation of potential new therapeutic goals is a priority. In this review, we explore the up-to-date knowledge of the neurobiology bases of bipolar disorders and the neurobiological mechanisms of dysfunction in a wide range of pathways, including: glutamatergic dysfunction, oxidative stress, mitochondrial dysfunction, impaired neurogenesis, increased inflammation, and apoptosis. The difficulty in the development of new drugs for bipolar patients is associated with a lack of knowledge about the precise pathophysiology of this disorder (Lund et al. 2012). Investigating these new pathophysiology systems purportedly related to the complexity of the neuronal systems in bipolar patients could improve our knowledge of the complex mechanisms related to the action of psychotropic drugs, as well as the neurobiology in bipolar patients (Fig. 1).

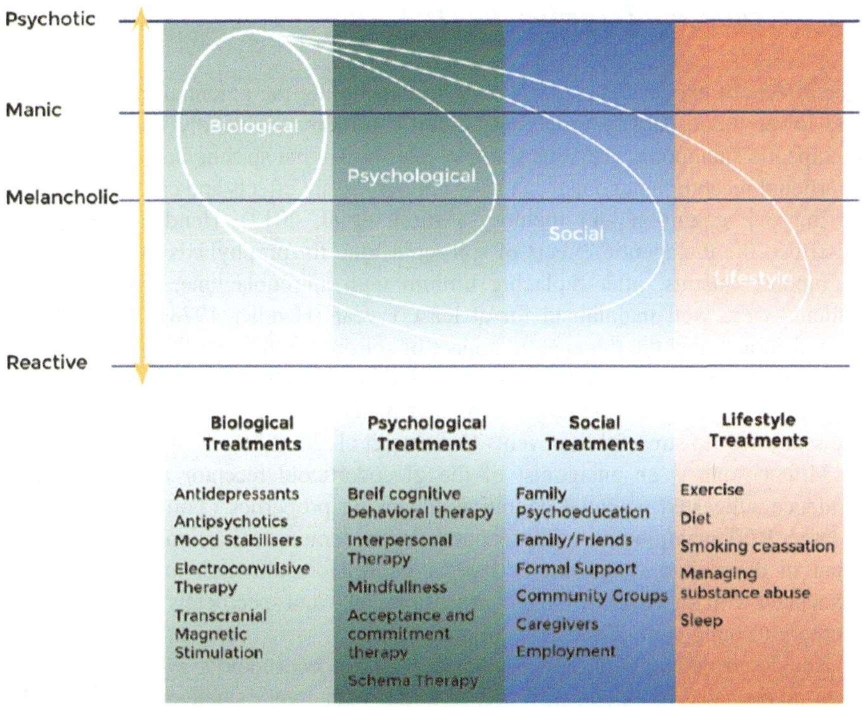

Fig. 1 The diagram provides a useful model for understanding the factors that contribute to planning clinical management. Therefore, a broad range of treatments is usually needed to treat bipolar disorders satisfactorily, adapted from Malhi et al. (2015). Illustrations are courtesy of Romayne Gadelrab

2 Hypothalamic-Pituitary-Adrenal Axis (HPA)

It is believed that disturbance of the hormonal stress system (the hypothalamic-pituitary-adrenal axis) can cause cognitive problems and worsen depressive symptoms in bipolar disorder (Watson et al. 2004) and can predict an inadequate clinical response to pharmacological treatments (Juruena et al. 2009a). One example of this is suggested in animal models, where a dysregulation in the HPA axis has an impact on serotonergic medications (Gartside et al. 2003).

The end product of the HPA axis is cortisol, which mediates its effects via two main receptors, the mineralocorticoid receptor (MR) and the glucocorticoid receptor (GR) (Juruena 2014). MR receptors modulate hormone release and complex behavior, such as emotion, memory, and sleep. In humans, the role of MR and GR receptors in the neurobiology of severe mental disorders has not been sufficiently characterized. However, recent research findings raise possibilities for new pharmacotherapies via modulation of the function of these receptors to normalize effects of cortisol and restore HPA axis balance (Geddes and Miklowitz 2013).

2.1 Treatments Targeting the HPA Axis

The effects of exogenous steroids on mood, including the potential to both elevate and lower mood, and to exacerbate mood disorders, are widely known. From a therapeutic standpoint, previous studies have shown that spironolactone, a mineralocorticoid receptor antagonist, can have antidepressant effects and can also maintain or improve cognition and memory (Murck et al. 2012). Hendler (1978) first described the therapeutic effects of spironolactone for prophylaxis in a case series of bipolar patients: after replacing lithium with spironolactone, five of the six patients were well maintained for at least 1 year (Hendler 1978). Some studies have demonstrated the therapeutic impact of spironolactone on affective symptoms in women with premenstrual syndrome (O'Brien et al. 1979; Wang et al. 1995) and eating disorders (Wernze 2000) and bipolar patients without remission, decreasing the sensitivity to stressful life events (Juruena et al. 2009).

Mifepristone is an antagonist of the glucocorticoid receptor and preliminary evidence suggested potential cognitive enhancing properties (Young et al. 2004). A larger RCT of adjunctive mifepristone failed to demonstrate a significant improvement in depressive symptoms in bipolar depression, although mifepristone was associated with improvement in cognitive function – and specifically spatial working memory (Young et al. 2004; Watson et al. 2012).

Ketoconazole and metyrapone, inhibitors of glucocorticoid synthesis, have also been investigated in bipolar disorder. In a trial with a very small sample in bipolar depression, with a treatment-resistant history, ketoconazole led to significant improvement in the severity of depression, but not mania (Brown et al. 2001). However, no RCTs have been performed.

A Cochrane review summarized the findings from antiglucocorticoid treatments for mood disorders (Gallagher et al. 2008). This comprised nine studies (three in psychotic major depression, five in non-psychotic major depression, and one in patients with bipolar depression); the authors described a significant reduction in the severity of depressive symptoms with antiglucocorticoid treatments.

3 Mitochondrial Dysfunction and Oxidative Stress in the Neurobiology of Bipolar

There is increasing data in the literature describing oxidative stress in the neurobiology of bipolar disorder. Corroborating the involvement of mitochondrial dysfunction in the pathophysiology of bipolar disorders, a study demonstrated that bipolar patients have an imbalance between the processes of mitochondrial fusion and fission (Scaini et al. 2017) observed by increased protein levels of fission protein (Fis-1) and decreased levels of fusion proteins (Mfn-2 and Opa-1), suggesting that the process of mitochondrial dynamics in patients with bipolar disorder is dysfunctional, which may increase mitochondrial fragmentation (Scaini et al. 2017). In addition, the same study indicated that the changes observed peripherally were directly correlated with functional decline in bipolar patients (Scaini et al. 2017). Effects of lithium in redox systems may have this effect in psychopathological disorders (Khairova et al. 2012).

3.1 Oxidative Stress and Pharmacological Approaches

Several clinical and preclinical studies have suggested the involvement of oxidative stress in bipolar disorders (Frey et al. 2006). Mitochondrial dysfunction has been described as the main triggering agent of this system with a consequent impairment in cellular energy metabolism (Berk et al. 2011b). An abnormal cellular energy state can lead to loss of neuronal function and plasticity and cognitive and behavioral changes characteristic of bipolar disorders (Kato 2007).

Several studies have found elevated levels of malondialdehyde (MDA) and carbonyl groups in the blood of patients with bipolar disorders (Descamps-Latscha et al. 2001). Postmortem studies have shown high levels of carbonyl, 4-HNE, and 8-ISSO groups in bipolar patients (Imai and Nakagawa 2003; Wang et al. 2009). In addition, the antioxidant defense system also appears to be altered in bipolar patients, even in the early stages of the disorder. A study detected an increase in glutathione (GST) in bipolar patients in the early stages of the disorder (Andreazza et al. 2009). The superoxide dismutase (SOD) enzyme has also been found to be increased in the blood of bipolar patients during episodes of mania and depression (Andreazza et al. 2010).

3.2 Alternative and Experimental Treatments Targeting Oxidative Stress

N-acetylcysteine (NAC) has been investigated in the treatment of acute and maintenance treatment of bipolar disorder. NAC is an immediate precursor to glutathione, an antioxidant protein with a significant role in the removal of numerous reactive oxygen species, levels of which are reduced in bipolar disorder (Rosa et al. 2014). In the first placebo-controlled RCT, adjunctive NAC treatment resulted in a significant improvement in depressive symptoms over placebo in bipolar patients at the maintenance treatment stage (Berk et al. 2008). Studying a subgroup analysis of bipolar patients in a depressive phase at baseline, the NAC treatment group showed significant improvements in depressive symptoms (eight of ten patients on NAC had treatment response at endpoint), alongside measures of executive function and QoL (Magalhaes et al. 2011). A further open-label study of 149 individuals with bipolar depression again showed adjunctive NAC treatment led to significant and robust reductions in depression scores and improvements in functioning and quality of life

(Berk et al. 2011a). However, more recently, a large RCT found no difference between adjunctive NAC and placebo in bipolar depression in terms of reduction of depressive symptoms at the 16-week endpoint (Berk et al. 2019).

Creatine is the precursor of phosphocreatine (PCr), which has an important role in brain energy homeostasis through the phosphocreatine circuit and has been shown to have antioxidant properties (Pereira et al. 2018). In a small sample of patients with unipolar and bipolar treatment-resistant depression, open label adjunctive treatment with creatine monohydrate led to significantly improved depressive symptoms; however, in both bipolar disorder patients, there was a momentary switch to elevated mood (Roitman et al. 2007).

In a placebo-controlled RCT evaluating creatine monohydrate as an added treatment in bipolar disorder in a depressive episode, the patients did not show improvement in depressive symptom scores. However, the study did find significant superiority of creatine augmentation over placebo using remission criteria (Toniolo et al. 2018).

Cytidine is a pyrimidine with the main function on phospholipid homeostasis and membrane balance, which potentially alleviates mitochondrial dysfunction and regulates dysfunctional glia-neuronal glutamatergic cycling in bipolar disorder. In a 12-week RCT of 35 individuals with bipolar depression, adjunctive cytidine resulted in an earlier decrease in depressive severity alongside a decrease in glutamate/glutamine levels as measured by proton magnetic resonance spectroscopy (^{1}H-MRS) (Yoon et al. 2009).

4 Immune System in Bipolar Disorder

It is known that multiple factors related to the immune system may be involved in the pathophysiology of bipolar disorder, including tumor necrosis factor-alpha (TNF-α) and interleukins (IL). These mechanisms, in turn, promote deleterious effects that contribute to exaggerated inflammation (Carvalho et al. 2013). Studies have suggested that the pathophysiology of bipolar disorder may be related to changes in the immune system, with mood episodes being characterized as pro-inflammatory states. A pro-inflammatory increase in cytokines has been described in patients with bipolar disorders, mainly TNF-α. In addition, IL-2, IL-1β, IL-4, IL-6, and IL-10 concentrations are elevated in bipolar patients (Modabbernia et al. 2013). Investigations about inflammation in bipolar disorder have significantly increased in the last years, showing the importance of the immune system in the neurobiology of bipolar disorder, but with variable results regarding the type of episode, the effect of treatments, and the progression of the disorder (Leboyer et al. 2012).

Horrobin and Lieb (1981) were the first to link immunological disorders with bipolar disorder; the authors hypothesize that relapses of mood episodes in bipolar patients would be driven by the immune system as well as in other inflammatory diseases such as multiple sclerosis (Maes 2011; Drexhage et al. 2010).

Specifically, alterations in the arachidonic acid (AA) signalling pathway and regulation of cyclooxygenase (COX)-generated metabolites including prostaglandins and thromboxanes have been implicated in the neurobiology of bipolar patients, and targeting AA pathway may be a potential therapeutic approach (Rao and Rapoport 2009). A postmortem brain study of bipolar patients showed changes in the AA metabolism cascade (Rao et al. 2010). As well as a significant increase in pro-inflammatory and anti-inflammatory cytokines, they have been consistently reported in patients with bipolar disorder (Modabbernia et al. 2013). The study suggests that pro-inflammatory cytokines are increased regardless of the mood episode of patients with bipolar disorder, as well as the occurrence of a reduction in anti-inflammatory cytokines (Barbosa et al. 2014).

Studies have highlighted TNF-α as an important marker of bipolar progression, with a significant increase in patients with a chronic history when compared to first-episode bipolar patients (Tatay-Manteiga et al. 2017). These studies reinforce the presence of changes in inflammatory biomarkers in bipolar patients (Modabbernia et al. 2013).

4.1 Inflammation and Treatment in Bipolar Disorders

The AA cascade has been the subject of some studies, and it may be the common route of action of several medications used to treat bipolar patients, for example, anticonvulsants (sodium valproate and carbamazepine) and lithium (Rao et al. 2008). Possible effectiveness of adjuvant celecoxib treatment in both poles of bipolar patients has also been reported, suggesting an essential impact of the inflammation pathway (Arabzadeh et al. 2015; Mousavi et al. 2017).

Reports of lithium's anti inflammatory properties may contribute to its therapeutic efficacy. However, data demonstrated in the literature are contradictory about the effects of lithium on pro- or anti-inflammatory markers (Rapaport and Manji 2001). In pre-clinical studies, chronic lithium treatment reduced levels of motor impulsivity, plasma levels of IL-1β, IL-10 and RANTES (CCL-5) were reduced selectively within the orbitofrontal cortex of lithium-treated rats (Adams et al. 2020). These findings demonstrate that lithium may improve impulse control deficits in clinical populations decreasing the pro-inflammatory signalling on neuronal activity, particularly within the orbitofrontal cortex (Adams et al. 2020). Studies evaluated the action of the medications utilized in bipolar patients and suggest that the AA cascade, as well as the decreased activity of cyclooxygenase 2 (COX-2) and prostaglandin E2 (PGE2), can be a common target in the mood-stabilizing action of these drugs (Bosetti et al. 2002). A further study demonstrated that bipolar patients who showed a poor response to lithium had increased levels of TNF-α compared to those with a good response (Guloksuz et al. 2012) (Fig. 2).

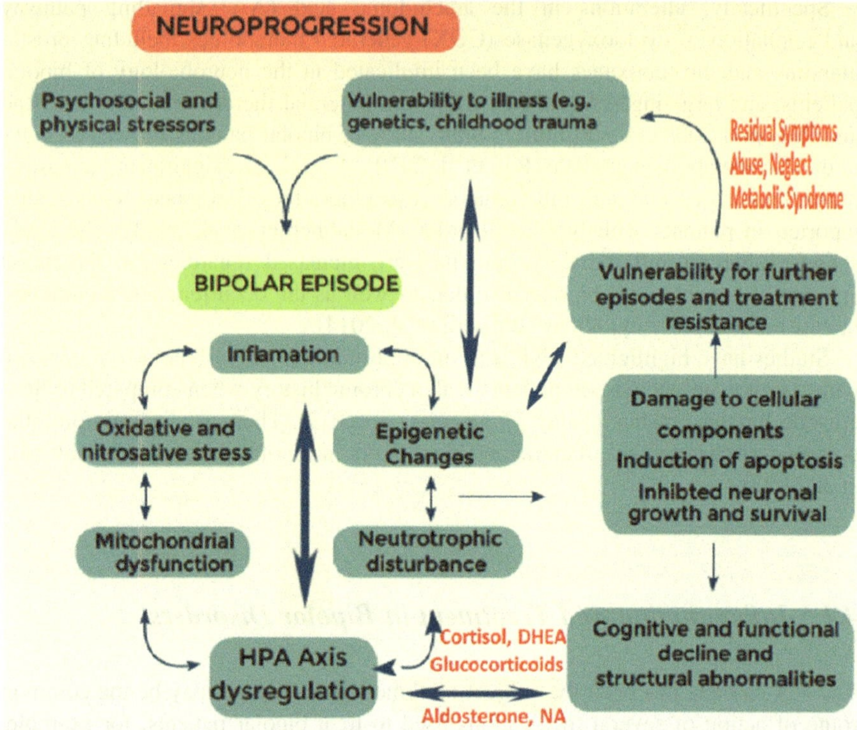

Fig. 2 Importance of early treatment: the impact on neuroprogression, Psycho-Neuro-Immune-Endocrine cascade. Adapted from Moylan et al. (2013). Illustrations courtesy of Romayne Gadelrab

4.2 *Alternative and Experimental Treatments and Inflammation*

Among non-steroidal anti-inflammatories (NSAIDs), there is the class of those specific inhibitors of COX-2; among these drugs is celecoxib. COX isoenzymes have a role as precursors of prostaglandin, thromboxanes, and prostacyclins. Prostaglandins are autacoid mediators with influence in most of the neurophysiological or pathophysiological progressions through the membrane receptors connected to the G-protein (Fitzpatrick 2004).

A meta-analysis examining the antidepressant effects of adjunctive anti-inflammatory agents in mood disorders, including MDD and bipolar disorder, found a moderate to large effect size (Husain et al. 2017). Interestingly, this meta-analysis also showed a significant reduction in manic symptoms when pooling the results of adjunctive anti-inflammatories (NSAIDs and NAC) in bipolar disorder.

The use of omega-3 fatty acids (naturally occurring anti-inflammatory molecules) in bipolar depression is restricted with one meta-analysis of five RCTs indicating a significant outcome in favor of omega-3 over placebo with moderate effect size (Sarris et al. 2012).

In a RCT with bipolar patients using celecoxib controlled with placebo, the authors found a significant decrease in TNF-α levels. However, in this study, there was no significant difference in the levels of IL-1β, IL-6, and high-sensitivity C-reactive protein (Kargar et al. 2014). This study evaluated inflammatory markers in the use of celecoxib. Although several studies are correlating bipolar disorder progression with inflammatory markers, the data are still controversial, highlighting the importance of new research in this area (Giridharan et al. 2019). Overall, these studies demonstrate strong evidence of inflammation in bipolar disorder leading to new therapeutic targets, such as the use of anti-inflammatory drugs, highlighting celecoxib and the AA pathway involved in the action of this drug (Husain et al. 2016).

A meta-analysis including three RCTs ($n = 121$) showed a significant effect on Young Mania Rating Scale scores from patients with bipolar disorder receiving adjunctive celecoxib treatment compared with placebo (Bavaresco et al. 2019). An RCT of adjunctive celecoxib in treatment-resistant bipolar depression found higher response and remission rates in the celecoxib than in the placebo arm (Halaris et al. 2020).

Minocycline is a second-generation tetracycline with anti-inflammatory properties. The effects of minocycline in combination with aspirin have also been examined in an RCT in bipolar depression that showed a significantly higher sustained clinical response rate in the minocycline and aspirin-treated group compared with placebo (Savitz et al. 2018).

A study investigating minocycline for bipolar depression patients demonstrated a significant effect size in reducing depressive symptoms (Soczynska et al. 2017). Remarkably, the clinical trials indicate that this second-generation tetracycline has more significant effects in bipolar disorder subjects with higher inflammatory indicators which highlights the need for inflammatory analysis +/− stratification in forthcoming RCTs of minocycline in bipolar patients. However, the largest and most recent RCT (*n*-266) in bipolar I and II and unipolar depression did not find a significant effect comparing placebo with celecoxib or minocycline (Husain et al. 2020). Husain et al. (2020) conclude that a possible therapeutic effect should not be considered in bipolar depression. Notwithstanding this negative finding, anti-inflammatory treatments are becoming a promising alternative for use in psychiatry.

5 Neurotrophins and Bipolar

In the middle of the twentieth century, the first neurotrophin, the nerve growth factor (NGF), was identified. This discovery broadened the horizon of neurobiology for the identification and elucidation of cellular functions. Almost 30 years after the identification of NGF, the prototype of neurotrophins for neurons of the autonomic nervous system was isolated, a homologue of NGF, which was called brain-derived neurotrophic factor (BDNF) (Lessmann et al. 2003).

Recent studies indicate that impairment of neuroplasticity and neuronal survival are the main events involved in the pathogenesis of bipolar disorders. These events are influenced by several factors, such as the harmonic action of neurotransmitters, hormones, neurotrophins, and inflammatory mediators, such as cytokines (Brietzke and Kapczinski 2008).

It has been shown that high levels of cortisol, released due to stress, can cause damage to the membranes of neuronal mitochondria, especially in patients with bipolar disorder, leading to the release of toxic compounds and culminating in changes in the structure of the DNA molecule in nucleus of these cells. All of this transformation triggers apoptosis mechanisms (Berk et al. 2011b).

5.1 Neurotrophins and Bipolar Treatment

The family of tyrosine kinase (Trk) receptors is composed of three receptors that can be activated by one or more neurotrophins: NGF, BDNF, NT-3, and NT-4/5. In studies on depression models, antidepressants increase TrkB signaling, which is dependent on the concentration of BDNF (Saarelainen et al. 2003). Antidepressants and mood stabilizers can increase serum BDNF levels (Frey et al. 2006). Chronic administration of antidepressants increases the expression of BDNF in the hippocampus, as well as in the CPF (Duman et al. 2000). Chronic treatment with lithium or sodium valproate demonstrated higher BDNF expression in preclinical studies (Fukumoto et al. 2001).

Several studies have suggested that BDNF/TrkB induction is one of the mechanisms responsible for the therapeutic effects of antipsychotics, mood stabilizers, and antidepressants (Coyle and Duman 2003; Nibuya et al. 1995). For example, it has been shown that the use of lithium modulates the phosphorylation of the TrkB and CREB receptors (Rantamaki et al. 2006). Studies show that the neuroprotective characteristics of lithium or sodium valproate may be responsible for its therapeutic effects, and one of the mechanisms involved would be the release of neurotrophins (Rosa et al. 2006; Laeng et al. 2004).

It has been proposed that the augmented expression of BDNF can trigger the neuroprotective properties of lithium and valproate (de Sousa et al. 2011; Hu et al. 2010).

Chronic treatment with lithium or sodium valproate produces protective effects against excitotoxicity and cell death induced by glutamate (Shao et al. 2005). Regarding BDNF, there is much evidence regarding its long-term role in synaptic plasticity in the hippocampus and neocortex. The application of exogenous BDNF enhances presynaptic efficacy by increasing the release of glutamate in excitatory synapses (Lessmann et al. 2003).

6 Bipolar Disorder and the Glutamatergic System

For a long time, the monoaminergic hypothesis explained both the neurobiology of bipolar disorders and the mechanisms of action of psychotropic drugs. Glutamate is related to a diversity of crucial functions including synaptic plasticity, learning, and memory (Riedel et al. 2003). This neurotransmitter acts on two classes of receptors: ionotropic and metabotropic. There are three types of ionotropic glutamate receptors: α-amino acid-3-hydroxy-5-methylisoxazol-4-proprionic (AMPA), N-methyl-D-aspartate (NMDA), and kainate. These three proteins are ion channels that, when activated, generate postsynaptic excitatory potential (Kew and Kemp 2005). Metabotropic receptors are not ion channels and are not exclusively located in the synapse region. When they bind to glutamate, they activate the G-protein, which is responsible for sending the message into the cell (Baudry et al. 2012).

Although all glutamate receptors respond to the same neurotransmitter, they perform very different functions (Jun et al. 2014). In physiological situations, ionotropic and metabotropic receptors regulate neurotransmission through excitatory synapses and modulate various neurophysiological functions in the brain, which include synaptic plasticity, mood, learning, and memory (Kew and Kemp 2005).

Several researchers have shown changes in glutamate receptors in the brain of patients with bipolar disorders, with decreased expression of NMDA receptor in the hippocampus and the prefrontal cortex of bipolar patients (McCullumsmith et al. 2007; Beneyto and Meador-Woodruff 2008). These functionally different receptors can be co-expressed in individual synapses, allowing precise temporal modulation of postsynaptic excitability and plasticity (Scheefhals and MacGillavry 2018). Recently, the NMDA receptor has been associated with the sociability and pathogenesis of autism spectrum disorder, which encourages further studies for the therapeutic exploration of the modulation of this receptor (Burket and Deutsch 2019).

6.1 Glutamate Neurotransmission and Bipolar Treatment

Several pharmacological approaches used in the treatment of psychiatric disorders act on the glutamatergic system. Significant evidence for the role of glutamate in bipolar disorder emerged from pharmacological interventions; for example, lamotrigine, an anticonvulsant also used in bipolar depression pharmacotherapy, indirectly inhibits the release of glutamate (Leach et al. 1991). Lithium is the gold standard mood stabilizer; it leads to an acute increase in the concentration of glutamate in the synapse and the chronic upregulation of transporter activity; its chronic use seems to promote the stabilization of excitatory neurotransmission (Li et al. 2002). Lithium attenuates the release of Ca^{2+} after acting on mGluR1 and mGluR5 receptors (Machado-Vieira et al. 2012; Sourial-Bassillious et al. 2009). As previously described, mitochondrial dynamics involve not only the processes of fission and fusion but also the movement of mitochondria through neurons, a process influenced by the concentration of ATP and Ca^{2+}, with lithium's influence (Curtis et al. 2011). In addition to changes in ATP levels, as already mentioned, studies have pointed out that bipolar patients show changes in intracellular Ca^{2+} signaling, causing an increase in cytosolic Ca^{2+} concentrations, related to excitotoxicity, decreased mitochondrial viability, and, ultimately, cell death (Mehta et al. 2013). The neuroprotective effects of lithium against the excitotoxicity of glutamate influences BDNF. Antipsychotics and anticonvulsants reduce glutamate in the central nervous system (Sitges et al. 2007; Juruena et al. 2009b; McLoughlin et al. 2009; Paraskevas et al. 2006; de la Fuente-Sandoval et al. 2013).

6.2 *Alternative and Experimental Treatments Targeting the Glutamatergic System*

6.2.1 Ketamine

Several studies have shown that glutamate-mediated synaptic plasticity is involved in bipolar pharmacotherapy. In addition, ketamine (an NMDA antagonist) has attracted attention for its rapid antidepressant activity, including in patients with bipolar depression (Zarate et al. 2012)

Ketamine is an N-methyl-D-aspartate (NMDA) receptor antagonist with significant evidence to treat both unipolar and bipolar depression, including in treatment-resistant cases (Coyle and Laws 2015). In an RCT crossover add-on study of 18 individuals with treatment-resistant bipolar depression (continued on lithium or valproate), following a single ketamine infusion (0.5 mg/kg over 40 min), depressive symptoms significantly decreased within 40 min compared with placebo, the effect lasting up to day 3 (Diazgranados et al. 2010). There was a notable response rate, with approximately 70% of subjects responding to ketamine. Acute dissociative symptoms were the most common side effect; otherwise ketamine at this dose was generally well tolerated. These findings were replicated in a similar clinical trial with treatment-resistant bipolar depression, which again showed a single ketamine infusion led to rapid, robust antidepressant effects, alongside significant improvements in suicidal ideation (Zarate et al. 2012). Although the antidepressant effects of ketamine are transient, the response may be maintained with repeated infusions (Diamond et al. 2014). Other forms of administration (oral, sublingual, intranasal, intramuscular, subcutaneous) may demonstrate more viable alternatives for repeated dosing. Nevertheless, additional research is required to fully assess and compare the safety and effectiveness of repeated dosing across different routes.

6.2.2 Memantine

Memantine is a non-competitive NMDA antagonist which, unlike ketamine, does not produce dissociative side effects at therapeutic doses. In a small RCT of memantine augmentation of lamotrigine in bipolar depression, although a primary benefit of memantine over placebo was demonstrated at 4 weeks, this effect was not maintained at the 8-week primary endpoint (Anand et al. 2012). A subsequent larger RCT exploring memantine augmentation to valproate in bipolar II depression again found no significant benefit over placebo in terms of response in the severity of depression (Lee et al. 2014). Of interest, this study found that TNF-α levels were decreased in the memantine arm, suggesting a potential anti-inflammatory mechanism. An open-label pilot trial in individuals with acute mania associated with bipolar I disorder found memantine monotherapy to have anti-manic effects at 3 weeks (Keck et al. 2009); however, larger RCTs are required to explore this further.

6.2.3 Riluzole

Riluzole is an NDMA modulator that is licensed for amyotrophic lateral sclerosis (ALS) treatment, acting to increase glutamatergic reuptake and enhance the production of neural growth factors, including BDNF (Bellingham 2011). Preliminary results in acute bipolar depression, in an open trial with riluzole added to lithium, found a significant treatment effect at 4 weeks with no switch observed and good tolerability (Zarate et al. 2005). However, in a subsequent RCT of 19 subjects with acute bipolar depression, there were no significant differences in depressive symptoms between riluzole monotherapy and placebo groups (Park et al. 2017).

7 Bipolar Disorder and the Cholinergic System

The cholinergic-adrenergic hypothesis of bipolar disorder was first proposed many decades ago (Janowsky et al. 1972) and suggested that mania was associated with an adrenergic predominance, while depression was associated with a central cholinergic predominance. Since this initial hypothesis, converging evidence has supported cholinergic dysfunction in bipolar disorder. A positron emission tomography (PET) study demonstrated decreased muscarinic type 2 (M2) binding in bipolar patients (Cannon et al. 2006); another study has reported reduced muscarinic receptor binding (M2 and M3) in the prefrontal cortex in bipolar patients (Gibbons et al. 2009).

7.1 Treatments Targeting the Cholinergic System

The short-acting acetylcholinesterase inhibitor physostigmine has been shown to lead to rapid but temporary reductions in manic symptoms in a few small case series (Davis et al. 1978). Augmentation with donepezil, a long-acting acetylcholinesterase inhibitor, was also shown to lead to marked reductions in mania in a small case series (Burt et al. 1999). However, a subsequent small RCT failed to show any significant benefit of adjunctive donepezil over placebo in treatment for refractory manic symptoms (Eden Evins et al. 2006).

Scopolamine, a muscarinic receptor antagonist and NMDA modulator, has also been examined for potential treatment of bipolar patients. In a double-blind RCT, including subjects with both major depressive disorder and bipolar depression, scopolamine administration led to a significant rapid and robust antidepressant response (Furey and Drevets 2006). An RCT (SCOPE-BD) is studying scopolamine in bipolar patients in a depressive episode (Hallahan n.d., ClinicalTrials.gov ID: NCT04211961).

8 Bipolar Disorder and the Melatonergic System

Melatonin is a vital hormone in the brain that is responsible for the regulation of circadian rhythms but also has essential roles in metabolic, immune, antioxidative, and mitochondrial functioning (Claustrat et al. 2005). Melatonin exerts its effects primarily through the MT_1 and MT_2 G-protein-coupled receptors.

There is accumulating evidence on the role of the melatonin system in bipolar disorder, with studies reporting changes in patters on melatonin secretion and supersensitivity of light-induced melatonin suppression (Lanfumey et al. 2013).

8.1 Treatments Targeting the Melatonergic System

Agomelatine is a potent MT_1 and MT_2 agonist and 5HT-2C antagonist and also acts to increase dopamine and noradrenaline levels (Guardiola-Lemaitre et al. 2014). A clinical trial in bipolar depression with agomelatine found that 81% of patients showed greater than 50% response in HAM-D score after 6 weeks (Calabrese et al. 2007); however, a larger RCT subsequently did not find a difference in depressive severity between agomelatine or placebo adjunctive therapy (Yatham et al. 2016).

Ramelteon is an MT_1 and MT_2 agonist, which has been shown to improve depressive symptoms in bipolar disorder with sleep problems (McElroy et al. 2011). Further work has shown ramelteon to be effective in maintaining mood stability in euthymic bipolar disorder individuals with sleep disturbances (Norris et al. 2013). Most recently a phase 3 RCT examining the efficacy and safety of adjunctive ramelteon as a maintenance treatment in bipolar disorder failed to show the efficacy of ramelteon in preventing relapse in bipolar subjects (Mahableshwarkar et al. 2017).

9 Conclusion

In the face of the complex symptomatology of bipolar disorder, there is still a hiatus in the understanding of its more complex neurobiology and, accordingly, in the treatment of bipolar disorder. Understanding new biological pathways will provide data with potential for clinical translation. This update of the neurobiological bases shows multiple factors probably interact in this complex equation of pathophysiology of bipolar disorder, such as genetic, biochemical, psychosocial, and environmental stress events, correlating with the development and severity of the bipolar disorder. Developing new pharmacological treatment for bipolar patients is crucial; however, new effective medications can only be developed when we understand better the neurobiology of bipolar disorder and find novel targets such as those described in this chapter.

References

Adams WK, Levesque DL, Cocker PJ, Kaur S, Bodnar TS, Young AH, Winstanley CA (2020) Decreased motor impulsivity following chronic lithium treatment in male rats is associated with reduced levels of pro-inflammatory cytokines in the orbitofrontal cortex. Brain Behav Immun 89:339–349

Anand A, Gunn AD, Barkay G, Karne HS, Nurnberger JI, Mathew SJ, Ghosh S (2012) Early antidepressant effect of memantine during augmentation of lamotrigine inadequate response in bipolar depression: a double-blind, randomized, placebo-controlled trial. Bipolar Disord 14:64–70

Andreazza AC, Kapczinski F, Kauer-Sant'Anna M, Walz JC, Bond DJ, Gonçalves CA et al (2009) 3-Nitrotyrosine and glutathione antioxidant system in patients in the early and late stages of bipolar disorder. J Psychiatry Neurosci 34(4):263–271

Andreazza AC, Shao L, Wang JF, Young LT (2010) Mitochondrial complex I activity and oxidative damage to mitochondrial proteins in the prefrontal cortex of patients with bipolar disorder. Arch Gen Psychiatry 67(4):360–368

Arabzadeh S, Ameli N, Zeinoddini A, Rezaei F, Farokhnia M, Mohammadinejad P, Ghaleiha A, Akhondzadeh S (2015) Celecoxib adjunctive therapy for acute bipolar mania: a randomized, double-blind, placebo-controlled trial. Bipolar Disord 17(6):606–614

Barbosa IG, Bauer ME, Machado-Vieira R, Teixeira AL (2014) Cytokines in bipolar disorder: paving the way for neuroprogression. Neural Plast 2014:360481

Baudry M, Greget R, Pernot F, Bouteiller JM, Xiaoning B (2012) Roles of group I metabotropic glutamate receptors under physiological conditions and in neurodegeneration. WIREs Membr Transp Signal 1(4):523–532

Bavaresco DV, Colonetti T, Grande AJ, Colom F, Valvassori SS, Quevedo J, da Rosa MI (2019) Efficacy of celecoxib adjunct treatment on bipolar disorder: systematic review and meta-analysis. CNS Neurol Disord Drug Targets 18:19–28

Bellingham MC (2011) A review of the neural mechanisms of action and clinical efficiency of riluzole in treating amyotrophic lateral sclerosis: what have we learned in the last decade? CNS Neurosci Ther 17:4–31

Beneyto M, Meador-Woodruff JH (2008) Lamina-specific abnormalities of NMDA receptor associated postsynaptic protein transcripts in the prefrontal cortex in schizophrenia and bipolar disorder. Neuropsychopharmacology 33(9):2175–2186

Berk M, Copolov DL, Dean O, Lu K, Jeavons S, Schapkaitz I, Anderson-Hunt M, Bush AI (2008) N-acetyl cysteine for depressive symptoms in bipolar disorder--a double-blind randomized placebo-controlled trial. Biol Psychiatry 64:468–475

Berk M, Dean O, Cotton SM, Gama CS, Kapczinski F, Fernandes BS, Kohlmann K, Jeavons S, Hewitt K, Allwang C, Cobb H, Bush AI, Schapkaitz I, Dodd S, Malhi GS (2011a) The efficacy of N-acetylcysteine as an adjunctive treatment in bipolar depression: an open label trial. J Affect Disord 135:389–394

Berk M, Kapczinski F, Andreazza AC, Dean OM, Giorlando F, Maes M et al (2011b) Pathways underlying neuroprogression in bipolar disorder: focus on inflammation, oxidative stress and neurotrophic factors. Neurosci Biobehav Rev 35(3):804–817

Berk M, Turner A, Malhi GS, Ng CH, Cotton SM, Dodd S, Samuni Y, Tanious M, McAulay C, Dowling N, Sarris J, Owen L, Waterdrinker A, Smith D, Dean OM (2019) A randomized controlled trial of a mitochondrial therapeutic target for bipolar depression: mitochondrial agents, N-acetylcysteine, and placebo. BMC Med 17:18

Bosetti F, Rintala J, Seemann R, Rosenberger TA, Contreras MA, Rapoport SI, Chang MC (2002) Chronic lithium downregulates cyclooxygenase-2 activity and prostaglandin E(2) concentration in rat brain. Mol Psychiatry 7(8):845–850

Brietzke E, Kapczinski F (2008) TNF-alpha as a molecular target in bipolar disorder. Prog Neuro-Psychopharmacol Biol Psychiatry 32:1355–1361

Brown ES, Bobadilla L, Rush AJ (2001) Ketoconazole in bipolar patients with depressive symptoms: a case series and literature review. Bipolar Disord 3:23–29

Burket JA, Deutsch SI (2019) Metabotropic functions of the NMDA receptor and an evolving rationale for exploring NR2A-selective positive allosteric modulators for the treatment of autism spectrum disorder. Prog Neuro-Psychopharmacol Biol Psychiatry 90:142–160

Burt T, Sachs GS, Demopulos C (1999) Donepezil in treatment-resistant bipolar disorder. Biol Psychiatry 45:959–964

Calabrese JR, Guelfi JD, Perdrizet-Chevallier C, Agomelatine Bipolar Study Group (2007) Agomelatine adjunctive therapy for acute bipolar depression: preliminary open data. Bipolar Disord 9:628–635

Cannon DM, Carson RE, Nugent AC, Eckelman WC, Kiesewetter DO, Williams J, Rollis D, Drevets M, Gandhi S, Solorio G, Drevets WC (2006) Reduced muscarinic type 2 receptor binding in subjects with bipolar disorder. Arch Gen Psychiatry 63:741–747

Carvalho LA, Torre JP, Papadopoulos AS, Poon L, Juruena MF, Markopoulou K, Cleare AJ, Pariante CM (2013) Lack of clinical therapeutic benefit of antidepressants is associated overall activation of the inflammatory system. J Affect Disord 148(1):136–140

Claustrat B, Brun J, Chazot G (2005) The basic physiology and pathophysiology of melatonin. Sleep Med Rev 9:11–24

Coyle JT, Duman RS (2003) Finding the intracellular signaling pathways affected by mood disorder treatments. Neuron 38:157–160

Coyle CM, Laws KR (2015) The use of ketamine as an antidepressant: a systematic review and meta-analysis. Hum Psychopharmacol 30:152–163

Curtis D, Vine AE, McQuillin A, Bass NJ, Pereira A, Kandaswamy R, Lawrence J, Anjorin A, Choudhury K, Datta SR, Puri V, Krasucki R, Pimm J, Thirumalai S, Quested D, Gurling HM (2011) Case-case genome-wide association analysis shows markers differentially associated with schizophrenia and bipolar disorder and implicates calcium channel genes. Psychiatr Genet 21(1):1–4

Davis KL, Berger PA, Hollister LE, Defraites E (1978) Physostigmine in mania. Arch Gen Psychiatry 35:119–122

de la Fuente-Sandoval C, León-Ortiz P, Azcárraga M, Stephano S, Favila R, Díaz-Galvis L, Alvarado-Alanis P, Ramírez-Bermúdez J, Graff-Guerrero A (2013) Glutamate levels in the associative striatum before and after 4 weeks of antipsychotic treatment in first-episode psychosis: a longitudinal proton magnetic resonance spectroscopy study. JAMA Psychiat 70 (10):1057–1066 de Sousa RT, van de Bilt MT, Diniz BS, Ladeira RB, Portela LV, Souza DO, Forlenza OV, Gattaz WF, Machado-Vieira R (2011) Lithium increases plasma brain-derived neurotrophic factor in acute bipolar mania: a preliminary 4-week study. Neurosci Lett 494(1):54–56

Descamps-Latscha B, Drüeke T, Witko-Sarsat V (2001) Dialysis-induced oxidative stress: biological aspects, clinical consequences, and therapy. Semin Dial 14(3):193–199

Diamond PR, Farmery AD, Atkinson S, Haldar J, Williams N, Cowen PJ, Geddes JR, McShane R (2014) Ketamine infusions for treatment resistant depression: a series of 28 patients treated weekly or twice weekly in an ECT clinic. J Psychopharmacol 28:536–544

Diazgranados N, Ibrahim L, Brutsche NE, Newberg A, Kronstein P, Khalife S, Kammerer WA, Quezado Z, Luckenbaugh DA, Salvadore G, Machado-Vieira R, Manji HK, Zarate CA Jr (2010) A randomized add-on trial of an N-methyl-D-aspartate antagonist in treatment-resistant bipolar depression. Arch Gen Psychiatry 67:793–802

Drexhage RC, van der Heul-Nieuwenhuijsen L, Padmos RC et al (2010) Inflammatory gene expression in monocytes of patients with schizophrenia: overlap and difference with bipolar disorder. A study in naturalistically treated patients. Int J Neuropsychopharmacol 13 (10):1369–1381

Duman RS, Malberg J, Nakagawa S, D'as C (2000) Neuronal plasticity and survival in mood disorders. Biol Psychiatry 48(8):732–739

Eden Evins A, Demopulos C, Nierenberg A, Culhane MA, Eisner L, Sachs G (2006) A double-blind, placebo-controlled trial of adjunctive donepezil in treatment-resistant mania. Bipolar Disord 8:75–80

Fitzpatrick FA (2004) Cyclooxygenase enzymes: regulation and function. Curr Pharm Des 10 (6):577–588

Frey BN, Valvassori SS, Réus GZ, Martins MR, Petronilho FC, Bardini K et al (2006) Effects of lithium and valproate on amphetamine-induced oxidative stress generation in an animal model of mania. J Psychiatry Neurosci 31(5):326–332

Fukumoto T, Morinobu S, Okamoto Y, Kagaya A, Yamawaki S (2001) Chronic lithium treatment increases the expression of brain-derived neurotrophic factor in the rat brain. Psychopharmacology 158:100–106

Furey ML, Drevets WC (2006) Antidepressant efficacy of the antimuscarinic drug scopolamine: a randomized, placebo-controlled clinical trial. Arch Gen Psychiatry 63:1121–1129

Gallagher P, Malik N, Newham J, Young AH, Ferrier IN, Mackin P (2008) Antiglucocorticoid treatments for mood disorders. Cochrane Database Syst Rev 1:CD005168

Gartside SE, Leitch MM, Young AH (2003) Altered glucocorticoid rhythm attenuates the ability of a chronic SSRI to elevate forebrain 5-HT: implications for the treatment of depression. Neuropsychopharmacology 28(9):1572–1578

Geddes JR, Miklowitz DJ (2013) Treatment of bipolar disorder. Lancet 381(9878):1672–1682

Gibbons AS, Scarr E, McLean C, Sundram S, Dean B (2009) Decreased muscarinic receptor binding in the frontal cortex of bipolar disorder and major depressive disorder subjects. J Affect Disord 116:184–191

Giridharan VV, Sayana P, Pinjari OF, Ahmad N, da Rosa MI, Quevedo J, Barichello T (2019) Postmortem evidence of brain inflammatory markers in bipolar disorder: a systematic review. Mol Psychiatry 25(1):94–113

Gitlin MJ, Swendsen J, Heller TL, Hammen C (1995) Relapse and impairment in bipolar disorder. Am J Psychiatry 152(11):1635–1640

Guardiola-Lemaitre B, De Bodinat C, Delagrange P, Millan MJ, Munoz C, Mocaer E (2014) Agomelatine: mechanism of action and pharmacological profile in relation to antidepressant properties. Br J Pharmacol 171:3604–3619

Guloksuz S, Altinbas K, Aktas Cetin E, Kenis G, Bilgic Gazioglu S, Deniz G, Oral ET, van Os J (2012) Evidence for an association between tumor necrosis factor-alpha levels and lithium response. J Affect Disord 143(1–3):148–152

Halaris A, Cantos A, Johnson K, Hakimi M, Sinacore J (2020) Modulation of the inflammatory response benefits treatment-resistant bipolar depression: a randomized clinical trial. J Affect Disord 261:145–152

Hallahan B (n.d.) Scopolamine in bipolar depression (SCOPE-BD). https://clinicaltrials.gov/ct2/show/NCT04211961

Hendler NH (1978) Spironolactone prophylaxis in manic-depressive disease. J Nerv Ment Dis 166 (7):517–520

Horrobin DF, Lieb J (1981) A biochemical basis for the actions of lithium on behaviour and on immunity: relapsing and remitting disorders of inflammation and immunity such as multiple sclerosis or recurrent herpes as manic-depression of the immune system. Med Hypotheses 7 (7):891–905

Hu Y, Yu X, Yang F, Si T, Wang W, Tan Y, Zhou D, Wang H, Chen D (2010) The level of serum brain-derived neurotrophic factor is associated with the therapeutic efficacy of modified electroconvulsive therapy in Chinese patients with depression. J ECT 26(2):121–125

Husain MI, Chaudhry IB, Hamirani MM, Minhas FA, Kazmi A, Hodsoll J, Haddad PM, Deakin JF, Husain N, Young AH (2016) Minocycline and celecoxib as adjunctive treatments for bipolar depression: a study protocol for a multicenter factorial design randomized controlled trial. Neuropsychiatr Dis Treat 13:1–8

Husain MI, Strawbridge R, Stokes PR, Young AH (2017) Anti-inflammatory treatments for mood disorders: systematic review and meta-analysis. J Psychopharmacol 31:1137–1148

Husain MI, Chaudhry IB, Khoso AB, Husain MO, Hodsoll J, Ansari MA, Naqvi HA, Minhas FA, Carvalho AF, Meyer JH, Deakin B, Mulsant BH, Husain N, Young AH (2020) Minocycline and celecoxib as adjunctive treatments for bipolar depression: a multicentre, factorial design randomized controlled trial. Lancet Psychiatry 7(6):515–527

Imai H, Nakagawa Y (2003) Biological significance of phospholipid hydroperoxide glutathione peroxidase (PHGPx, GPx4) in mammalian cells. Free Radic Biol Med 34(2):145–169

Janowsky DS, el-Yousef MK, Davis JM, Sekerke HJ (1972) A cholinergic-adrenergic hypothesis of mania and depression. Lancet 2:632–635

Jun C, Choi Y, Lim SM, Bae S, Hong YS, Kim JE, Lyoo IK (2014) Disturbance of the glutamatergic system in mood disorders. Exp Neurobiol 23:28–35

Juruena MF (2014) Early-life stress and HPA axis trigger recurrent adulthood depression. Epilepsy Behav 38:148–159

Juruena MF, Gama CS, Berk M, Belmonte-de-Abreu PS (2009) Improved residual symptoms in bipolar affective disorder with adjunctive spironolactone (mineralocorticoid receptor antagonist): case series. J Psychopharmacol 23(8):985–987

Juruena MF, Pariante CM, Papadopoulos AS, Poon L, Lightman S, Cleare AJ (2009a) Prednisolone suppression test in depression: prospective study of the role of HPA axis dysfunction in treatment resistance. Br J Psychiatry 194:342–349

Juruena MF, Ottoni G, Machado-Vieira R, Carneiro R, Weingarthner N, Fleig SS, Broilo L, Busnello ED (2009b) Bipolar I and II disorder residual symptoms: Oxcarbazepine and carbamazepine as add-on treatment to lithium in a double-blind, randomized trial. Prog Neuro-Psychopharmacol Biol Psychiatry 33:94–99

Kargar M, Yousefi A, Mojtahedzadeh M, Akhondzadeh S, Artounian V, Abdollahi A, Ahmadvand A, Ghaeli P (2014) Effects of celecoxib on inflammatory markers in bipolar patients undergoing electroconvulsive therapy: a placebo-controlled, double-blind, randomised study. Swiss Med Wkly 144:w13880

Kato T (2007) Mitochondrial dysfunction as themolecular basis of bipolar disorder: therapeutic implications. CNS Drugs 21(1):1–11

Keck PE Jr, Hsu HA, Papadakis K, Russo J Jr (2009) Memantine efficacy and safety in patients with acute mania associated with bipolar I disorder: a pilot evaluation. Clin Neuropharmacol 32:199–204

Kew JN, Kemp JA (2005) Ionotropic andmetabotropic glutamate receptor structure and pharmacology. Psychopharmacology 179(1):4–29

Khairova R, Pawar R, Salvadoran G, Juruena MF, De Sousa RT, Soeiro-de-Souza MG, Salvador M, Zarate CA, Gattaz WF, Machado-Vieira R (2012) Effects of lithium on oxidative stress parameters in healthy subjects. Mol Med Rep 5:680–682

Laeng P, Pitts RL, Lemire AL, Drabik CE, Weiner A, Tang H, Thyagarajan R, Mallon BS, Altar CA (2004) The mood stabilizer valproic acid stimulates GABA neurogenesis from rat forebrain stem cells. J Neurochem 91:238–251

Lanfumey L, Mongeau R, Hamon M (2013) Biological rhythms and melatonin in mood disorders and their treatments. Pharmacol Ther 138:176–184

Leach MJ, Baxter MG, Critchley MA (1991) Neurochemical and behavioral aspects of lamotrigine. Epilepsia 32(2):4–8

Leboyer M, Soreca I, Scott J, Frye M, Henry C, Tamouza R, Kupfer DJ (2012) Can bipolar disorder be viewed as a multi-system inflammatory disease? J Affect Disord 141(1):1–10

Lee SY, Chen SL, Chang YH, Chen PS, Huang SY, Tzeng NS, Wang YS, Wang LJ, Lee IH, Wang TY, Yeh TL, Yang YK, Hong JS, Lu RB (2014) The effects of add-on low-dose memantine on cytokine levels in bipolar II depression: a 12-week double-blind, randomized controlled trial. J Clin Psychopharmacol 34:337–343

Lessmann V, Gottmann K, Malcangio M (2003) Neurotrophin secretion: current facts and future prospects. Prog Neurobiol 69:341–374

Li X, Ketter TA, Frye MA (2002) Synaptic, intracellular, and neuroprotective mechanisms of anticonvulsants: are they relevant for the treatment and course of bipolar disorders? J Affect Disord 69(1–3):1–14

López-Muñoz F, Shen WW, D'Ocon P, Romero A, Álamo C (2018) A history of the pharmacological treatment of bipolar disorder. Int J Mol Sci 19(7):E2143

Lund C, Tomlinson M, De Silva M, Fekadu A, Shidhaye R, Jordans M, Petersen I, Bhana A, Kigozi F, Prince M, Thornicroft G, Hanlon C, Kakuma R, McDaid D, Saxena S, Chisholm D, Raja S, Kippen-Wood S, Honikman S, Fairall L, Patel V (2012) Prime. A programme to reduce the treatment gap for mental disorders in five low- and middle-income countries. PLoS Med 9 (12):e1001359

Machado-Vieira R, Ibrahim L, Henter ID, Zarate CA Jr (2012) Novel glutamatergic agents for major depressive disorder and bipolar disorder. Pharmacol Biochem Behav 100(4):678–687

Maes M (2011) Depression is an inflammatory disease, but cell-mediated immune activation is the key component of depression. Prog Neuro-Psychopharmacol Biol Psychiatry 35(3):664–675

Magalhaes PV, Dean OM, Bush AI, Copolov DL, Malhi GS, Kohlmann K, Jeavons S, Schapkaitz I, Anderson-Hunt M, Berk M (2011) N-acetylcysteine for major depressive episodes in bipolar disorder. Braz J Psychiatry 33:374–378

Mahableshwarkar AR, Calabrese JR, Macek TA, Budur K, Adefuye A, Dong X, Hanson E, Sachs GS (2017) Efficacy and safety of sublingual ramelteon as an adjunctive therapy in the maintenance treatment of bipolar I disorder in adults: a phase 3, randomized controlled trial. J Affect Disord 221:275–282

Malhi GS, Bassett D, Boyce P, Bryant R, Fitzgerald PB, Fritz K, Hopwood M, Lyndon B, Mulder R, Murray G, Porter R, Singh AB (2015) Royal Australian and New Zealand College of Psychiatrists clinical practice guidelines for mood disorders. Aust N Z J Psychiatry 49 (12):1087–1206

McCullumsmith RE, Kristiansen LV, Beneyto M, Scarr E, Dean B, Meador-Woodruff JH (2007) Decreased NR1, NR2A, and SAP102 transcript expression in the hippocampus in bipolar disorder. Brain Res 1127(1):108–118

McElroy SL, Winstanley EL, Martens B, Patel NC, Mori N, Moeller D, McCoy J, Keck PE Jr (2011) A randomized, placebo-controlled study of adjunctive ramelteon in ambulatory bipolar I disorder with manic symptoms and sleep disturbance. Int Clin Psychopharmacol 26:48–53

McLoughlin GA, Ma D, Tsang TM, Jones DN, Cilia J, Hill MD, Robbins MJ, Benzel IM, Maycox PR, Holmes E, Bahn S (2009) Analyzing the effects of psychotropic drugs on metabolite profiles in rat brain using 1H NMR spectroscopy. J Proteome Res 8(4):1943–1952

Mehta A, Prabhakar M, Kumar P, Deshmukh R, Sharma PL (2013) Excitotoxicity: bridge to various triggers in neurodegenerative disorders. Eur J Pharmacol 698(1–3):6–18
Modabbernia A, Taslimi S, Brietzke E, Ashrafi M (2013) Cytokine alterations in bipolar disorder: a meta-analysis of 30 studies. Biol Psychiatry 74(1):15–25
Mousavi SY, Khezri R, Karkhaneh-Yousefi MA, Mohammadinejad P, Gholamian F, Mohammadi MR, Zeinoddini A, Akhondzadeh S (2017) Randomized double-blind placebo-controlled trial on effectiveness and safety of celecoxib adjunctive therapy in adolescents with acute bipolar mania. J Child Adolesc Psychopharmacol 27:494–500
Moylan S, Maes M, Wray N, Berk M (2013) The neuroprogressive nature of major depressive disorder: pathways to disease evolution and resistance, and therapeutic implications. Mol Psychiatry 18:595–606
Murck H, Schüssler P, Steiger A (2012) Renin-angiotensin-aldosterone system: the forgotten stress hormone system: relationship to depression and sleep. Pharmacopsychiatry 45(3):83–95
Nibuya M, Morinobu S, Duman RS (1995) Regulation of BDNF and trkB mRNA in brain by chronic electroconvulsive seizure and antidepressant drug treatments. J Neurosci 15:7539–7547
Norris ER, Karen B, Correll JR, Zemanek KJ, Lerman J, Primelo RA, Kaufmann MW (2013) A double-blind, randomized, placebo-controlled trial of adjunctive ramelteon for the treatment of insomnia and mood stability in patients with euthymic bipolar disorder. J Affect Disord 144:141–147
O'Brien PM, Craven D, Selby C, Symonds EM (1979) Treatment of premenstrual syndrome by spironolactone. Br J Obstet Gynaecol 86(2):142–147
Paraskevas GP, Triantafyllou NI, Kapaki E, Limpitaki G, Petropoulou O, Vassilopoulos D (2006) Add-on lamotrigine treatment and plasma glutamate levels in epilepsy: relation to treatment response. Epilepsy Res 70(2–3):184–189
Park LT, Lener MS, Hopkins M, Iadorola N, Machado-Vieira R, Ballard E, Nugent A, Zarate CA Jr (2017) A double-blind, placebo-controlled, pilot study of riluzole monotherapy for acute bipolar depression. J Clin Psychopharmacol 37:355–358
Pereira C, Chavarria V, Vian J, Ashton MM, Berk M, Marx W, Dean OM (2018) Mitochondrial agents for bipolar disorder. Int J Neuropsychopharmacol 21:550–569
Rantamaki T, Knuuttila JEA, Hokkanen ME, Castrén E (2006) The effects of acute and long-term lithium treatments on TrkB neurotrophin receptor activation in the mouse hippocampus and anterior cingulate cortex. Neuropharmacology 50:421–427
Rao JS, Rapoport SI (2009) Mood-stabilizers target the brain arachidonic acid cascade. Curr Mol Pharmacol 2:207–214
Rao JS, Lee HJ, Rapoport SI, Bazinet RP (2008) Mode of action of mood stabilizers: is the arachidonic acid cascade a common target? Mol Psychiatry 13(6):585–596
Rao JS, Harry GJ, Rapoport SI, Kim HW (2010) Increased excitotoxicity and neuroinflammatory markers in postmortem frontal cortex from bipolar disorder patients. Mol Psychiatry 15 (4):384–392
Rapaport MH, Manji HK (2001) The effects of lithium on ex vivo cytokine production. Biol Psychiatry 50(3):217–224
Riedel G, Platt B, Micheau J (2003) Glutamate receptor function in learning and memory. Behav Brain Res 140:1–47
Roitman S, Green T, Osher Y, Karni N, Levine J (2007) Creatine monohydrate in resistant depression: a preliminary study. Bipolar Disord 9:754–758
Rosa AR, Frey BN, Andreazza AC, Cereser KM, Cunha AB, Quevedo J, Santin A, Gottfriend C, Gonçalves CA, Vieta E, Kapczinski F (2006) Increased serum glial cell line-derived neurotrophic factor immunocontent during manic and depressive episodes in individuals with bipolar disorder. Neurosci Lett 407(2):146–150
Rosa AR, Singh N, Whitaker E, de Brito M, Lewis AM, Vieta E, Churchill GC, Geddes JR, Goodwin GM (2014) Altered plasma glutathione levels in bipolar disorder indicates higher oxidative stress; a possible risk factor for illness onset despite normal brain-derived neurotrophic factor (BDNF) levels. Psychol Med 44:2409–2418

Saarelainen T, Hendolin P, Lucas G, Koponen E, Sairanen M, MacDonald E, Agerman K, Haapasaio A, Nawa H, Aloyz EP, Castren E (2003) Activation of the TrkB neurotrophin receptor is induced by antidepressant drugs and is required for antidepressant induced behavioral effects. J Neurosci 23:349–357

Sarris J, Mischoulon D, Schweitzer I (2012) Omega-3 for bipolar disorder: meta-analyses of use in mania and bipolar depression. J Clin Psychiatry 73:81–86

Savitz JB, Teague TK, Misaki M, Macaluso M, Wurfel BE, Meyer M, Drevets D, Yates W, Gleason O, Drevets WC, Preskorn SH (2018) Treatment of bipolar depression with minocycline and/or aspirin: an adaptive, 2x2 double-blind, randomized, placebo-controlled, phase IIA clinical trial. Transl Psychiatry 8:27

Scaini G, Fries GR, Valvassori SS, Zeni CP, Zunta-Soares G, Berk M et al (2017) Perturbations in the apoptotic pathway and mitochondrial network dynamics in peripheral blood mononuclear cells from bipolar disorder patients. Transl Psychiatry 7(5):e1111

Scheefhals N, MacGillavry HD (2018) Functional organization of postsynaptic glutamate receptors. Mol Cell Neurosci 91:82–94

Shao L, Young LT, Wang JF (2005) Chronic treatment with mood stabilizers lithium and valproate prevents excitotoxicity by inhibiting oxidative stress in rat cerebral cortical cells. Biol Psychiatry 58:879–884

Sitges M, Chiu LM, Guarneros A, Nekrassov V (2007) Effects of carbamazepine, phenytoin, lamotrigine, oxcarbazepine, topiramate and vinpocetine on Na+ channel-mediated release of [3H]glutamate in hippocampal nerve endings. Neuropharmacology 52:598–605

Soczynska JK, Kennedy SH, Alsuwaidan M, Mansur RB, Li M, McAndrews MP, Brietzke E, Woldeyohannes HO, Taylor VH, McIntyre RS (2017) A pilot, open-label, 8-week study evaluating the efficacy, safety and tolerability of adjunctive minocycline for the treatment of bipolar I/II depression. Bipolar Disord 19:198–213

Sourial-Bassillious N, Rydelius PA, Aperia A, Aizman O (2009) Glutamate-mediated calcium signaling: a potential target for lithium action. Neuroscience 161(4):1126–1134

Tatay-Manteiga A, Balanzá-Martínez V, Bristot G, Tabarés-Seisdedos R, Kapczinski F, Cauli O (2017) Clinical staging and serum cytokines in bipolar patients during euthymia. Prog Neuropsychopharmacol Biol Psychiatry 77:194–201

Toniolo RA, Silva M, Fernandes FBF, Amaral J, Dias RDS, Lafer B (2018) A randomized, double-blind, placebo-controlled, proof-of-concept trial of creatine monohydrate as adjunctive treatment for bipolar depression. J Neural Transm 125:247–257

Wang M, Hammarbäck S, Lindhe BA, Bäckström T (1995) Treatment of premenstrual syndrome by spironolactone: a double-blind, placebo-controlled study. Acta Obstet Gynecol Scand 74 (10):803–808

Wang JF, Shao L, Sun X, Young LT (2009) Increased oxidative stress in the anterior cingulate cortex of subjects with bipolar disorder and schizophrenia. Hron 11(5):523–529

Watson S, Gallagher P, Ritchie JC, Ferrier IN, Young AH (2004) Hypothalamic-pituitary-adrenal axis function in patients with bipolar disorder. Br J Psychiatry 184:496–502

Watson S, Gallagher P, Porter RJ, Smith MS, Herron LJ, Bulmer S, North-East Mood Disorders Clinical Research Group, Young AH, Ferrier IN (2012) A randomized trial to examine the effect of mifepristone on neuropsychological performance and mood in patients with bipolar depression. Biol Psychiatry 72:943–949

Wernze H (2000) Surprising effect of spironolactone on eating behavior and mood in bulimia nervosa. Psychopharmakotherapie 7:33–39

Yatham LN, Vieta E, Goodwin GM, Bourin M, de Bodinat C, Laredo J, Calabrese J, Agomelatine Study Group (2016) Agomelatine or placebo as adjunctive therapy to a mood stabilizer in bipolar I depression: randomized double-blind placebo-controlled trial. Br J Psychiatry 208:78–86

Yoon SJ, Lyoo IK, Haws C, Kim TS, Cohen BM, Renshaw PF (2009) Decreased glutamate/glutamine levels may mediate cytidine's efficacy in treating bipolar depression: a longitudinal proton magnetic resonance spectroscopy study. Neuropsychopharmacology 34:1810–1818

Young AH, Gallagher P, Watson S, Del-Estal D, Owen BM, Ferrier IN (2004) Improvements in neurocognitive function and mood following adjunctive treatment with mifepristone (RU-486) in bipolar disorder. Neuropsychopharmacology 29:1538–1545

Zarate CA Jr, Quiroz JA, Singh JB, Denicoff KD, De Jesus G, Luckenbaugh DA, Charney DS, Manji HK (2005) An open-label trial of the glutamate-modulating agent riluzole in combination with lithium for the treatment of bipolar depression. Biol Psychiatry 57(4):430–432

Zarate CA Jr, Brutsche NE, Ibrahim L, Franco-Chaves J, Diazgranados N, Cravchik A, Selter J, Marquardt CA, Liberty V, Luckenbaugh DA (2012) Replication of ketamine's antidepressant efficacy in bipolar depression: a randomized controlled add-on trial. Biol Psychiatry 71 (11):939–946

Correction to: Inflammation as a Mechanism of Bipolar Disorder Neuroprogression

Tatiana Barichello, Vijayasree Vayalanellore Giridharan, Gursimrat Bhatti, Pavani Sayana, Tejaswini Doifode, Danielle Macedo, and Joao Quevedo

Correction to: Chapter "Inflammation as a Mechanism of Bipolar Disorder Neuroprogression" in: T. Barichello et al., Curr Topics Behav Neurosci, https://doi.org/10.1007/7854_2020_173

The original version of this chapter is updated with the below mentioned updates as per the author's request:

1. The author name ''Vijayasree V. Giridharan" is now updated with full name as Vijayasree Vayalanellore Giridharan, so that in the PubMed the name appears as Giridharan VV.
2. Figure 1 is now included in the content in Sect. 3.1 Mechanisms of inflammation and their contribution to BD neuroprogression.

The original article has been updated.

The updated online version of this chapter can be found at https://doi.org/10.1007/7854_2020_173

Curr Topics Behav Neurosci (2021) 48: 325–326
https://doi.org/10.1007/7854_2020_214
Published Online: 2 February 2021

The manufacturer's authorised representative in the EU is Springer Nature Customer Service Centre GmbH, Europaplatz 3, 69115 Heidelberg, Germany. If you have any concerns regarding our products, please contact ProductSafety@springernature.com

Printed and bound by CPI Group (UK) Ltd, Croydon, CR0 4YY
28/09/2026
02224090-0016